I0477038

Gemstones and Precious Stones of North America

by G.F. Kunz

with an introduction by Kerby Jackson

This work contains material that was originally published in 1890.

This publication is within the Public Domain.

This edition is reprinted for educational purposes
and in accordance with all applicable Federal Laws.

Introduction Copyright 2015 by Kerby Jackson

Introduction

It has been years since the G.F. Kunz released their important publication "Gems and Precious Stones of North America". First released in 1890, this work has been unavailable to the mining community since those days, with the exception of expensive original collector's copies and poorly produced digital editions.

It has often been said that *"gold is where you find it"*, but even beginning prospectors understand that their chances for finding something of value in the earth or in the streams of the Golden West are dramatically increased by going back to those places where gold and other minerals were once mined by our forerunners. Despite this, much of the contemporary information on local mining history that is currently available is mostly a result of mere local folklore and persistent rumors of major strikes, the details and facts of which, have long been distorted. Long gone are the old timers and with them, the days of first hand knowledge of the mines of the area and how they operated. Also long gone are most of their notes, their assay reports, their mine maps and personal scrapbooks, along with most of the surveys and reports that were performed for them by private and government geologists. Even published books such as this one are often retired to the local landfill or backyard burn pile by the descendents of those old timers and disappear at an alarming rate. Despite the fact that we live in the so-called "Information Age" where information is supposedly only the push of a button on a keyboard away, true insight into mining properties remains illusive and hard to come by, even to those of us who seek out this sort of information as if our lives depend upon it. Without this type of information readily available to the average independent miner, there is little hope that our metal mining industry will ever recover.

Though this volume may not at first seem to be of great importance to gold miners, I feel that those miners with an interest in smelting and refining their finds, especially those recovered from lodes, will find the processes outlined to be of great value.

This important volume and others like it, are being presented in their entirety again, in the hope that the average prospector will no longer stumble through the overgrown hills and the tailing strewn creeks without being well informed enough to have a chance to succeed at his ventures.

Please note that at times it is necessary to rearrange illustration plates in these texts. Any illustrations not found in their original sequence may be found following the index.

Kerby Jackson
Josephine County, Oregon
June 2015

www.goldminingbooks.com

CONTENTS

———

Reeless 12-3-25 P. K. L

INTRODUCTION

NEARLY all the known varieties of precious stones are found in the United States, but there is very little systematic exploration for them, as the indications seldom justify the investment of much capital in such search. The daily yield from the coal and iron mines, or from the South African diamond mines, or a week's yield of the granite quarries, would exceed in value the entire output of precious stones found in the United States during a year. Systematic search for gems and precious stones has been carried on in only two States—Maine and North Carolina. Otherwise, the gems are found accidentally, in connection with other substances that are being mined, or in small veins which are only occasionally met with, as the turquoise in Mexico. They are often gathered on the surface, as is the case with garnet or olivine from Arizona and New Mexico; or in sluicing for gold, as the sapphires from Montana; or in connection with mica mining, as the beryl from Connecticut and North Carolina; or from the beds of streams and decomposing rocks, as the moss-agate from Wyoming; or on the beaches, as the agate, chlorastrolite, and thomsonite from the shores of Lake Superior.

Nearly all the gems found in these various ways are sent to the large cities in small parcels, or sold in the neighborhood to tourists, or sent to other places to be disposed of as having been found in their vicinity. Many of them are only known locally, some to mineralogists, while others, mentioned in the following pages, are accepted by the few gem collectors of the United States, whose only object is to find something possessing the qualities of a gem or precious stone, wherewith to enrich their cabinets. A list of such gem stones will be of interest to many who have not known of their existence in this country, and to others this knowledge may have a commercial value, should some of these·minerals be met with in sufficient quantities and of good quality; and, in directing attention to valuable deposits where a small amount has already been realized, it may stimulate the interest in and search for gems, and aid in developing what may become an important industry.

It is known that the Indians worked the turquoise mines of New Mexico more than two centuries ago; that they made arrow and spear points of rock crystal, smoky quartz, agatized and opalized woods, agates, jaspers, and obsidian, and buried crystals of quartz with their dead; that the richly colored fluorite of Hardin County, Ill., was worked by them into ornaments. Some of the most beautiful of their arrow-points are now used for decoration by the white man, paralleling the proverbial conversion of swords into plowshares. Mention will be made of a few localities where gem specimens have been found which are remarkable as such, and which have a special claim on the collector; some notes will also be given concerning specimens that have been of value to the finder, such, for example, as the Pike's Peak amazonstone, or the spinels found at Monroe, N. Y., which have little or no gem value. Many of these stones are as beautiful, in their native form, as they are after having undergone the grinding process. The cutting of such material, therefore, for its money value, is really vandalism and should be discouraged by all scientists, although we may not all be willing to accept more broadly Ruskin's opinion that "gems should not be cut, but worn in the natural state."

In 1882 the writer was invited to prepare a paper on precious stones of the United States for the first annual report of the Division of Mining Statistics, and since then has prepared similar annual reports. From that beginning the present work has grown, and it contains much additional information obtained from studying the collections of the United States as well as from a personal examination of many of the localities where gems are found. Its object is to present, in convenient form, as many of the facts as possible regarding the precious stones peculiar to the United States, Canada, and Mexico, so that they may be available, not only to the mineralogist, the miner, the mineral and gem collector, the archæologist and the jeweler, but also to the public, the conditions under which they occur, the methods by which the mining and search for them are conducted, the value and production of different stones, and also an account of the collections in these countries.

A brief general description of each important gem will be found at the beginning of the article and a series of analyses indicating the composition of each precious stone, from the latest or most reliable authority, and for comparison a typical analysis is generally included. Full reference to the literature of the subject is given in the foot-notes.

The chapter on Canadian precious stones is based on a report prepared for " The Mining and Mineral Statistics of Canada " for 1887, and its use is permitted through the courtesy of the authorities of the Canadian Geological Survey. One chapter is devoted to pearls, with a full account of their mention by the early explorers, their occurrence in mounds, Indian graves, and similar remains ; another devoted to the imports and values, and to the cutting of gem stones, with mention of some remarkable gems owned in the United States, and a brief description of the best known collections in this country.

A number of minerals are enumerated that are not only below 7 in the scale of hardness, but that are even below 6, and apparently too soft for cut gems; yet cups, vases, and other objects may be made of these stones, such as serpentine and catlinite, which could be successfully used where transparent apatite could not, because they are opaque, do not show scratches, and always present an even, good color.

During recent years a number of items have appeared in the newspapers relative to the finding of alleged valuable gems, which have proved on investigation to be without foundation. As newspaper statements are sometimes copied into special literature, it may be well to refer briefly to them.

The "Blue Ridge sapphire," or "Georgia marvel," as it was called in the reports, was found in 1883 in a brook in the Blue Ridge Mountains in Georgia. It was estimated to be worth about $50,000 by the owner, he having been assured of its genuineness as a sapphire by two Southern jewelers, who had arrived at its valuation by computing its weight. Anything scratched by a file is sure to be pronounced glass, whether it is glass, topaz, or some other equally hard stone; while, on the other hand, the common fallacy prevails that anything that a file cannot scratch is a genuine stone, even though it may be only glass. In this instance the gem proved to be a piece of rolled blue bottle-glass, of which fact its owner could be convinced only when he saw a platinum wire coated with a melted fragment.

Another wonder was a stone weighing 9 ounces, plowed up near Gibsonville, Guilford County, N. C., which was pronounced a genuine emerald by some local expert, who tested it, and with the microscope showed that it contained various small diamonds. Its value was estimated up in the thousands, and $1,000 was reported to have been refused for it by its owner, who, as it was believed to be the largest known emerald, expected that it would bring him a fortune. Being, therefore, too valuable to be entrusted to an express company, he put himself to the expense of a trip to New York, where his prize proved on examination to be a greenish quartz crystal, filled with long hair-like crystals of green byssolite or actinolite, on which were series and strings of small liquid cavities that, glistening in the sun, had led to the included diamond theory. The best offer that he received for the stone was $5.

The "Wetumpka Ruby," from Elmore County, Ala., was supposed to be a ruby of six ounces weight, "after cutting away all the roughness." Owing to its assumed value, it was deposited in the Wetumpka bank vault, and on no consideration would be sent to any one on approbation. A small fragment sent to New York City proved the stone to be only a garnet and from its quality, of no gem value, even if a ruby.

Another is a crystal found near Danbury, N. C., which was examined and pronounced to be a genuine diamond by the local jewelers, and valued at $7,000.

A number of the supposed diamond discoveries will be found at the end of the first chapter.

It is not intended to make this volume either a complete treatise on precious stones or on the science of mineralogy, but to confine it more particularly to the occurrence of precious stones in North America, and for comparison, occasional reference is made to foreign sources and authorities.

The beautiful colored plates contained in this volume are the work of the eminent art lithographers, Messrs. L. Prang & Co., of Boston, Mass., and are unquestionably the finest work of the kind ever published. The writer's thanks are extended to Messrs. Tiffany & Co. for the facilities afforded by their corps of artists in the preparation of the original designs used in the production of these plates.

During the preparation of this work, much valuable assistance has been received from the following gentlemen, to whom the author begs to tender his most sincere thanks.

Maj. John W. Powell, William H. Holmes, Prof. Frank W. Clarke, Joseph S. Diller, and Dr. David T. Day, of the United States Geological Survey; Lester F. Ward, Frank H. Knowlton, William H. Dall, George H. Merrill, Dr. Thomas Wilson and Dr. Robert E. C. Stearns, of the United States National Museum; Prof. Edward S. Dana, Dr. Samuel L. Penfield and Mr. O. H. Drake, of Yale University; Dr. Augustus C. Hamlin, Bangor, Me.; Dr. Robert Lilley, Dr. Marcus Benjamin, Prof. Daniel S. Martin, Mr. James D. Yerrington, Prof. Oliver P. Hubbard and Mr. C. J. Cottier, of New York City, also Dr. E. Hamy, of Paris, France.

CHAPTER I.

Diamonds.

THE diamond crystallizes in the isometric system, and is usually found as an octahedron or as some modification of that form. It is 10 in the scale of hardness and the hardest of all known substances. Its composition is pure carbon. It has a greater refractive and dispersive power on light than any other gem, and is the only one that is combustible. Its specific gravity is about 3·525. In color its range is extensive, and it is found in almost all the shades of the spectrum, more commonly, however, white, yellow, brown ; rarely rose-red, red, blue, and green.

· Ninety-five per cent. of the diamonds at present obtained are from the Kimberley Mines, Griqua Land West, South Africa. The remainder come from Brazil, India, and Borneo. A few have been found recently in New South Wales, and they are known to exist in the Ural Mountains. Since the discovery of South African mines in 1867, and the opening a short time afterward of 3,143 claims that are now consolidated into a small number of large companies, all within a radius of a mile and a half, more diamonds have been found than during the two preceding centuries throughout the whole world. Over 9 tons (40,000,000 carats) of diamonds, valued in the rough at $250,000,000, and after cutting at over $500,000,000, have been taken from these mines. Diamonds are sold by the carat. The

International carat weighs 205 grams, equivalent to 3·168 grains troy.   The proportions according to quality of the entire South African yield are as follows : First quality, eight per cent., second quality, twenty-five per cent., third quality, twenty per cent., and the balance bort, which is used for slitting gems, polishing diamonds, more recently for saws, and ground into powder for use in the arts.   An impression seems to prevail that a diamond will not break if struck with a hammer on an anvil, and several that were supposed to be good specimens were broken in this way.   While the diamond is hard, it is also very brittle, and can be easily broken, and although every substance from the hardness of feldspar up, including a cleavage or cut diamond, will scratch glass, nothing but the natural edge of a diamond crystal will cut it.   To determine whether a given specimen is a diamond, the best test is to try if it will scratch corundum.   If no mark is produced, and if the specimen cannot be scratched by a diamond, it is safe to assume that it is a diamond. It is well to make the trial on a smooth or polished surface, otherwise the scratch will not be perceptible.

The occurrence of diamonds in the United States is chiefly confined to two regions, geographically very remote and geologically quite dissimilar.   The first is a belt of country lying along the eastern base of the southern Alleghanies, from Virginia to Georgia, while the other extends along the western base of the Sierra Nevada and Cascade ranges in northern California and southern Oregon.   In both cases the mode of occurrence has several marked resemblances.   The diamonds are found in loose material, among deposits of gravel and earth, and are associated with garnets, zircons, iron sands, monazite, anatase, and particularly with gold, in the search for which they have usually been discovered.   This resemblance is due altogether to the fact that these loose deposits, in both regions, are merely the débris of the crystalline rocks of the adjacent mountains, and therefore present a general similarity, while the ages of the rocks themselves are widely different.   In the case of the South Atlantic States, the rocks of the Blue Ridge and eastern Alleghanies are of ancient Archaen and Cambrian ages, while in the western belt, the Sierra Nevada was not elevated and metamorphosed until the middle or later Mesozoic. From the general resemblance of conditions above referred to, the details of discovery in the two regions are remarkably similar, and in both occasional diamond crystals are found, accidentally picked up on the surface, or more frequently encountered in the search for gold, sometimes in placer-mining and sometimes in the flumes and sluices of hydraulic workings. They have sometimes been overlooked, unrecognized, or destroyed by rude and ignorant methods of testing, and at other times have been made the basis of fabulous estimates and exaggerated tales, but they have not as yet been found in sufficient quantities to justify an attempt at diamond-mining, nor have the specimens obtained been of more than local interest and moderate value.

With regard to the finding of diamonds in other parts of the country, there have been various reports, but little or no positive evidence. The supposed diamond field of central Kentucky has been the subject of much interesting study and discussion on account of the striking resemblance of the rock to that of the diamantiferous region of South Africa; but the conditions are found, upon closer examination, to present important differences, and the diamonds are yet to be discovered. The formations in the eastern portions of the United States where diamonds have been found are entirely different from those of South Africa. They resemble more nearly those of the diamond fields of Brazil and of parts of India. The diamonds found in the United States are much older than those of South Africa, and if they have ever occurred in rock similar to that in Kimberley, there is nothing to indicate it now, since the rocks in American diamond-bearing localities are mainly granitic. It may be said that, while diamonds are found to some extent within the limits of the United States, there is no reason as yet to believe that they will ever be numbered among our important mineral products. Their local and scientific interest is of course very great; and this fact will justify the somewhat detailed account of their occurrence given in this volume as an important part of a work on precious stones in the United States. Prof. H. Carvill Lewis paid much attention to this subject, visiting many of the localities where diamonds had been found in the eastern part of the United States, and personally in-

vestigating the history of such stones as he could trace. His conclusions are to be published in a final memoir on the " Genesis of
the Diamond," the completion and editing of which has been
undertaken, since his death, by his friend and associate, Prof.
George H. Williams, of the Johns Hopkins University. ⁒ A
moderate number of well-authenticated diamonds have been found
in Georgia and in North Carolina, a very few are reported from
South Carolina, and one or two are known from southern Virginia. These are all apparently derived from the detritus of the
crystalline metamorphic rocks that extend through these States
as the eastern ranges of the Appalachian system. Of these, the
great continuous Blue Ridge is, of course, the leading member ;
eastward of it lies the so-called Atlantic or Tidewater gneiss, by
many regarded of later age; and another belt, perhaps distinct,
extends from Richmond, Va., to the vicinity of Raleigh, N. C.
It is much to be regretted that the geology of these crystalline
belts is thus far so little known. At some points they appear to
be well distinguished, while at others they merge into one another
and have not been clearly defined. The names Laurentian, Huronian, Montalban, and Cambrian are variously applied to different portions of them by different geologists. The Blue Ridge
proper is generally admitted to be chiefly true Laurentian ; and it
is certain that Cambrian beds appear at some points in the area.
The remarkable itacolumite rock, popularly associated with the
occurrence of diamonds, is found at many points on the flanks of
the Blue Ridge, but its geological age is not yet clearly established.

Beginning with the account of the one well-known Virginia
diamond, we shall pass on southward, taking up the States in order. The Dewey diamond (see Colored Plate No. 1), was found at
Manchester, Va., in 1855, and John H. Tyler, Sr., of Richmond,
Va., who was the first to see it, says : " This diamond was found
just opposite Richmond, by a laborer engaged in grading one of
the streets. It was brought to me to ascertain its character and
value. I pronounced it at once a valuable diamond, and recommended the finder to keep it carefully and to see me about it again.
I did not know his name, and have not seen him since, but after-
·wards learned that he sold it." The first record that we have of

it is from the "New York Evening Post" of April 28, 1855, where it says : " We were shown yesterday, on board the steamship 'Jamestown,' what is said to be the largest diamond ever discovered in North America. It was found several months ago by a laboring man at Manchester, Va., in some earth which he was digging up. It was put into a furnace for melting iron, at Richmond, where it remained at red heat for two hours and twenty minutes. It was then taken out and found to be uninjured and brighter than ever. It was valued in Richmond at $4,000." This stone next passed into the possession of Samuel W. Dewey, of Jackson, now of Philadelphia, and by him was named the Oninoor, or "Sun of Light," though it has more generally been known as the Dewey diamond. It was for a time on exhibition in New York, at the store of Ball, Black & Co., and was cut at an expense of $1,500 by Henry D. Morse, of Boston. Having passed out of Mr. Dewey's hands, through his failure to redeem it on a loan, it was then sold to J. Anglist, who received from John A. Morrissey a loan of $6,000 on it, and as he failed to redeem it, it became part of the Morrissey estate and was known as the Morrissey diamond. It had a large flaw on one side (see Colored Plate No. 1), and was an octahedron with slightly rounded faces. Its original weight was 23¾ carats, and after cutting it weighed 11¼ carats. As it is off-color and imperfect, it is to-day worth not more than from $300 to $400. Exact copies of it in glass, as well as copper electrotypes of it as it was found, and as cut, were deposited by Mr. Dewey in the United States Mint at Philadelphia, and also at the Peabody Museum in New Haven and in a number of cabinets.

North Carolina, so rich and varied in mineral resources, has long been known to yield a certain amount of gold ; and in the same region have been found some diamonds, either loose in the soil, or taken from the washings of auriferous gravel. The portion of the State which has yielded these valuable substances is that known as the Piedmont region,—a belt of country lying, as its name indicates, at the foot of the mountains, along the eastern base of the Blue Ridge. The rocks here are metamorphic and crystalline, with some Cambrian beds a little farther west. There runs throughout much of this region a belt or belts of itacolumite, the so-called " flexible sandstone," which is sometimes found in Brazil and in the Ural Mountains, and is generally supposed to be the matrix of diamond crystals.   The presence of this peculiar rock and the occasional discovery of diamonds in adjacent districts have led to the idea that the itacolumite belt of North Carolina might prove to be a valuable diamantiferous region ; but as yet no diamonds have actually been discovered there, and but few have been found in the loose débris of the crystalline beds. Prof. Frederick A. Genth, of the University of Pennsylvania, describes [1] the occurrence of the two crystalline varieties of carbon in that State,—the graphite in beds interstratified with mica schist or gneiss ; the diamond in the débris of such rocks, associated with gold, zircon, garnet, monazite, and other minerals, and after speaking of this occurrence in connection with rocks of identical age, as a very interesting circumstance, he says : " The diamond has not been observed in North Carolina in any more recent strata, and in the itacolumite regions no diamonds have ever been found, as in Brazil ; from which it appears that the itacolumite of Brazil is either simply a quartzose mica slate of similar age with the North Carolina gneissoid rocks, or, if it be contemporary with the North Carolina itacolumite, the diamonds were not produced in the same but came from the older rocks, and were redeposited with the sands resulting from the reduction to powder of these and are now found imbedded in the same, their hardness having prevented their destruction.   Seven or eight diamonds have thus far been found.   They occur distributed over a wide area of surface in the counties of Burke, Rutherford, Lincoln, Mecklenburg, and Franklin, and I have no doubt if a regular search were to be made for them, they would be more frequently found."   To the counties named by Professor Genth, must now be added McDowell, and these all form, with the exception of Franklin, a group lying together in the line of the general drainage of the country, southeast of the Blue Ridge.   Franklin County is far to the northeast of the others ; and any diamonds occurring there must be derived from the disintegration of another belt of crystalline rocks, that traverses the eastern portion of the State, near Weldon, in Halifax County, or else have been

[1] Mineral Resources of North Carolina, p. 28, Philadelphia, 1871.

transported to a great distance by rivers. The same is doubtless true of the diamond found at Manchester, Chesterfield County, Va.

Historically the North Carolina diamonds are reported as follows : The first specimen was picked up at the ford of Brindletown Creek, Burke County, in 1843, by Dr. F. M. Stephenson. It was an octahedral crystal, and was estimated to be worth $100. Another was found in the same neighborhood by Prof. George W. Featherstonhaugh; but there seems to be no account of its characters. The third found, but the first to attract much attention, was obtained in 1845, from the gold-washings of D. J. Twitty's mine, in Rutherford County. It was owned by Gen. Thomas L. Clingman, of Asheville, and was described by Prof. Charles U. Shepard.[1] It was a curved and remarkably distorted octahedron, clear, almost flawless, and faintly tinged with yellow. The weight was about 1$\frac{1}{4}$ carats (4·12 grains). Professor Shepard had announced the existence of itacolumite in the gold-bearing region of North Carolina, at the meeting of the American Association of Geologists and Naturalists in 1845, and under the impression that the itacolumite is their matrix, had predicted the further discovery of diamonds in that region, as in Brazil. For this reason, diamonds, when found, were naturally submitted to him. C. Leventhorpe, of Patterson, Caldwell County, N. C., reports a small and poor specimen found in a placer-mine on his property in Rutherford County, and states that he presented it to Professor Shepard, who retained it in his cabinet. The fourth important specimen was found in gold-washings in 1852, by Dr. C. L. Hunter, near Cottage Home, Lincoln County. It is said to have been an elongated octahedron of a delicate greenish tint, transparent, and about half a carat in weight. Another, said to be a very handsome white crystal of 1 carat, was obtained in the summer of the same year, at Todd's Branch, Mecklenburg County; and a beautiful black stone " as large as a chinquapin," was afterwards found by some gold-washers in the same locality. This specimen, unfortunately, was crushed with a hammer, sharing the fate of several American diamonds when submitted to the mistaken test which con-

[1] Am. J. Sci. II, Vol. 2, p. 253, Sept. 1846.

founds hardness with strength. The fragments of the black diamond scratched corundum with ease, thereby proving its genuineness. The next discovery reported is that mentioned by Professor Genth,—two diamonds, one a beautiful octahedron, from the Portis Mine in Franklin County. These specimens, as before remarked, came from localities remote from all the others, and must have been either transported a long distance by river action, or else derived from the belt of gneissic rocks that extends from Richmond to Raleigh. McDowell County has yielded two specimens, one a small crystal found some years ago on the head waters of Muddy Creek, and a much larger one, picked up on the surface in 1886, at Dysortville. This is a somewhat distorted and twined hex-octahedron of 4½ carats' weight, 10 millimeters in height and 7 millimeters in diameter, transparent, but with a grayish-green tinge of color, and is valued, for gem purposes alone, at from $100 to $150. The circumstances of its discovery are thus related : Willie Chrystie, the twelve-year-old son of Grayson Chrystie, was sent for a pail of water to a spring on the Alfred Bright farm, in Dysortville. While sitting at the spring, he saw a glistening object among the gravel, and picking it up as a "pretty trick," brought it home. It lay on a shelf almost unnoticed for a fortnight, and was then shown at the store of the village grocer. Here it became an object of general curiosity, and elicited various opinions, until the idea grew that it was probably a diamond. It was sent to Tiffany & Co., of New York, and its real character at once determined. A year later the present writer visited the spot, and fully authenticated all the facts of the discovery. The sediment in the bed of the spring was taken out and examined, and also the small hollows on the adjacent hillside. None of the ordinary associations of the diamond were observed, and hence it is probable that the crystal was washed down with decomposing rock-soil from higher ground, perhaps during some freshet ; or possibly it may have been carried to the spring by miners, and left unobserved or unrecognized among the "wash-up" of the gold-bearing sand from some neighboring placer. There are gold mines in McDowell County, worked chiefly by hydraulic sluicing, but as a rule the stones that remain in the sluices are carefully examined, as the miners know that gems are sometimes thus found. The value of the Dysortville diamond as a jewel will hardly represent the interest that attaches to it as a local specimen of large size and fine appearance.

The foregoing list includes all the authentic diamonds thus far discovered in North Carolina. A number of small stones, exhibited as diamonds, have been found at Brackettstown. They are similar to supposed diamonds found by J. C. Mills at his mine at Brindletown, but these were transparent zircon or smoky-colored quartz, the former of which has a lustre readily mistaken by an inexperienced person for that of a diamond. A number of pieces of rough diamond, exhibited as from the same section, have been decided to be of South African, not Carolinian, origin. It is to be hoped that the few legitimate discoveries actually made in this locality will not lead to deceptions, which would greatly retard any natural development of interest. It is quite possible that diamonds may be found widely distributed throughout the auriferous belt of the Carolinas and northern Georgia ; and that, in the often rude and hurried methods of gold-washing employed, they may have been overlooked in the past, and now lie buried in the piles of sand that stretch for miles along the water-courses.

It would naturally be expected that in the extension of the Piedmont region through the extreme northwestern part of South Carolina, the same possibilities of diamond discoveries would exist. The reports are few and uncertain. Mr. Leventhorpe, already referred to, has stated, in writing to the " New York Sun," that in 1883 D. J. Twitty, of Spartanburg, had a fine diamond valued at some $400, that was obtained from a place in South Carolina. He had it cut and mounted as a stud ; but it was unfortunately stolen from him while riding in a car in New York City. The loss of so interesting a specimen is much more than that of an ordinary diamond of the same gem value.

On passing into Georgia the same metamorphic belt, with its localities for gold, itacolumite, and to some extent diamonds, extends across the State to the Alabama line. The counties in which diamonds are claimed to have been found are Habersham, White, Banks, Lumpkin, Hall, Forsyth, Gwinnett, Cobb, Clayton, Bartow, Carroll, and Haralson.   Dawson, Cherokee, Milton, and Paulding, lying in the same line, and very possibly other counties adjacent to the metamorphic belt, should perhaps be included in the list.   The mode of occurrence is similar to that of North Carolina, as previously described, a few real diamonds, and many supposed ones, having been found in connection with mining for gold, in the detritus of the crystalline rocks spread along streams and placers.   From time to time glowing accounts have been published, in which Georgia is announced as the future diamond-field of the continent; but up to the present the specimens actually obtained have been few and small, and it has not been considered worth while to mine for them.   Of these diamonds interesting stories are told.   An Atlanta lady wears in a ring one of the best specimens ever found in Georgia.   Another Georgia lady would not marry until her prospective husband gave her a ring with a Georgia diamond for an engagement ring.   Several stones have been lost, and it has been found that they were destroyed by ignorant people who attempted to test them. The earliest discoveries reported were by gold-washers in Hall County over forty years ago and later in White County. Most of the specimens were found near Gainesville, in the troughs and sluices of the Hall County placers.   Two small crystals, less than ⅓ carat each, are in the cabinet of Samuel R. Carter, of Paris, Me. They are opaque and without definite form.   They were found in 1866, in the Racoochee Valley, White County, at the Horshaw placer gold-mine.   One was discovered by Dr. Augustus C. Hamlin, of Bangor, Me., and the other by H. Ashbury.   Another specimen from the same region is thus described by C. Leventhorpe, of Patterson, Caldwell County, N. C., in a letter to the "New York Sun," in August, 1883.   He says : " Numerous diamonds have been discovered in Georgia.   After the war, during the prevalence of a mining fever, a company was formed, I believe, for exploring and diamond washings.   I heard nothing further of this enterprise, and if dividends were declared the announcement escaped my notice."   There is in the writer's possession, a rough diamond taken from a " Long Tom " in White County, Ga.   It is of very perfect water and crystallization, and weighs almost a carat.   The " Long Tom " is a narrow plank trough set with a steep pitch.   An iron grating at its lower end closes it so as to form an obtuse angle.   The detritus from the gold-bearing streams is shovelled into this box, and a second operator stirs it with a shovel under a small stream of water.   The coarser gravel is thrown out, and the gold, and such small gravel as may possess a superior gravity, do not pass off with the current.   It was thus that this diamond was detained.   In April, 1887, Lewis M. Parker, a tenant of Daniel Light, found a diamond on his farm, situated three-quarters of a mile northeast of Morrow's Station, Clayton County.   The stone afterwards came into the possession of W. W. Scott, of Atlanta, who sent it to me for examination.   It proved to be an octahedral crystal weighing $4\frac{1}{11}$ carats (12·672 grains), $\frac{2}{5}$ of an inch long and $\frac{1}{4}$ of an inch wide (9 x 10 x 7 millimeters), is slightly yellow and has one small black inclusion.   It would afford a stone from $1\frac{1}{2}$ to 2 carats in weight.   Its specific gravity is 3·527.   Its surface is curiously marked with long, shallow pittings.   L. O. Stevens, of Atlanta, Ga., has communicated to the writer that a negro called on him during the past year with a 2-carat diamond, defective and of very poor color, which he had found in his garden a few miles from Atlanta.   He showed no desire to sell the stone or loan it for examination.

A book by Dr. M. F. Stephenson [1] records some of the exaggerated accounts of Georgia diamonds that have been given in good faith, but upon mere hearsay evidence, and often after years have passed.   Although diamonds have been found in Georgia, and the smaller ones mentioned are doubtless genuine, yet it is certain that in some instances Dr. Stephenson was unable to discriminate between a paste imitation and a genuine stone, and his enthusiasm may have overreached his judgment in other cases. The large specimens described were evidently quartz crystals and not diamonds.   This is almost certain as to the one mentioned which was used for a marker in a game of marbles and bore considerable concussion, as a diamond could not withstand this concussion without cleaving, whereas a rolled quartz pebble would bear a good deal of such treatment.   It is possible that quartz crystals without any prismatic faces, like those found in Arizona (hexag-

[1] Geology and Mineralogy of Georgia, Atlanta, 1871.

onal dodecahedrons), may have misled the Georgia prospectors.

Many notices have from time to time appeared, both in local
newspapers and in scientific journals, of the occurrence of diamonds in California. After making due allowance for errors and
unfounded rumors, the fact of such occurrence in certain localities
is well established ; but the number and size of the diamonds
found have not been such as to render the search for them profitable. The fact of their discovery is highly interesting, and some
of the specimens possess both elegance and value ; but as a rule
they are small and rare. The mode of their occurrence seems to
be in all cases that they are imbedded in the auriferous gravels,
and thence washed out in the search for gold. These gold-bearing gravels of California present two types of distribution : first, as
loose material in the valleys and bars of the modern streams, and,
second, as great accumulations of gravel occupying the valleys of
much larger ancient streams, and now covered with masses of lava
or compact volcanic tufa. The sides of the Sierra Nevada are
trenched with cross-valleys running down into the great, trough-
like valley of California, which lies between the Sierras on the
east and the Coast Range on the west. Along this great depres-
sion, the drainage from the mountains on both sides finds its way
to the sea through the Sacramento and San Joaquin rivers, the
former flowing from the north and the latter from the south into
the Bay of San Francisco, where a break in the Coast Range, at
Golden Gate, allows a passage between the ocean and the bay.
In the northern part of the State, where the streams from the
Sierras run down to the Sacramento, this remarkable system of
"buried river gravels" is found. In and before the tertiary
period of geology these streams had worn valleys on the slopes
of the Sierras, and made extensive deposits of gravel, by the erosion of the mountain-sides. Then came a period, or a succession
of volcanic disturbances and outflows, which made the great
"lava-beds" of northern California and Oregon. In many cases
the lavas flowed down and filled up the river-beds from side to
side, covering the gravel deposits, and in some instances hard-
ening and compacting them. The rivers have since then worn
down a new series of channels between these hard lava-streams,

and the old river gravels, with their protecting caps of volcanic rock, are now seen running out as spurs from the Sierras and forming the divides between the modern streams.   The latter have formed their own more recent gravels, from the wear, partly of the old deposits and partly of the mountain sides, as at first. The surface-diggings and placers of the early prospecting days of California were, of course, in these modern gravels and bars. , The older gravels, equally rich, are worked either by the hydraulic process, or when compacted into what are called " cement-beds," by stamp-mills.   It is in these deposits that the diamonds have been found, picked from the sluices and flumes.   , In the case of the cement-beds, only fragments are obtained, as the diamond-crystals have been crushed under the stamps.   There is much in the mode of their occurrence that recalls, at first sight, the diamond mines of Brazil and South Africa.   In Brazil the matrix is also a gravel, and is frequently cemented into a conglomerate (" cascalho ") by oxide of iron.   , In Africa the diamond gravels contain associated minerals similar to those found in some of the California placers, notably in those of Butte County, where zircons, garnets, and rutile are met with.   But these are not important relations, and afford no ground for assuming either a similar richness of yield or an identity of geological origin.

' The first recognition of diamonds in the State goes back to the early gold-seeking days of 1850.   In that year, Mr. Lyman, a clergyman from New England, was shown a crystal about the size of a small pea, with convex faces, and of a straw-colored tint. He saw it for a moment only, yet its general aspect was enough to identify it as a true diamond, and the interesting fact was published.[1] The first diamond from the Cherokee district, Butte County, was obtained in 1853.   This has since proved one of the principal localities in the State.   In 1854 Melville Attwood called attention, in a newspaper article, to the general similarity of the California deposits to the diamantiferous gravel and conglomerate of Brazil, with which he had become familiar by a residence there of some years.   He advised that search be made and care exercised, lest diamonds should pass unheeded in the gold-washings.   Since then diamonds have been reported from

---

[1] Am. J. Sci., II, Vol. 8, p. 294. Sept. 1849.

a number of points, and at present, according to Henry G. Hanks, the State Mineralogist, five counties, Amador, Butte, El Dorado, Nevada, and Trinity, are known to have yielded them. Other localities and larger numbers are yet, in his judgment, to be discovered. The hydraulic mining is in some respects a wasteful and unfortunate process, as the force of the current sweeps away the greater part of any material that does not amalgamate with the mercury; and thus many valuable substances are probably lost, such as iridosmine, platinum, and diamonds. Moreover, whatever diamonds occur in the hard cement are crushed into fragments by the stamps, and such fragments and particles are not infrequent in the tailings and sluices.

  The following is a brief summary of the principal diamond discoveries in California up to the present time, arranged by localities. At Indian Gulch, near Fiddletown, and Jackass Gulch, near Volcano, Amador County, numerous diamonds have been found. In 1867, the younger Silliman of Yale College exhibited several specimens before the California Academy of Sciences:[1] one of these, a little over 1 carat in weight (3·6 grains) was from near Fiddletown; and four others from the same region were at that time known. These stones occurred in a compact volcanic ash or tufa, forming a gray "cement-gravel." At Volcano the rock is similar, and some sixty or seventy diamonds have been reported thus far. This is one of the places where the cement-rock is worked by stamping, and the tailings show pulverized diamonds. The crushed gravel pays well in gold; and it has not been thought desirable to change the present method and break up the rock in other ways more costly and troublesome, in order to save the diamonds that it may contain. In August, 1887, Mr. Hanks exhibited before the San Francisco Microscopical Society, a beautiful stone of 1·57 carat weight (4·97 grains), found at Volcano in 1882, and belonging to J. Z. Davis, a member of the society. It is a modified octahedron, about $\frac{4}{10}$ inch in diameter, transparent and nearly colorless, though slightly flawed. The curvature of the faces gives the crystal a subspherical form, but the edges of the pyramids are channels instead of planes. Closer examination shows that the channeled

[1] See Proc. Cal. Acad. Sci., Vol. 3, p. 354.

edges, the convex faces, and the solid angles are caused by an apparently secondary building up of the faces of a perfect octahedron, and for the same reason the girdle is not a perfect square, but has a somewhat circular form. These observations were well shown by enlarged drawings. The faces seem to be composed of thin plates overlying each other, each slightly smaller than the last. These plates are triangular, but the lines forming the triangles are curved, and the edges of the plates are beveled. Mr. Hanks remarked that it could be seen by the enlarged crystals shown under the microscope, and by drawings exhibited, that each triangular plate was composed of three smaller triangles and that all the lines were slightly curved. The building up of plate upon plate caused the channeled edges and the somewhat globular form of this exquisite crystal. A close examination of the crystal revealed tetrahedral impressions, as if the corners of the minute cubes had been imprinted on the surface of the crystal while in a plastic state. These are the result of the law of crystallography, as was shown by the faint lines forming a lace-work of tiny triangles on the faces when the stone was placed in a proper light. Mr. Hanks concluded with the remark that it would be an act of vandalism to cut this beautiful crystal, which is doubly a gem, and he protested against its being defiled by contact with the lapidary's wheel. The Cherokee district, in Butte County, has been, from as early as 1853, one of the most prolific diamond localities in the State. Cherokee is near the North Fork of Feather River, and the geological relations of the diamonds and gold are essentially the same as those in Amador County, a hundred miles to the northwest, both districts lying among the western foot-hills of the Sierras, as previously described. Mr. Hanks calls attention to included leaf-impressions in the volcanic beds, as proving them to be tufas and not lavas. In number, the Cherokee diamonds obtained are about equal to those from Volcano. One was shown by Professor Silliman, on the occasion already mentioned, in 1867; and others were then known to be from that locality. William Bradreth obtained a crystal in the same year which he afterward had cut into a fine white stone of $1\frac{3}{16}$ carats. In 1873, several were obtained from the ground of the Spring Valley and Cherokee Mining Company, in cleaning up the sluices. One of these was described as large and straw-colored, while others were smaller, but very pure. Various stones, white, yellow and pink, have from time to time been reported, and some have been cut and set. A fine crystal was presented to the State Museum by Mr. Williams, Superintendent of the Spring Valley Mining Company. Two others, found at the same place in the summer of 1881, by Lucinda Voght, were shown by the present writer before the New York Academy of Sciences in 1886. Professor Silliman made the concentrations from the sluices of these Cherokee mines the subject of a minute investigation, the results of which were published in two papers.[1] In the first he describes his treatment of the material, both chemical and mechanical; and in the second he gives additional particulars, with results. He finds here the following association of interesting minerals; light-colored zircons, crystals of topaz, fragments of quartz, rutile, epidote, pyrite, and limonite, with some platinum, iridium, iridosmine, and gold, and a large quantity of black grains, which are proved by the magnet to consist about equally of chromite and titanite. At first he could find but little of the platinum and iridosmine, but this was due, as above stated, to the force of the hydraulic streams, which sweep away all small particles that do not amalgamate. Mr. Hanks adds that platinum minerals have been found rather abundantly in Butte County. At St. Clair Flat, near Pentz, they were found in quantity in the early days of placer-mining. They are found, also, at the Corbier Mine, near Magalia (Dogtown). In 1861, a diamond was found one and a half miles northwest of Yankee Hill, Butte County, in cleaning up a placer-mine. The stone was taken from the sluice with the gold, and sold to M. H. Wells, to whom I am indebted for this information. Mr. Wells presented the gem to John Bidwell of Chico, who had it cut in Boston. It weighed 1¼ carats (4·75 grains). Mr. Bidwell gave the diamond to his wife, who now wears it in a ring. This was the only diamond found in this locality. In all the northern counties of California, drained by the Trinity River, in the vicinity of Coos Bay, in Oregon, and on the banks

[1] See Mineralogical Notes on Utah, California, and Nevada, in The Eng. and Min. J., Vol. 17, p. 148, March 11, 1873, and the Am. J. Sci. III., Vol. 6, p. 127, Aug., 1873.

of Smith River, Del Norte County, diamonds are very likely to be found in the flumes and sluices. Diamonds have been found at a few points in El Dorado County. In 1867, Professor Silliman, at the meeting of the California Academy of Sciences, before mentioned, showed a crystal of 1⅜ carats (4·75 grains), of good color, though a little defective, from Forest Hill. It was found at great depth, in a tunnel run into the auriferous gravel. W. P. Carpenter, of Placerville, gives the following account of it in a letter to Mr. Hanks, in 1882: "In 1871, W. A. Goodyear, Assistant State Geologist, while examining the deposits of auriferous gravels in the ancient river bed, about three miles east of Placerville, found several specimens of itacolumite, and expressed the opinion that diamonds should be found in the gravels. I assisted him in searching for them, and we found several in the hands of the miners. Mr. Goodyear bought one of them as a geological specimen. None of the parties who had them knew what they were, but kept them as curiosities. The gravel in the channel is capped with lava from 50 to 450 feet in depth. Of late years the gravel is worked by stamp gravel mills, and I know of instances where fragments of broken diamonds have been found in panning out the batteries."

He goes on to give the particulars of about fifteen diamonds obtained at different times in the neighborhood, some yellow and some white. One of these was a nearly spherical crystal, over ¼ of an inch in diameter, that was sold in San Francisco for $300, and another was sent to England to be cut. Professor Silliman also showed to the California Academy of Sciences a very clear and symmetrical crystal from French Corral, Nevada County. It was thrown out of the cement-rock of deep gold washings, as usual, and weighed 1⅜ carats (5·11 grains). The color was slightly yellowish; but this was perhaps due to its having been exposed to a red heat, as a test of its authenticity. Prof. Josiah D. Whitney of Harvard College stated, at the same meeting, that diamonds had been found in some fifteen or twenty localities in the State, and that the largest that he had seen was from French Corral, and weighed 7¼ carats. Some small ones are reported from Trinity County; and their mode of occurrence, similar to that of the diamonds of Cherokee County and of Oregon, is described in a letter to Dr. Charles F. Chandler, of the Columbia College School of Mines.[1] Prof. Fredrich Wöhler, of Göttingen, mentions having observed in the native platinum sands of the Trinity River, Oregon, transparent zircons associated with laurite, sulphide of ruthenium and osmium, iridosmine, chromic iron, etc., and microscopic rounded crystals which he supposed were diamonds. In a subsequent communication, dated Göttingen, August 8, 1869, Professor Wöhler continues: "On examination under the microscope the mineral powder which had been freed from platinum, gold, chromic iron (in part), silica, iron and tin, and from which the ruthenium, etc., had been removed by aqua regia, besides many grains of chromic iron and beautiful hyacinth crystals, colorless and transparent grains resembling quartz were observed; but besides these, grains resembling rounded diamond crystals were detected." He then describes in full his methods of testing these grains, and expresses his conviction that they were true diamonds./

A few small diamonds have been found in the placer diggings of Idaho, of about the same quality and occurring under the same conditions as those in California. In neither region have they been made the object of special search, those found having been picked up by miners while washing the gravel for gold. Fragments of diamonds have been noticed in the tailings from the quartz mills, being the remains of stones broken under the stamp. About twenty years ago, quite an excitement prevailed for a time over Idaho diamonds. Local and mining papers, during the latter part of 1865 and the spring of 1866, had many references to the reported or anticipated diamond-yield of that territory. Small crystals, answering to all the usual tests, were said to be abundant in a tract of country some forty miles square, between Boise City and Owyhee. After a few months the excitement subsided, and the ordinary quartz-crushing industry resumed its sway over the attention of the people and the press.

In the latter part of 1883, an octahedral diamond is said to have been taken from a placer claim called Nelson Hill, near

[1] Chemical News, Am. Ed., Nov., 1869, and Am. J. Sci., II., Vol. 48, p. 44, Nov., 1869.

Blackfoot, Deer Lodge County, Mont. It was panned out by a Chinaman, who handed it to Edward Mason, one of the owners of the claim. The latter did not regard it as of any particular value and left it lying about his cabin. Afterward, while on a visit to Helena, he showed the stone to a jeweler, who examined it and made several offers to purchase it. These Mr. Mason declined, suspecting that the stone was of greater value than he had imagined. He subsequently came to New York, and submitted it to a diamond-broker, who pronounced it a true diamond. According to a recent article in the Butte "Inter-Mountain," the stone is retained in its natural state by Mr. Mason.

A few years ago reports were started of the finding of diamonds in central Kentucky. Prof. Edward Orton, the State Geologist of Ohio, made a visit to that district, and found that it presented certain resemblances to the diamond-bearing region of South Africa. He found dykes of trap-rock (peridotite) breaking through fissures in shale, and spreading to some extent over the adjacent country. Garnets and other associated minerals derived from the decomposition of the peridotite were found, suggesting the possibility of a diamond yield, from the similarity of the conditions to those of Africa. Similar investigations and results were reported by A. R. Crandall.[1] It had been previously suggested by E. J. Dunn, E. Cohen, H. Huddleston, and Rupert Jones that the South African diamonds were formed in a sort of volcanic mud. Mr. Huddleston thought that the action was hydrothermal rather than igneous, the diamonds resulting from the action of steam in contact with magnesian mud, under pressure, upon carbonaceous shales, and compared the rock to boiled plum-pudding.

At a meeting of the Manchester Literary and Philosophic Society, held in October, 1884, Sir Henry E. Roscoe presented a paper on the diamond-bearing rocks of South Africa, in which he said that he had noticed a peculiar odor, somewhat like that of camphor, which was evolved on treating the soft "blue" diamond earth with hot water. He powdered a quantity of this earth and digested it with ether ; and after filtering and evaporating, he ob-

[1] Note on the Peridotite of Elliott County, Ky., Am. J. Sci. III., Vol. 32, p. 121, Aug. 1886.

tained a small quantity of strongly aromatic crystalline body, volatile, burning easily with a smoky flame, and melting at about 50°C. Unfortunately the quantity obtained was too small to admit of a full investigation of its composition and properties. He suggested that perhaps the diamond was formed from hydrocarbon simultaneously with this aromatic body. Prof. H. Carvill Lewis, at the meeting of the British Association at Birmingham, in September, 1886, in a paper "On the Genesis of the Diamond," stated that from the De Beers' Mine, in South Africa, at a depth of 600 feet, there had been sent him specimens of unaltered rock which proved to be peridotite containing carbonaceous shale. He added that information received from New South Wales, Borneo, and Brazil led him to believe all diamonds to be the result of the intrusion of a peridotite through carbonaceous rocks and coal seams. The similarity of the South African peridotite to that described by Joseph S. Diller in Kentucky [1] led Professor Lewis to suggest interesting possibilities as to the occurrence of diamonds there; and on the invitation of Prof. John R. Proctor, State Geologist of Kentucky, in the summer of 1887, Mr. Diller and the writer were sent, by Major John W. Powell, the Director of the United States Geological Survey, to make an investigation. The locality is easily reached by way of the East Kentucky Railroad, which ends in Carter County, at Willard, where conveyances may be obtained of the farmers for the remaining ten miles. The best exposures of the peridotite occur along Ison's Creek, in Elliott County. The peridotite alters and disintegrates readily, but because the declivity of the surface here is considerable, the transportation of material almost keeps pace with disintegration, and there is no great accumulation of residuary deposits upon the narrow divides and hillsides. The specific gravity and durability of the gems found in connection with peridotite are generally greater than of serpentine and other products of its alteration. On this account they accumulate upon the surface, and in favorable positions along adjacent lines of drainage. The plan followed was to search by sifting and carefully panning the beds, receiving the drainage directly from the surface of the peridotite, and to enlist the services of the people

[1] Am. J. Sci. III., Vol. 32, p. 121, Aug., 1886.

in the neighborhood to examine the steep slopes where gems that had weathered out of the peridotite might be exposed. Particular attention was directed also to the study of the solid rock and residuary deposits which so closely resemble the diamantiferous material of South Africa. Till this time the actual contact of the peridotite and shale had not been observed. It is exposed in the bed of a branch of Ison's Creek, within 100 yards of Charles Ison's house. The intrusion of the peridotite has displaced and greatly fractured the shale, besides locally indurating it and enveloping a multitude of its fragments. The latter are dark-colored, like the peridotite, and are strongly contrasted with the light-colored dolomite nodules of secondary origin. Besides pyrope garnets, a few of which are suitable for cutting, several fairly good specimens of pyroxene were found here, resembling the same transparent mineral from Arizona. The South African mineral is a little more opaque, but of a richer green color. When suitably prepared, they will make worthy additions to the gem collection of the United States National Museum. An altered biotite also occurs, identical with the South African vaalite. During a careful search over a small area for nearly two days, no diamonds were found, but this by no means demonstrates that they are not there. The similarity between the peridotite here and that of the Kimberley and other diamond mines of South Africa is very striking; and when this fact alone is considered, the probability of finding diamonds in Kentucky seems correspondingly great; but when it is noted that the carbonaceous shale, and not the peridotite itself, is the source of the carbon from which the diamond is formed, and that the shale in Kentucky is much poorer in carbon than that of South Africa, the probability is proportionally diminished. Recent excavations have shown that large quantities of this shale surround the South African mines, and that they are so highly carbonaceous as to be combustible, smouldering during long periods of time when accidentally fired. In the chemical laboratory of the United States Geological Survey, J. Edward Whitfield found 37·52 per cent. of carbon in the shale from near the Kimberley Mine, while in the blackest shale adjoining the peridotite of Kentucky, he found only ·68 per cent. of carbon. The peridotite at the time of its intrusion must have been forced up through a number of coal beds, and at a greater depth it penetrated the Devonian black shale, which is considerably richer in carbon than the shale now exposed at the surface. It is quite possible, if the theory of the origin of diamonds proposed by Professor Roscoe and independently advanced by Professor Lewis be true, that a number of diamonds may have been formed in the Kentucky peridotite; but the general paucity of carbon in the adjacent rock is certainly discouraging to the prospector. The best time to search for gems in that locality is immediately after a heavy rain, when they are most likely to be exposed upon the surface. It is proposed to continue the search economically, by furnishing to responsible persons living in the vicinity a number of rough diamonds mounted in rings, for comparison, that they may know what to look for under the most favorable circumstances.

The " Jewelers' Review " for June, 1888, gave an account of a diamond from Russell County. It is described as a small octahedron, with curved faces, lustrous and nearly white, though with a yellow tinge, and weighing $\frac{1}{16}$ of a carat. It was found in a gravelly field on the top of a hill some 300 feet above Cabin Fork Creek. The country rock is said to be composed of granite dykes, slates, and some floating rocks, such as quartz, feldspar, magnetic iron ore, flint, garnet, etc., mingled in clayey hills. The rocks near Montpelier, Adair County, Ky., belong either to Keokuk or to the St. Louis group, probably to the former. From the absence of any direct geological information concerning the two counties, they have been referred to these groups by Professor Proctor.

Various reports of the discovery of diamonds in different parts of the country are from time to time published by local papers; but they generally prove to have been written without exact information as to the character of the stone, or for speculative purposes. A few of these reported diamonds will be referred to, of which only the following are known to be genuine. Two diamonds have been on exhibition for several years at the store of Frederick N. Herron, Indianapolis, and are reported by him to have been found at some locality in Indiana. They are perfect elongated hexoctahedrons of 2 carats each. The stones are genuine diamonds, but the particulars of their occurrence and discovery have not been obtained, and therefore nothing definite can be stated regarding them. J. D. Yerrington of New York city has had a brown diamond weighing 1 carat, that will yield, when cut, a gem weighing ⅓ carat, which was found near Philadelphos, Ariz. Two pieces of blue bottle-glass that had been rolled so as to lose all form, were naturally supposed by the finder to be sapphires, being in the same locality with the diamond. It is stated that three diamond crystals were obtained many years ago on Koko Creek, at the headwaters of the Tellico River, in East Tennessee, on the "bench lands" of the Smoky or Onaka Mountains. If this statement be correct, it probably points to a western extension of the diamond-belt of North Carolina, or to the transportation of the stones thence by streams.

In 1884, quite an excitement was aroused in Wisconsin by a reported diamond-discovery at Waukesha, in that State. A Milwaukee jeweler purchased for $1 from a lady, a stone which he stated was a topaz. It was said to have been found at a considerable depth, in digging a well on the property of the lady's husband, at Waukesha, some years before. Subsequently it was thought to be a diamond, and as the first ever found in Wisconsin was valued at a high price and made the basis of much local excitement and speculation. The land where it was found was purchased at an increased price and two other small diamonds were produced as from the same locality. The gravel in which they were claimed to occur was simply the ordinary glacial drift of the whole region, and the diamonds have the aspect of being African stones. In 1888, it was announced that a fine and large diamond of over 80 carats had been found by a laborer while attending a bowlder-crushing machine in Cincinnati. The theory was advanced that it might be the stone lost in 1806, at Blennerhassett Island, by Mrs. Clark, and described by Aaron Burr in a letter to his daughter. The story lacks foundation. Another instance is that of a stone, supposed to be a diamond, found in working for coal a few years since at Ponca,

Neb. Great excitement was aroused, but the stone proved not to be a diamond.

The well-known "Arizona diamond swindle" was a clever one, and its locality could hardly have been better selected; but it should not have received so much credence, since gem minerals are so readily recognized through their local characteristics by all collecting mineralogists. A few words in regard to this celebrated swindle may not be amiss. Twenty years ago fabulous stories were circulated about the richness of New Mexico and Arizona. Companies with high-sounding titles were organized to collect not only the diamonds, but the rubies, sapphires, emeralds, and other gems, which were said to abound there. In 1870, a large scheme for this purpose was originated. It was represented in San Francisco that Philip Arnold and John B. Slack had found diamonds and other precious stones in great quantities in a certain Territory of the United States. Among some of the objects shown in confirmation were 80,000 carats of so-called rubies and a large number of diamonds, one of 108 carats weight. These gems were all displayed for the inspection of those interested in the scheme, and were deposited with the Bank of California for safe-keeping. Subsequently the same parties again visited the fields, which were reported to lie somewhere in New Mexico, and returned with another lot of stones, not so large as the former ones, however. It was remarked at the time that one could scarcely expect to pick them up by the bushel. Heavy capitalists on both coasts soon became deeply interested, and on May 10, 1872, a bill was passed by Congress in the interests of the diamond-miner. Finally, a party composed of representatives of both the East and the West, with a mining expert, a graduate of the Royal School of Mines, Freiberg, Saxony, chosen by the investors, started out prospecting, equipped for a sixty days' expedition. They left Rawlins, Wyo., May 28th, first taking a southwestern course, then a northwestern course, until some of the party thought that they had missed their way, and began to doubt the truth of the discovery. But when the mountain was reached, the promised Golconda, every one picked up gems, and hope rose correspondingly. In a week 1,000 carats of diamonds and 6,000 to 7,000 carats of rubies were gathered, and the party returned, well pleased with their success.   Another expedition, setting out late in the season, failed to reach the fields, and was abandoned.   On hearing of the failure of the third party, Clarence King, Director of the United States Geological Survey, started on his famous expedition, which proved that the whole affair was a humbug and that the mine had been "salted."   The "rubies" were shown to be ordinary garnets, and the 108-carat diamond proved to be a piece of quartz.   It was ascertained that an American had purchased a large quantity of rough diamonds in London, regardless of their value, and so plentiful was the salting that some years afterward diamond crystals were still found there.   A number gathered by a shoemaker are still in the cabinet of Prof. Joseph Leidy, of the University of Pennsylvania, in Philadelphia.   So carefully was the swindle planned that an eastern expert was only shown a paper of cube diamonds, a form quite rare, and peculiar to Brazil.   About $750,000, taken principally from capitalists on the California coast, was realized by the promoters of this gigantic fraud.   Had the company employed a competent gem-expert or a gem-collecting mineralogist, no such swindle could have occurred.   The expert retained by the investors was himself deceived, and this fact, of course, greatly facilitated the fraud.

To insure the finding of diamonds in a new district, one of the best methods is to familiarize the searchers with their lustre. This can readily be accomplished, and was once partly carried out by Dwight Whiting, of Boston.   He has suggested selling to the miners small, imperfect diamond crystals (bort), mounted in a very inexpensive manner, so that the entire ring or charm could be sold at from $5 to $10.   Several thousand searchers thus prepared would soon ascertain whether diamonds really existed, and the crystal would also serve for testing the hardness as well as the lustre of the stone.

A geologist of North Carolina conceived the happy idea of interesting the children of his vicinity in the search for minerals. A trifling reward was sufficient to awaken a keen interest, so that healthful exercise certainly, and often valuable specimens, were the result of his plan.   Some of the series of modified quartz crystals described by Prof. Gerhard von Rath, as well as the beautiful rutiles, emeralds, and other minerals that we are now familar with, we owe to the industry and sharp sight of these children.   It would aid much in the development of new mineralogical fields if this plan of Mr. Stephenson's could be widely introduced. ιOne of the minerals most likely to be mistaken for the diamond is a form of small quartz crystal found principally at Santa Fé and Gallup, N. M.; Fort Defiance, Ariz.; Deadwood, Dak.; and Shell Creek, Nev.   These crystals range in size from 1 to 5 millimeters and the prism is nearly or entirely obliterated. In addition to this, as a rule, the surface is slightly roughened, and by an inexperienced person such a crystal is easily mistaken for an octahedron, which is almost universally considered to be the only diamond shape.

CHAPTER II.

Corundum, Sapphire, Ruby, Oriental Topaz, Oriental Emerald, Diaspore, and Spinel.

CORUNDUM is nearly pure alumina (Al₂ O₃), and is found in almost all colors of the rainbow. The transparent varieties rank among the most valuable of gems. The names, ruby, sapphire, oriental amethyst, oriental emerald, and oriental topaz are given to the transparent red, blue, purple, green, and yellow varieties of the mineral. These colors are due to the addition of minute quantities of metallic oxides to the alumina. Its specific gravity varies from 3·97 to 4·05, and its hardness is 9; that of the ruby is generally about 8·8. The finest pigeon's-blood-colored rubies are found at Mandalay in Burmah, where mines have recently been leased by a London syndicate. Fine rubies, which are generally small, sometimes of a pink color, and often with a currant-wine or purplish tint, are found at Ratnapoora in Ceylon ; likewise in Siam, where, however, the color is most commonly a dark red, almost that of a garnet, often with a tinge of brown. The finest sapphires are found in Burmah or Ceylon. Some of the finest cornflower blue varieties are from Ceylon. Many of the rich velvety blue, as well as the lighter-colored stones, are from the Simla Pass in the Himalayàs. Fine sapphires have recently been found in Siam and in Australia, the latter generally of an opaque, milky-blue color.

Corundum is a mineral of great importance, though not of frequent occurrence; in combination, however, especially with silica, alumina enters into a vast number of mineral species and varieties. ⅃ Its great hardness gives it value as a polishing material, and as such it has no substitute. It is found in the United States, chiefly in the crystalline rocks along the Appalachian Mountains, from Chester, Mass., to northern Georgia, and also in Montana. At Chester, where the deposits have long been known and worked, the mineral consists chiefly of emery, which is corundum mixed with magnetite, and somewhat softer than corundum alone. No gems have been found here. At Pelham, Mass., corundum in small quantities has been recognized, and Prof. Charles U. Shepard[1] found asteriated crystals in nodules of cyanite at Litchfield, Conn., also at Norwich, Conn., where he found small blue crystals enclosed in fibrolite. It is likewise found in the metamorphic rocks of the Highlands of New York and northern New Jersey. At Vernon, N. J., forty years ago, crystals of sapphire and ruby corundum were found, but always opaque, so that, while many specimens were obtained from this locality, some of which have been cut, it is probable that none of them has furnished a transparent gem.

It is of interest to know that rubies from Mandalay, Burmah, occur in similar association with limestone; hence they are generally found detached and separated from their original matrix. Some handsome cabinet specimens, showing asterism, have been obtained from Delaware, Chester, and Lancaster Counties, Pa.; few, however, were suitable for cutting. Crystals have been found in Virginia, in Louisa County, and near Staunton, Augusta County.

The great corundum region is in the crystalline rocks of North Carolina, where in Madison, Buncombe, Haywood, Jackson, Macon, and Clay Counties, numerous localities are known. A second and a third line of localities are recognized, but they are of slight importance. According to Thomas M. Chatard,[2] of the United States Geological Survey, the corundum region extends from the Virginia line through the western part of

[1] Report on the Geological Survey of Connecticut, p. 64, New Haven, 1837.
[2] Mineral Resources of the United States, p. 714, 1883–1884.

South Carolina, and across Georgia as far as Dudleyville, Ala. Its greatest width is estimated to be about one hundred miles. This belt is frequently referred to as the chrysolite or chromiferous series, owing to the presence of chrysolite and chromium, from which corundum is believed, by certain authorities, to have been derived by alteration.[1] In this decomposed and altered chrysolite, throughout the Southern States, corundum is found in place. Dr. J. Lawrence Smith says that "outside of serpentine it has not been found," while Professor Shepard says that it occurs "only in a single formation, which may be designated as chrysolite rock, though from its color and some other peculiarities it has often been confounded with serpentine." Charles W. Jenks gives the following account of the Culsagee locality, which is typical of most Southern deposits. He says : "The aspect of the ridge is somewhat barren, like that of all the corundum and emery localities with which I am acquainted in any part of the world. The granite rocks which make up the principal masses of the mountains have been fissured with a large dyke of chrysolite and serpentine, in which the corundum-carrying veins are found. These veins traverse the dyke, and are mainly composed of chlorite and chloritic minerals, carrying with them corundum in massive and crystal forms. The veins are five in number, dip to the northwest at an angle of 45°, and contain the mineral, in size from microscopic crystals to those of from 1 to 500 pounds. The two varieties of chlorite known as ripidolite and jefferisite form the usual vein-gangue or matrix of the mineral. Some gem masses were in their native matrix of ripidolite between hanging walls and foot walls of serpentine ; others, from the size of a hen's egg to a fifty-pound shot, were found locked up in geodes of chlorite ; others still in pockets of partially changed or decaying schists of mica or talc."[2] It is believed by some that corundum is derived from the breaking up of alumina compounds, especially hydrates, like the minerals diaspore and bauxite. Professor Genth, in his monograph on corundum, refers to a locality near Friendship, in Guilford

[1] See, Corundum : Its Alterations and Associated Minerals, by Frederick A. Genth, in Contributions from the Laboratory of the University of Pennsylvania, No. 1. Philadelphia, 1873.
[2] Corundum and Its Gems. A Lecture before the Society of Arts, Boston, 1876.

County, where corundum is found associated with titaniferous iron ore. In other localities, in Gaston and Rutherford Counties, the corundum was found in a series of slates, and was regarded by Prof. Ebenezer Emmons, Chief of the North Carolina Geological Survey, as belonging to the Taconic system. At these places it is found associated with pyrophyllite, rutile, damourite and lazulite. Professor Genth says: "There are reasons to believe that the pyrophyllite beds in Orange, Chatham, Moore and Montgomery Counties are analogous to the corundiferous strata of Gaston County, and the same appears to be true for those at Graves' Mountain, Lincoln County, Ga." At this locality there is also to be found lazulite with rutile as well as at Crowder's Mountain in Gaston County, N. C. The earliest reference to corundum in this country is found in Silliman's Journal for 1819,[1] in an article on the mineralogy and geology of parts of South and North Carolina, by John Dickson, who sent a number of specimens to illustrate the paper. Among these was one nearly an inch in length and very like the East Indian specimens, which Prof. Benjamin Silliman, Sr., of Yale College, recognized as a very perfect hexagonal crystal of blue corundum. The locality from which it came was subsequently found to be near Andersonville, Laurens District, S. C., and it has lately yielded a large amount of corundum mingled with zircon. The Massachusetts emery deposits near Chester were first described by Dr. Charles T. Jackson[2] and later by Professor Shepard[3] and Dr. Smith.[4] The Connecticut localities were described by Professor Shepard, and that at Pelham, Mass., by J. H. Adams, a few years later; meanwhile the Pennsylvania corundum, and that of Vernon, N. J., and Orange County, N. Y., had been found. Dr. Smith writes[5] that this mineral was first discovered in North Carolina in 1846, but does not specify where or by whom. Professor Shepard, in 1872, states[6] that he had received an hexagonal prism

[1] Am. J. Sci. I., Vol. 3, p. 4, 1819.

[2] Am. J. Sci. II., Vol. 39, p. 88, Feb., 1865, and the Proc. Boston Soc. Nat. His. for 1864.

[3] Am. J. Sci. II., Vol. 40, p. 112, Aug., 1865; Vol. 42, p. 42, Nov., 1866; Vol. 64, p. 256, Oct., 1868.

[4] Am. J. Sci. II., Vol. 42, p. 83, Aug., 1866.   [5] Am. J. Sci. III., Vol. 6, p. 180, Sept., 1873.

[6] Am. J. Sci. III., Vol. 4, p. 175, Sept., 1872.

of ruby-red color, "upwards of twenty years ago," from a gentle-man of Macon, Ga., who said that it came from a gold mine in Habersham County of that State. The specimen was apparently a loose crystal that had been washed down to the placers east of the Blue Ridge. About the same time Gen. Thomas L. Cling-man sent him several pounds of a coarse blue sapphire broken from a large crystal " picked up at the base of a mountain on the French Broad River in Madison County, N. C."

This is probably the same discovery as that in 1846 or 1847, for at that time Madison County was part of Buncombe County. Dr. C. L. Hunter discovered the Gaston County corundum, and Professor Emmons refers to it in his report on the midland coun-ties of North Carolina in 1853.[1] The civil war began soon after, putting a stop to further research, and it was not until its close that investigations were resumed.

Rev. C. D. Smith, of Franklin, N. C., who had served as an assistant to Professor Emmons on the State Geological Survey, discovered most of the important localities in North Carolina. In 1865 a specimen was brought to him from a point west of the Blue Ridge which he recognized as corundum; he visited the lo-cality, found the mineral, collected specimens, and announced the occurrence. This was the origin of the mining industry now so valuable. These discoveries led to further exploration, and many localities were found in the same region which have since been more or less developed. The principal deposits that are now worked are the Jenks, Lucas, or Culsagee Mine; Corundum Hill Mine, near Franklin, Macon County, N. C.; the Buck Creek or Cullakenee Mine in Clay County, also at Laurel Creek in Rabun County, Ga., and near Gainesville, Hall County, Ga. The Jenks Mine is on the Culsagee or Sugartown fork of the Ten-nessee River. Its two names are derived from the locality and from the name of its first operator, Charles W. Jenks, of Boston, Mass. Prof. Washington C. Kerr, State Geologist of North Carolina, placed the mica-bearing rocks in the upper part of the Laurentian series, identifying them provisionally with those called by Dr. T. Sterry Hunt, Montalban. Thomas M. Chatard, of the United States Geological Survey, has described quite fully the occur-

[1] Am. J. Sci. II., Vol. 15, p. 373, May, 1853.

rence of corundum[1] at the Culsagee and Laurel Creek localities, both of which are now operated by the Hampden Emery Company, of Chester, Mass. The Culsagee outcrop, covering some thirty acres, consists of chrysolite (dunite) mingled with hornblende. The corundum is enclosed among various hydromicaceous minerals, commonly grouped, under the term chlorite, between the gneiss and the dunite, from the alteration of which they have evidently been formed. It occurs chiefly in crystalline masses, often of considerable size, and sometimes suitable for gems. At other parts of the mine it is found in small crystals and grains mingled with scales of chlorite, forming what is called the "sand vein." This is so loose and incoherent that it is worked by the hydraulic process; and the small size of such corundum is the saving of much labor in the next process of pulverizing. The Laurel Creek Mine is similar in character. At Buck Creek the chrysolite rocks cover an area of over 300 acres, and from that point southward the hornblende rocks assume greater proportions, being associated with albite instead of the ordinary feldspar and forming an albitic cyanite rock. There is also found here the beautiful green smaragdite, called by Professor Shepard chrome arfvedsonite, which, with red or pink corundum, forms a beautiful and peculiar rock curiously resembling the eclogite or omphacite of Hoff, in Bavaria, Germany. At Shorting Creek in Clay County and in Towns County, Ga., there are also corundum localities. The resemblance in the occurrence of the North Carolina corundums to that of Mramorsk in the Ural Mountains, as described by Prof. Gustav Rose of the University of Berlin, has been shown by Professor Genth.[2] There the associated species are serpentine and chlorite schist, sometimes with emery, diaspore, and zoisite, very similar to the chrome serpentine corundum belt of the Southern States. The emery deposits of Asia Minor and the Grecian Archipelago, according to Dr. J. Lawrence Smith,[3] yield that substance in marble or limestone, overlying gneissic rocks; while with it are associated many of the same hydromicaceous and chloritic species that accompany both the New England emery and the Southern corundum.

[1] Mineral Resources of the United States, p. 714, 1883–1884.
[2] Contributions to the Laboratory of the University of Pennsylvania, No. 1, 1873.
[3] Am. J. Sci. II., Vol. 10, p. 355, Nov., 1850; and Vol. 12, p. 53, Jan., 1851.

With more particular reference now to the actual gems yielded at these various localities, we may note that they occur in two distinct forms : first, as crystals, of which the usual forms for sapphire are doubly terminated hexagonal pyramids, often barrel-shaped by the occurrence of a number of pyramidal planes of successively greater angle; and second, as nodules of purer and clearer material, in the midst of larger masses of ordinary cleavable corundum. The latter, when broken or falling out, are sometimes taken for rolled pebbles, which they resemble.

In 1886, a London periodical made the statement that any one who found the sapphire or the ruby in its original matrix would be called the "King of Rubies," and that his fortune would be assured. This recalls the fact that Charles W. Jenks, of Boston, was the original finder of the true corundum or sapphire gems in place in the Jenks Mine at Franklin, N. C., and that he obtained from this locality nearly all the fine crystals of the best American collections. One of the most interesting of these is a piece of blue crystal with a white band running across it and a place in the center where a nodule had dropped out. This piece was cut and put back in its place, and the white band can be seen running across both gem and rock. (See Colored Plate No. 1). Nearly all the fine gems from Franklin, N. C., were brought to light by Mr. Jenks' mining; but although found here in their original matrix, they were of such rare occurrence that it was found unprofitable to mine for them alone. The work was suspended for some time in consequence of the financial crisis of 1873, but has lately been resumed by the Hampden Emery Company, as mentioned, who now own the mines, and are operating them for corundum under the direction of Dr. S. F. Lucas, whose name has been given to the mine at Culsagee, formerly called after Mr. Jenks. What success in gem-discovery is at present attained, it is not easy to learn. Certainly but few gems have appeared in the market of late from that locality.

The largest crystal ever found, which is five times larger than any other known crystal, is one early discovered by Mr. Jenks and described by Professor Shepard. It is now in the cabinet at Amherst College; but much injured by the disastrous fire of 1882, which destroyed so many fine specimens of the Shepard Collection there. \ This crystal weighs 312 pounds, is perfectly terminated, partly red and partly blue in color, but opaque. (See Illustration). Another large crystal, also obtained by Mr. Jenks and purchased by Professor Shepard, weighed 11¼ pounds. These two specimens are more fully described as follows: The largest is red at the surface, but of a bluish-gray color within. The general figure is pyramidal, showing, however, more than a single six-sided pyramid, whose summit is terminated by a rather uneven and somewhat undefined hexagonal plane. The smaller crystal is a regular hexagonal prism, well terminated at one of its extremities, the other being drusy and incomplete. The general color is a grayish-blue, though there are spots, particularly near the angles, of a pale sapphire tint. Its greatest breadth is 6 inches and its length over 5. Some of the lateral planes are coated in patches with a white, pearly margarite. Only the smaller crystals found at Franklin furnish material suitable for use in jewelry. They are frequently transparent near their extremities, so that small gems can be cut from them; but scarcely any of those thus far obtained are worth $100 and not 100 have been found in all.

\ In variety of color the North Carolina corundum excels; it is gray, green, rose, ruby-red, emerald-green, sapphire-blue, dark-blue, violet, brown, yellow of all intervening shades and colorless. Many specimens have been cut and mounted, especially of the blue and red shades, and make good gems, though not of the choicest quality. The two finest rubies are in the collection of Clarence S. Bement, of Philadelphia, in a suite of the choicest crystals found at the Culsagee Mine. Among these is probably the finest known specimen of emerald-green sapphire (oriental emerald). It is the transparent part of a crystal of corundum, 4 x 2 x 1¼ inches, from which could be cut several pieces that would together furnish from 80 to 100 carats of very fine, almost emerald-green gems (not too dark, as in the Siamese), the largest possibly 20 carats or more in weight. As its color is one of the rarest known, it makes this specimen a very valuable one. There is in this collection a beautiful crystal of yellow and blue in consecutive bands (see Colored Plate No. 1), from which it is estimated that at least $1,000 worth of gems could be cut. A dark-blue stone of

1 carat weight is in the United States National Museum at Washington, and a series of fine red and blue crystals have been deposited there by S. F. Lucas. In the collection disposed of by Prof. Joseph Leidy, of Philadelphia, a few years ago, were several gems from the same mine, including a wine-yellow sapphire of 3¼ carats (660 milligrams); a violet-blue stone of a little over 1 carat (215 milligrams); and three dark-blue ones weighing respectively about 1½ (320 milligrams), 1¼ (250 milligrams) and ¾ (145 milligrams) carats each. In Professor Genth's suite of corundums are some from North Carolina and Pennsylvania that would afford opalescent stones with fixed stars and other interesting forms. Many fine examples of corundum from Pennsylvania are in the cabinets of W. W. Jefferis, now of Philadelphia, Lewis W. Palmer, of Media, and Dr. Cardesa, of Claymont. Specimens from Pennsylvania and North Carolina are also to be found in the cabinets of Joseph Wilcox and Dr. Isaac Lea, and in the William S. Vaux cabinet at the Philadelphia Academy of Natural Sciences. Near the Franklin, N. C., locality there has been obtained a considerable amount of a brown variety of corundum,[1] which shows distinct asterism, both by artificial light and in the sunlight, when the stone is cut en cabochon. A similar variety, though of darker brown, with a bronze-like reflection, has also been found, some twelve miles from Franklin, by Mr. Chatard. These all show a slight bronze play of light on the dome of the cabochon in ordinary light, and under artificial light they show well-defined stars, being really asterias or star-sapphires, and not cat's-eyes, as might seem at first sight to be the case. Similar light-brown corundums, showing asteriation and cleavage faces of the crystals, are found in Delaware County, Pa. A fine opalescent variety of deep indigo color is reported by E. A. Hutchins, as obtained by him from near Franklin and elsewhere in Macon County. Red and pink corundum is found at the Cullakenee Mine, in Buck County, and also at Penland's, on Shooting Creek, in Clay County. From the former locality there is a fine ruby-colored specimen in the cabinet of the Philadelphia Academy of Natural Sciences, and in the Vaux Collection a remarkable black crystal, the locality given for which is Buncombe County.

[1] Transactions New York Academy of Sciences, p. 52, Jan., 1884.

Among other varieties found at various points in North Carolina, the following may be noted : Two miles northeast of Pigeon River, near the crossing of the Asheville road, in Haywood County, and two miles north of this, on the west fork of Pigeon River, at Presley Mine, are found some of the finest colored specimens of blue and grayish-blue corundum. Twenty miles northeast of this, at the Carter Mine, fine white and pink corundum occurs in crystals and in a laminated form. Blue, bluish-white and reddish varieties occur at Swannanoa Gap, Buncombe County. J. A. D. Stephenson found fine hexagonal prisms of a pale brownish corundum at Belt's Ridge and more recently some very fine, fair colors from several new localities near Statesville. Fine crystals have also been found in the Hogback Mine, Jackson County.

↖ The chief locality for gem-sapphires in the United States is near Helena, Mont., where they occur as loose crystals, usually small, but often transparent and of good colors. They are found on bars in the Upper Missouri River, more or less rolled among gravel, and in the riffles and sluices of the gold-washers, with the gold, garnets and other heavy minerals of the placer mines. Dr. J. Lawrence Smith was the first to describe these Montana sapphires, as follows : " These pebbles are found on the Missouri River near its source, about sixty-one miles above Benton ; they are obtained from bars on the river, of which there are some four or five within a few miles of each other. In the mining region of this territory considerable gold is found on these bars, it having been brought down the river and lodged there, and the bars are now being worked for gold. The corundum is scattered through the gravel (which is about 5 feet deep) upon the rock bed. Occasionally it is found in the gravel and upon the rock bed in the gulches, from 40 to 50 feet below the surface, but it is very rare in such localities." [1]

It is most abundant upon the Eldorado bar, situated on the Missouri River about 20 miles from Helena, where, at one time, a man could collect from 1 to 2 pounds a day. Some of these have been cut, and one very perfect stone of 3½ carats and of good green color, almost equal to the best oriental emerald, has

[1] Am. J. Sci. III., Vol. 6, p. 185, Sept., 1873.

been obtained. The Montana specimens rarely exceed ⅓ inch to ½ inch in length. (See Colored Plate No. 1.) They are brilliant but usually of pale tints. Two gems are in the Amherst College Collection, which weigh about ⅓ carat each. One is a true ruby-red, and the other a sapphire-blue, colors rarely met with here. The gems are usually of a light-green, greenish-blue, light-blue, bluish-red, light-red and red, and the intermediate shades. They are usually dichroitic, and often blue in one direction and red in another, or when viewed through the length of the crystal, and frequently all the colors mentioned will assume a red or reddish tinge by artificial light. A fine one of 9 carats was found of a rich steel-blue. A very beautiful piece of jewelry, in the form of a crescent, was made of these stones by Tiffany & Co., in 1883; at one end the stones were red, shaded to bluish-red in the center, and blue at the other end; by artificial light the color of all turned red. Perfect gems of from 4 to 6 carats each are frequently met with. Occasionally crystals are found which would afford ruby and sapphire asterias of a poor quality. The value of the gems cut from material found in this district amounted at one time to fully $2,000 a year. Many are found that are never cut, for it requires greater skill, involving much higher cost, to cut sapphire, than gems which are less hard. In the latter part of 1889 specimens were shown to the writer of a trachyte rock, imbedded in which were well-defined crystals of sapphire, similar to those found on the Eldorado bar, from a dyke on the Missouri River near and above that locality. The sapphire on Eldorado bar evidently came from this rock, and, on its disintegration, was washed down the river.

CORUNDUM

Silica	3.28	Lime	1.99
Alumina	85.75	Water	1.37
Ferric Oxide	4.26	Color	red to gray
Titanic Oxide	2.74	Locality	Shimerville, Pa.
Magnesia	trace	Analyst	Edgar F. Smith [1]

Sapphires are obtained to a limited extent in Colorado. William B. Smith states, in the "Proceedings of the Colorado Scientific Society," that near Calumet, about twelve or fourteen miles from Salida, corundum is found in what has proved to be a corundum schist. The crystals are in flat hexagonal plates,

[1] Am. Chem. J., Vol. 5, p. 272.

have a bluish tinge, in some cases quite deep, and are from 1 to 5 millimeters (·039 to ·196 inch) in thickness. Hoffman mentions corundum occurring in fragments near Silver Peak, Nev.[1] Rubies and sapphires have been erroneously reported to be found in the surface sands and gravels of Arizona and New Mexico, associated with the pyrope garnet.

SPINEL

crystallizes in the isometric system, and is generally found in the form of octahedrons. Its hardness is 8 and its specific gravity about 3·65. Following the order of the rainbow, it exists in all shades of red, orange, green, blue, and indigo, as well as white and black. The crimson and flame-red colored varieties are exceedingly beautiful. The red is called ruby spinel, and fine stones command high prices. Spinel is found associated with ruby in Burmah, Ceylon, and Siam. Its composition consists of one molecule each of alumina and magnesia, equivalent to 72 per cent. of alumina and 28 of magnesia.

Spinel fine enough to cut into gems has been only occasionally met with in the United States. The Rev. Alfred Free of Toms River, N. J., had in his possession at one time cut gems of a smoky blue or velvety green and a dark-tinted claret color, from the locality near Hamburgh, Sussex County, N. J. They were all good specimens, weighing about 2 carats each. Some half dozen from San Luis Obispo, Cal., of very good quality and weighing about 2 carats each, were brought to the notice of the writer by James W. Beath, of Philadelphia, Pa. Silas C. Young, who, for over twenty years has collected minerals in Orange County, N. Y., writes that in his extensive working for minerals he has found small ruby spinels, also others of a smoky and purple tint, sufficiently clear to cut. The locality at Hamburgh, N. J., was discovered by his father over fifty years ago.

The region of granular limestone and serpentine in which spinels abound extends from Amity, N. Y., to Andover, N. J., a distance of thirty miles. Monroe, Norwich, and Cornwall, N. Y., are well-known localities. The finest crystals from the locality known as Monroe, N. Y., are in the Vaux and Bement Collec-

[1] Mineralogy of Nevada.

tions, Philadelphia, and in the Amherst College Collection. The place that furnished the monster spinel crystals so well-known to collectors of twenty years ago, is probably somewhere between Monroe and Southfield. Its exact situation was known only to the two collectors, Silas Horton and John Jenkins, both now deceased, who secretly worked the locality some years by moonlight, and from it took crystals that realized for them over $6,000, although many fine crystals were ruined in blasting and breaking out. Since the death of these workers the location has been lost.

The gahnite green spinel from the Deak Mine, Mitchell County, N. C., is of a very dark-green color, translucent on the edges, and appears to be compact enough for cutting. The localities of Franklin and Sterling, N. J., have afforded some of the finest known crystals of this mineral, some of which would cut into mineralogical gems. At the lead mine at Canton, Ga., some fine crystals were found implanted on galenite. Professor Genth mentions in his " Contributions to Mineralogy " large, rough crystals, 3½ inches (9 centimeters) long, from the Cotopaxi Mine, Chaffee County, Col. In a specimen of gahnite sent the writer from a lead mine in New Mexico the crystals were bright, polished octahedrons, from ⅛ to ¼ inch across, translucent on the edges, imbedded in galenite. This most interesting and curious association was accompanied with massive garnet. The locality may rightfully be regarded as one of the most interesting for this variety, and it is to be regretted that more exact information cannot be obtained regarding it. Gahnite is found in the pyrite mines, associated with iron pyrite quartz, at Rowe, Mass.,[1] the larger crystals having a diameter of ¼ inch.

DIASPORE

This is an aluminium hydrate, with a hardness of about 7 and a specific gravity of 3·4.

Probably the finest known diaspores are those which were found with corundum near Unionville, Chester County, Pa. There crystals have been obtained from ⅛ to 1½ inches in length, and ¼ inch in thickness, the color varying from white to a fawn

[1] Am. J. Sci. III., Vol. 29, p. 455, June, 1885.

TRUE SPINEL

	COLOR.	LOCALITY.	Silica.	Alumina.	Ferric Oxide.	Chromic Oxide.	Ferrous Oxide.	Zinc Oxide.	Lime.	Magnesia.	Cupric Oxide.	Nickel Oxide.	Hardness.	Specific Gravity.	Analyst.
Theoretical Composition.				72.03						27.97					
Typical Analysis.	Greenish	Madison Co., N.C.		66.74	1.34	trace	11.94	0.22		19.34	0.09	0.33		3.751	F. A. Genth.[1]
Other Analyses.	Black	Culsagee Mine, N.C.		54.32	1.31	3.06	11.10			19.05					Koenig.[2]
"	Black	"		50.58	9.31	2.28	14.60			16.88	0.11	0.25			" [3]
"	Dark green.	"		66.63	1.80	trace	11.35			19.86					" [4]
"	Green.	Franklin, N.J.	5.62	73.31					7.42	13.63					Thomson.[5]
"	Green.	Amity, N.Y.		61.79					10.56	17.87					" [6]

GAHNITE (ZINC-SPINEL)

	COLOR.	LOCALITY.	Silica.	Alumina.	Ferric Oxide.	Ferrous Oxide.	Manganous Oxide.	Zinc Oxide.	Magnesia.	Cupric Oxide.	Moisture.	Hardness.	Specific Gravity.	Analyst.
Theoretical Composition.				55.98				44.02						
Typical Analysis.	Blackish green	Deak Mica Mine, Mitchell Co., N.C.		54.86	4.50		0.29	38.05	0.79		0.30	7.5	4.576	F. A. Genth.[6]
Other Analyses.		Franklin, N.J.		57.09		1.74	trace	34.85	2.22			7.5–8.0	4.1–4.6	Abich.[7]
"		Cantor Mine, Ga.	1.22	53.37	6.68	4.55	0.20	30.27	3.22			7.5–8.0	4.0–4.6	F. A. Genth.[8]
Variety Dysluite		Stirling, N.J.	2.37	39.49	41.93	3.02	7.60	16.80						Thomson.[10]
"	Greenish	Rowe, Mass.	2.97	54.67	3.22	3.25	trace	36.91	2.01				4.53	A. G. Dana.[11]
"	Greenish	Rowe, Mass.	0.57	55.04	2.79	3.49	trace	36.93	1.86				4.53	A. G. Dana.[12]
"	Blackish green	Mine Hill, Franklin, N.J.	0.48	49.78	8.58		1.13	39.62	0.13	und't.				I. S. Adams.[13]
"	Blackish green	Cotopaxi Mine, Chaffee Co., Col.	0.57	66.76	0.58			23.77	10.33				4.91	H. F. Keller.[14]
"	Dark green.	Gilmore's Mine, Montgomery Co., Md.		55.46	2.77	4.56		40.07	0.59	1.93	0.30			T. M. Chatard.[15]
"	D'k leek green.	Cantor Mine, N.J.	0.57	53.37	6.68	3.01	0.20	30.27	3.22				4.59	Abich.[17]
"	Deep green.	Franklin, N.J.	2.37	57.80				31.93	2.25					F. A. Genth.[18]
"	Deep green.	Cantor Mine, Ga.		55.36	3.40	4.60	0.21	31.40	3.34					
"	Deep green.	Mine Hill, Franklin Furnace, N.J.		50.07	7.14	1.29	1.13	39.85	0.13					G. J. Brush.[19]

HERCYNITE (IRON-SPINEL)

	COLOR.	LOCALITY.	Silica.	Alumina.	Ferric Oxide.	Ferrous Oxide.	Magnesia.	Specific Gravity.	Analyst.
Theoretical Composition.				58.86		41.14			
Analysis.	Iron black.	Shimerville, Lehigh Co., Pa.	2.62	56.42	13.17	22.95	4.04	4.056	G. M. Lawrence.[20]

[1] F. A. Genth, Am. Phil. Soc., Aug. 18, 1882.
[2,3] Koenig, Am. Phil. Soc., Sept. 19, 1873.
[4,5] Thomson, Thomson's Min. I., 244.
[6] F. A. Genth, Am. Phil. Soc., Aug. 18, 1882.
[7] Abich, Ak. H. Stockh, 840, 6.
[8] F. A. Genth, Am. J. Sci. (2), 33, 196.
[9] Thomson, Thomson's Min. I., 221 | Albert Seile,
[10] Cour. de Min., 1876, 5m.
[11,12] A. G. Dana, Am. J. Sci. (2), 9, 455.
[13] I. S. Adams, Am. J. Sci. (2), 1, 28.
[14] H. F. Keller, Am. Phil. Soc., Aug. 18, 1882.
[15] T. M. Chatard, U. S. Geol. Sur., Bull. No. 9.
[16] F. A. Genth, Am. J. Sci. (2), 33, 196.
[17] Abich, Rammelsberg's Mineralchemie, p. 139.
[18] F. A. Genth, Am. J. Sci. (2), 33, 196. Ram. Min., p. 139.
[19] G. J. Brush, Am. J. Sci. (9), 1, 28.
[20] G. M. Lawrence, Am. Phil. Soc., Aug. 18, 1882.

color, inclining to a topaz-yellow, while some are of a slightly brownish tint. They closely approach topaz in appearance, and would afford gems as fine as any yet obtained. The best of these specimens are in the cabinets of Dr. Isaac Lea and Joseph Wilcox, of Philadelphia, Pa. At the emery mines of Chester, Mass., have been found masses of small crystals, which might be cut into minute cabinet gems. Joseph C. Trautwine, of Philadelphia, obtained some minute acicular crystals in a cavity of massive corundum at the Culsagee Mine in North Carolina. Gen. Thomas L. Clingman also observed the mineral associated with blue corundum near Marshall, Madison County, N. C.

CHAPTER III.

Turquoise.

TURQUOISE is a hydrated phosphate of alumina sometimes containing small quantities of copper, iron, or manganese. Its hardness is 6, and specific gravity 2·75. The finest varieties, which generally do not lose their color easily, have been for centuries found in small veins in a clay slate in the vicinity of Nishapoor, Persia. Large quantities are brought from Egypt, but this variety, although dark-blue when found, often changes in a short time to a verdigris green.

This mineral is found at Los Cerrillos, N. M.; Turquoise Mountain, Cochise County, Ariz.; Mineral Park, Mohave County, Ariz.; near Columbus, Nev.; Holy Cross Mountain, Col.; and Taylor's Ranch, Fresno County, Cal. The first-named locality is part of a group of conical mountains situated about twenty-two miles southeast of Santa Fé, N. M., and north of the Placer or Gold Mountains, from which they are separated by the valley of the Galisteo River. The rocks of which they are composed are yellow and gray quartzite sandstones and porphyry dykes. Probably the sandstones are of the Carboniferous period, and they are so much uplifted and metamorphosed that the sedimentary character is partly obliterated. William P. Blake describes the locality as being an immense pit, with precipitous sides of angular rock, projecting in crags, sustaining in the fissures a growth of pines and shrubs.   On one side, the rocks' tower into a precipice, and so overhang as to form a cave, at another place the side is low, and formed by the broken rocks that were removed from the top of the cliff.   The excavations, which appear to be about 200 feet in depth and 300 or more in width, were made in the solid rock, and thousands of tons of rock have been broken out. The lower part of the working is funnel-shaped, and is formed by the sloping banks of the débris or fragments of the side walls.   On the débris, at the bottom of the pit, and on the bank of the refuse rock, pine trees are now growing.   There are several other pits in the vicinity more limited in extent, and some of them, apparently, more recently excavated. Prof. Benjamin Silliman, Jr., who visited this locality in 1880, states: " The age of eruption of these volcanic rocks is probably tertiary.   The rocks which form Mount Chalchihuitl are at once distinguished from those of the surrounding and associated ranges of the Cerrillos by their white color and decomposed appearance, closely resembling tufa and kaolin, and giving evidence of extensive alteration, due probably to the escape through them, at this point, of heated vapors of water and perhaps of other vapors or gases, by the action of which the original crystalline structure of the mass has been completely decomposed or metamorphosed with the production of new chemical compounds.   Among these, the turquoise is the most conspicuous and important.   In this yellowish-white and kaolin-like tufaceous rock the turquoise is found in thin veinlets and little balls or concretions called nuggets, covered with a crust of the nearly white tuff, which within consists generally, as shown on a cross fracture, of the less valued varieties of this gem, but occasionally affords fine sky-blue stones of higher value for ornamental purposes.   Blue-green stains are seen in every direction among the decomposed rocks, but the turquoise in mass is extremely rare, and many tons of the rocks may be broken without finding a single stone that a jeweler or collector would value as a gem.   The waste or débris excavated in the former workings covers an area which extends over twenty acres at least.   On the slopes and sides of these great piles are large cedars and pines,

---

[1] The Chalchihuitl of the Ancient Mexicans : Its Locality and Association, and Its Identity with Turquoise.   Am. J. Sci. II., Vol. 25, p. 227, March, 1858.

the age of which, judging from their size and the slowness of growth in this very dry region, must be reckoned by centuries."[1]

It is well known that in 1680 a large section of the mountains suddenly fell in from the undermining of the mass by the Indian miners, killing a number of them, and that this accident was the immediate cause of the uprising of the Pueblos, which resulted in the expulsion of the Spaniards. On both the east and west side of the mountain, shafts have been sunk, which were intended to be connected at their base by a subterranean tunnel. The entrance to the main mining shafts on the west side is 194 feet below the spot where the Indians originally began their excavations. (See Illustration.) Recently several caves have been unearthed extending from the level of the long-abandoned mine. Some of the most curious of these openings, named the Wonder Caves, are about 75 feet northwest of Shaft No. 1, on the east side of the mountain, and appear to have been hermetically sealed by the Indian peons on abandoning the mine; their discovery was purely accidental. The Wonder Caves are almost 25 feet from the surface and run 100 feet from the apex of the mountain, being about 30 by 25 feet in width and from 6 to 8 feet in height above the débris. The group resembles in shape the five fingers with the hand. Here were found numerous veins of turquoise from $\frac{1}{8}$ inch to 2 inches in thickness, and strips of gold-bearing quartz cover the walls of the central cave. The bottom is composed of loose rock, almost 20 feet deep, which is supposed to have been thrown there by the Indians when the mine was sealed. The roof is supported by pillars from 10 to 20 feet thick. It is presumed that further explorations would bring to light openings through these walls, showing that the entire mountain was honeycombed by the ancients, and the pillars left by them to support the roof. This information was obtained in 1880 by the efforts of the mining company under J. B. Hyde, who supposed that the mine could be worked for gold and turquoise; but the effort, after the expenditure of thousands of dollars, proved unsuccessful. The only work that is carried on at present at the Los Cerrillos Mines is done in a very desultory manner by either the local lapidaries, poor whites, or Indians. It

[1] Eng. and Min. J., Vol. 31, p. 169, Sept. 10, 1881.

consists in building large fires against the base of the rock which becomes heated, whereupon water is thrown over it. The sudden change of temperature cracks off large pieces, and much of the turquoise is ruined in the process. After cracking off the rock, the turquoise is picked out of the exposed seams with pieces of pointed iron, such as old harrow-teeth, or any other sharp-pointed instrument. Only occasionally is there a blast put in. The turquoise is sold in Santa Fé, or along the line of the railroad in the vicinity of the mines, by the Indians of the San Domingo pueblo, N. M. The specimens are ground into round or heart-shaped ornaments, which are pierced with a crude form of bow-drill, called by them "malakates." The drilling point is either quartz or agate, and the wheel to give velocity was in one instance made of the bottom of a cup. The selling price of the ornaments is now very low, the Indians disposing of their specimens at the rate of twenty-five cents for the contents of a mouth, where they usually carry them. A string made up of many hundreds of stones, they value at the price of a pony. Comparatively little of the American turquoise finds sale except as cabinet specimens, or as mementos of travel. Still, for ornamental or inlaying work, were it properly introduced, it might have a large sale, as the green and blue-green tints would contrast favorably with many stones or with dark wood. It is possible that deeper workings will develop finer stones, perhaps of such material as will maintain a more permanent color. Concerning the origin of the turquoise veining rock, both Prof. John S. Newberry and Prof. Benjamin Silliman, Jr., regard it as eruptive. According to Prof. Frank W. Clarke, the very small size of the veins and their limited distribution show that the turquoise is of local origin, and he emphasizes the idea that it has resulted from the alteration of some other mineral. In addition to the facts tending to show its derivation from apatite, there is also the fact that epidote containing lime is present as a secondary product. The existence of the pyrite in the gold-bearing veins may have had something to do with initiating the process of alteration, and the alumina of the turquoise was probably derived from decomposing feldspar. During the summer of 1885 a very full suite of specimens was collected by Maj. John W. Powell, and placed for analysis in the hands of Professor Clarke, chief chemist of the United States Geological Survey, and are now deposited in the United States National Museum Collection.

This mineral varies in color from a fine sky-blue through many shades of bluish-green and apple-green to dark-green, showing no blue whatever. The dark-green nodules pass to white at the center, sometimes resembling in structure certain varieties of malachite. Many of the specimens obtained by Major Powell, which are seamed or streaked by limonite, Professor Clarke suggests have been derived from the accompanying pyrite; and the latter mineral is occasionally found, bright and unaltered, enclosed completely in masses of clear blue turquoise. Three samples, selected as representing as nearly as possible the most definite types of the mineral, may be briefly described as, A. Bright blue, faintly translucent in thin splinters. B. Pale blue with a slight greenish cast, opaque and earthy in lustre, and having a specific gravity of 2·805. C. Dark green in color and opaque. These were analyzed, with the following results :

	A[1]	B	C
Alumina } .... Ferric Oxide }	39·53	36·88 2·40	37·88 4·07
Phosphorus Pentoxide..	31·96	32·86	28·63
Copper Monoxide......	6·30	7·51	6·56
Lime.................	·13	·38	
Silica...............	1·15	·16	4·20
Water...............	19·80	19·60	18·49
	98·87	99·79	99·83

In Professor Silliman's paper there is reported 3·81 of copper, which corresponds to 4·78 of copper monoxide. On account of the value of this gem, attempts have been made to color it by artificial means. The discovery of this deception was made by the writer, who saw numerous parcels of turquoise sent to New York from New Mexico, and among them several small lots with an exceptionally fine color for American specimens. This color did not appear to be natural, although the stones

[1] Analysis A was not completed, as material enough could not be obtained without the destruction of two valuable specimens. The silica in it was due to traces of admixed rock from which the material could not be perfectly freed. C, however, was free from rock, and the silica in it must be otherwise accounted for.

were found to have the same specific gravity as others from New Mexico, and when tried with a knife cut with the characteristic soapy, ivory feel. It was only after the back had been scraped off to some depth that the fact was revealed that they were artificially stained. The coloring matter used was the same as that employed in Germany to make the breccia agate that resembles lapis-lazuli, and is often sold as such to tourists. In this case, however, the Prussian blue is only a superficial stain, and the intensity of the blue is modified by the green. It can readily be removed, without injury to the stone, by scraping the back with a knife. Prussian blue dissolves readily in ammonium hydroxide, so that the simplest test is to wash the stone in alcohol, and after wiping it, to remove any grease, and lay it in the ammonia solution for a moment, when the blue color will partially or wholly disappear, and the gem resume its natural greenish hue. If it is desired to examine the stone without destroying the color, the face should be covered with wax, which should be allowed to project above the back, and a little strong ammonium hydroxide poured into this groove. If artificial, the difference of the shades of the two sides will be apparent at once. If stones thus stained are worn in rings, their color is soon affected by the water used in washing the hands. Ammonia does not affect the color of true Persian turquoises, although washing the hands with them on usually does. By artificial light the color of this stained turquoise is rather gray-blue, and appears duller instead of lighter, as is the case with the genuine turquoise. A stone costing $100 to $200, if found to be stained, would depreciate to only a hundredth part of its original cost. The deception is to be regretted, since it will cast suspicion on any fine turquoise that may be found in this country hereafter; but the test is so simple that any one can satisfy himself as to the genuineness of the specimen. A few stones cut from New Mexico turquoise, which had at the time of cutting a very good color, changed to the characteristic green within a few days. William P. Blake[1] also describes a second locality in Cochise County, Ariz., about twenty miles from Tombstone and not far

[1] New Locality of the Green Turquoise known as Chalchihuitl. By William P. Blake. Am. J. Sci. II., Vol. 25, p. 227, March, 1858.

from the stronghold of the Apache chief, Cochise.   This locality, likewise worked by the ancients, is now known as Turquoise Mountain, and as there are several deposits of silver ores in the vicinity, a mining district has been formed known as the Turquoise District.   At the place itself, there are two or more ancient excavations upon the south face of the mountain, and large piles of waste or débris thrown out are overgrown with vegetation.   The place has been worked only for a short time, and probably never by the Apaches.   The excavations are not so extensive as those at Los Cerrillos, and the mineral is more difficult to find; but, though it is less abundant here, its identity with the New Mexican chalchihuitl has been satisfactorily established.   The rock is all similar, and the turquoise occurs in seams and veinlets rarely more than ⅛ or ¼ inch in thickness. In color it is light apple-green or pea-green, rather than blue. The specific gravity of two different fragments gave 2·710 and 2·828, of which the first was slightly porous and earthy and the second dense, hard, and homogeneous.

In 1883 the author saw a series of finely colored specimens, which had been obtained at Mineral Park, Ariz., and brought to New York city.   They had been taken from three veins, varying in thickness from 1 to 4 inches, about 100 yards apart, running almost parallel, and traceable for nearly half a mile.   This deposit showed evidences of having been mined by the Spaniards, and a large number of stone hammers was found, indicating that it had also been worked by the Indians.   Hoffmann, in the " Mineralogy of Nevada," states that turquoise is also found in a locality situated in the Sierra Nevada Mountains, five miles north of Columbus.   This locality was visited by J. E. Clayton, who reports that, on a sharp ridge, about half a mile southwest of the Northern Bell Mine, in the Columbus District of southern Nevada, he found turquoise in seams and bunches in a metamorphic sandstone of a brownish color, not vitreous enough to be classed as a quartzite.   The best specimens were in small, roundish pebbles in clusters, imbedded in the brown sandstone, in size from that of a duckshot up to a third of an inch in diameter.   Some fine ones have been obtained, equal in color and hardness to the best standard.   Those which occurred in seams were higher colored and softer. The principal sale is in San Francisco, where the sandstone is cut with the turquoise in it, making a rich mottled stone for jewelry. Although the nodules are small, this is the finest turquoise for color and quality found on the continent. At Taylor's Ranch, Chowchillas River, Fresno County, Cal., several hexagonal crystals of bluish-green turquoise have been found, each about 1 inch in length. They were identified as turquoise by Dr. Gideon E. Moore, and are of great interest as to the origin of turquoise. The crystalline characters were such that V. von Zepharovich believed them to be pseudomorph after crystals of apatite.[1] (See Fig. 1.)

That the ancient Mexicans held the turquoise in high esteem is well known, and that the Los Cerrillos Mines were extensively worked prior to the discovery of America, is proved by fragments of Aztec pottery—vases; drinking, eating, and cooking utensils; stone hammers, wedges, mauls, and idols—discovered in the débris found everywhere. While Major Hyde was exploring this neighborhood, in 1880, he was visited by several Pueblo Indians from San Domingo, who stated that the turquoise he was taking from the old mine was sacred, and must not go into the hands of

FIG. 1.
PSEUDOMORPH OF
TURQUOISE AFTER
APATITE.

those whose saviour was not Montezuma, offering, at the same time, to purchase all that might come from the mine in the future. In the Mystery Cave, there was found a stone hammer weighing 13$\frac{4}{16}$ pounds, with its handle attached. Additional evidence of the antiquity of the turquoise workings of New Mexico and Arizona has been gathered by the Hemenway Expedition, sent out by Mrs. Hemenway, under the direction of Lieut. Frank H. Cushing. There was found a prairie dog cut out of white marble, with turquoises for eyes (see Illustration); also, about ten miles from Tempe, Ariz., enclosed in asbestos, in a decorated Zuni jar, a sea shell coated with black pitch, in which were incrusted turquoises and garnets in the form of a toad, the sacred emblem of the Zuni. (See Colored Plate No. 2.) The Christy Collection in London contains two human skulls which

[1] Kallait pseudomorph nach Apatit aus Californien, Zeitschrift für Krystallographie, Vol. 10, p. 240, 1885.

are inlaid with turquoise and have eyes made of iron pyrites (see Illustration), and a finger-ring made of the central whorl of a cone-like shell (see Colored Plate No. 2), in which triangular-shaped pieces of turquoise and red spondylus shell were inlaid. Pieces of dark wood were also inlaid with turquoise.

Bernal Diaz, who came over with Cortez, mentions that on the landing of the explorers at San Juan de Ulloa, the ambassador from Montezuma brought various rich presents, including four chalchihuitls, each of which the ambassador claimed was worth more than a load of gold. Diaz states that the chalchihuitls were green stones of uncommon value, and held in higher estimation among the Indians than the smaragdus or emerald was among the Spaniards. Torquemada, who regarded chalchihuitl as a species of emerald, states that the Mexicans gave the name "Chalchihuitl" to Cortez, intending thus to show their respect for him as a captain of great valor, "for chalchihuitl is of the color of the emerald, and emeralds were held in great esteem." Offerings of this stone were made by the Indians in the temple of the goddess Matlalcueye, and it was their custom to place a fragment in the mouths of distinguished chiefs when buried. Torquemada, in recording this fact, says that these stones were emeralds, but that they were called chalchihuitl by the Indians. When Alvarada and Montezuma played together at games of chance, Alvarada paid, if he lost, in chalchihuitl stones, but received gold if he won.

The Indians claimed that the art of cutting and polishing chalchihuitl was taught them by the god Quetzalcohvatl. Bernardino de Sahagun considered chalchihuitl to be a jasper of a very green color, or a common smaragdus. He states that they are green and opaque, and are much worn by the chiefs strung on a thread around their wrists, being regarded as a badge of distinction. (See Illustration.) Friar Marco de Nica in 1539 made a journey among the Indians of New Mexico, and in his narrative frequently mentions green and bluish stones, which were worn as ornaments by them, pendant from the ears and nose. He also mentions seeing many "turqueses," which there is little doubt he considered the green stones to be. These turquoises were worn, not only in the ears and nose, but as necklaces and girdles. They were called Cacona by the Indians, and were obtained from the kingdom. On arriving at this place De Nica observes that "the people have emeralds and other jewels, although they esteem none so much as turquoises, wherewith they adorn the walls of the porches of their houses and apparel and vessels, and they use them instead of money through all the country." Coronado, who visited Civola in 1540, denies De Nica's statement respecting the turquoises upon the porches of the houses, but he obtained turquoise ear-rings and tablets set with the stones. The turquoise has always been the favorite jewel of the western tribes of Indians and was extensively in use at the time of the Conquest by Coronado, in 1541. Fra Saverio Claverigo,[1] alluding to the minor kingdom states tributary to the main kingdom, says: "Among articles of tribute annually required from these natives, mention is made of ten small measures of fine turquoises and one carga of ordinary turquoises," and elsewhere the first present from Montezuma to Charles V. of Spain, through Cortez, is thus referred to: "The present of the Catholic king consisted of various works of gold, ten bales of most curious rolls of feathers and fair gems, so highly valued by the Mexicans that, as Tehuitlile himself, the ambassador of Montezuma to Cortez, affirmed, each gem was worth a load of gold." According to the Mexican system of weights, 240 pounds constituted a load of gold. Estimating gold at $20 an ounce, the value of these gems was over $57,000. It is a well authenticated fact that these gems referred to were turquoises, and it is believed that they are now among the crown jewels of Spain. In the memoir on ancient turquoise mosaics, recently published by Luigi Pigoni, director of the Ethnographic Museum in Rome,[2] it is stated that the objects of this kind known as Mexican are distributed as follows: five in the Museum in Rome; seven in the Christy Collection in London; one in a private collection in England; two in the Ethnographic Museum in Berlin; and one in Gotha. Those in the Christy Collection have been described by E. B.

[1] History of Mexico, Cesena, 1780-1881.
[2] Gli Antichi Oggette Messicani Incrostati di Mosaico Isistenti Nel Museo Preistorico-Etnografico di Roma. Roma, 1885.

Tylor in his "Anahuac; or, Mexico and the Mexicans, Ancient and Modern," p. 337; also in the "British Museum Guide to the Christy Collection" (1868), p. 20; and by Brasseur de Bourbourg in his "Recherches sur les ruines de Palenque et sur les origines de la civilization du Mexique" with drawings by M. de Waldeck (Paris, 1866). The specimens in the Copenhagen Museum have been described in "Congrès International d'anthropologie préhistorique, Compte Rendu de la 4me Session" (Copenhagen, 1869), p. 462, and by Steinhauer in "Das königliche Ethnographische Museum zu Copenhagen" (1881), p. 19. The three in Berlin have been described in a lecture before the Anthropological Society of Berlin. Adolph Bastian claimed that one had originally been the property of Alexander von Humboldt,

TURQUOISE

CHEMICAL COMPOSITION AND PROPERTIES.	Theoretical Composition	LOCALITY Los Cerrillos, New Mexico. Analyst, F. W. Clarke.[1]	LOCALITY Los Cerrillos, New Mexico. Analyst, F. W. Clarke.[2]	LOCALITY Los Cerrillos, New Mexico. Analyst, F. W. Clarke.[3]	LOCALITY Taylor's Ranch, Colorado. Analyst, G. E. Moore.[4]
Color..............	.....	Bright Blue.	Pale Blue.	Dark Green.	Blue Green.
Phosphoric Acid...	32·60	31·96	32·86	28·63	33·21
Alumina	46·90	} 39·53 {	36·88	37·88	35·98
Ferric Oxide.......	.....		2·40	4·07	2·99
Copper Oxide	.....	6·30	7·51	6·56	7·80
Lime	.....	0·13	0·38	.....	.....
Silica..............	.....	1·15	0·16	4·20	.....
Water ....·...	20·50	19·80	19·60	18·49	19·98
Specific Gravity....	.....	.....	2·805	.....	2·806

[1], [2], [3] F. W. Clarke, Am. J. Sci. III., Vol. 32, p. 211, Sept., 1886.
[4] Gideon E. Moore, Zeit. für Kryst. u. Min. 10, 240.

while the other two were from the Ducal Museum of Brunswick. See "Verhandlungen der Berliner Gesellschaft für Anthropologie" (1885), p. 201. The exact ownership of the one in Gotha does not appear to be known. Illustrations of these objects are to be found in the works of E. B. Tylor and Brasseur de Bourbourg, and notices of them appear in various books of the seventeenth century, among which are "Pyranarcha sive de fulminum natura" by Liceti (Padua, 1643), p. 143, and "Musæum Metallicum," by Aldrovandi (Bologna, 1647), p. 550; "Museo Cospiano," by Legati (Bologna, 1677), p. 477; and in Clavigero "Storia antica del Messico" (Vol. II., Book 7, Chap. 52). These mosaics are made with pieces of broken shells. The art is still practised in Guatemala. Pigoni's pamphlet is specially devoted to a description of the masks of the Museum in Rome. Of these, three are mentioned in the books of the seventeenth century, the first having been the property of Aldrovandi, while the other two are from the Museo Cospiano. The mask shown in the plate of the pamphlet as No. 4 is the one mentioned by Aldrovandi in his "Musæum Metallicum." It is made of wood, one side of which is left natural and carved out so as to fit the human face, while parts of the front side are painted, and these are incrusted in mosaic. Among the materials composing the incrustations are turquoise, white, pearly, red, and black sea shells, also small garnets, with several minute square pieces of metal. This mask was in the Archæological Museum of Bologna until 1878, and its history is well known, as it originally belonged to the Aldrovandi Collection. The mask designated on the plate as No. 5 is well preserved, and was acquired in 1880 from Florence. The mosaic is formed of red shell and turquoise. In the ethnographic collection of the College of the Propaganda in Rome, there are also two masks, differing from the others in not being incrusted with mosaic, but tinted red, and engraved with lines that are filled in with white material. These have been described and illustrated by Dr. Guiseppe A. Colini in the "Bulletino della Societa geografica Italiana," Vol. 19, p. 324, 325.

CHAPTER IV.

Topaz and Tourmaline (Rubellite, Indicolite, and Achroite).

TOPAZ crystallizes in the orthorhombic system, and occurs in prisms with one end regularly terminated, and has a very perfect cleavage transverse to the prism. Its hardness is 8, and specific gravity 3·53. It is a silicate of alumina containing fluorine. A blue crystal weighing 20 pounds is in the Imperial Mining School at St. Petersburg, Russia. Fine blue and sherry colored crystals have been found in Siberia, blue ones in Scotland and Ireland, yellow in Minas Geraes, Brazil; white in Villa Rica, Brazil; and blue and white in Ceylon and Australia. Brazilian or true mineralogical topaz is often confounded with two other minerals, namely, citrine and Spanish or Saxon topaz, the color of which is made by heating and so decolorizing smoky quartz to various shades of yellow or brown. Yellow sapphire is called Oriental topaz. The specific gravities of the three varieties are given for comparison.

	SPECIFIC GRAVITY.	HARDNESS.	COMPOSITION.
Oriental Topaz............	4·01	9	Alumina.
True or Brazilian Topaz.....	3·55	8	Fluo silicate of Alumina.
False or Saxon Topaz......	2·65	7	Silica.

True yellow topaz, if heated for a time, becomes pink, and continued heating renders it colorless.

The gem topaz has been found in Huntington and Middletown, Conn.; Stoneham, Me.; North Chatham, N. H.; Sevier Lake, Utah; at Nathrop, Chalk Mountain, Crystal Park, Florissant and Devil's Head Mountain, Col.; and at Ruby Mountain, Nev. The first discovery of topaz in the United States was that of Trumbull, Conn. Specimens of it, found there in a vein of fluorite, associated with a chlorophane variety of fluorite, were sent to Prof. Benjamin Silliman, who determined it to be topaz. Six different determinations of its specific gravity gave results varying from 3·42 to 3·47, with a mean of 3·45. In their modification and color, the crystals afforded by this locality very strikingly resemble those from Saxony, but are generally of larger dimensions, and scarcely any of them would afford a gem, since they are nearly all opaque. This same authority, in 1838, in a " Notice of a Second Locality of Topaz in Connecticut," says:[1] " Among specimens which I obtained at China Stone Quarry, in Middletown, two years ago, I find one that contains above fifty crystals of topaz. They measure from $\frac{1}{4}$ to $\frac{1}{2}$ of an inch in length, are very slender and perfectly transparent, being attached by a lateral plane to crystals of albite." Probably the most beautiful and brilliant crystals of topaz known in the United States are those found forty miles north of Sevier Lake, Utah, and the same distance north of the town of Deseret on the Sevier River. This locality, known as Thomas Mountain, is an isolated and arid elevation about six miles long, and is described by Henry Engelman, geologist of the expedition that, under Capt. James Simpson, crossed Utah in 1859. He found crystals loose on the surface. James E. Clayton, of Salt Lake City, visited the place in June, 1884, and obtained a large number of beautiful crystals, larger than those from Nathrop, Col., and equally as brilliant as those from San Luis Potosi, Mexico, which they closely resemble. Mr. Clayton states that still larger crystals are found, and he says: " They are evidently not secondary products, like zeolites, but primary, and produced by sublimation or crystallization from presumably heated solutions, contemporaneous, or nearly so, with the final consolidation of the rocks."[2] Prof. J. Alden Smith refers

[1] Am. J. Sci. I., Vol. 34, p. 329, Oct., 1838.
[2] Am. J. Sci. III., Vol. 31, p. 432, June, 1886.

to beautiful topazes occurring in the lithophyses of rhyolite, which is the first noted occurrence of this gem in an eruptive rock.[1] This rock was, however, first identified by Whitman Cross, and its exact locality is directly opposite Nathrop, Col., on a ridge a quarter of a mile in length and about 200 feet in height. Here the topaz is found in more or less rounded cavities, partially filled by its curved walls, which by concentric arrangement and an overlapping often produce a roselike form. These cavities are often lined with minute, glassy quartz crystals, and on them are found the topazes, which are prismatic in form, and, being attached to

TOPAZ

CHEMICAL COMPOSITION AND PROPERTIES.	Locality Stoneham, Oxford Co., Maine. Analyst, Chatard.[1]	Locality Stoneham, Oxford Co., Maine. Analyst, F. A. Genth.[2]	Locality Stoneham, Oxford Co., Maine. Analyst, C. M. Bradbury.[3]	Locality Florissant, Colorado. Analyst, W. F. Hillebrand.[4]	Locality Turnbull, Connecticut. Analyst, Rammelsberg.[5]	Locality Huntington, Connecticut. Analyst, Silliman & Hitchcock.[6]
Silica.....................	31·92	32·03	21·37	33·15	32·38	29·74
Alumina	57·38	57·18	51·26	57·01	55·32	47·46
Fluorine.................	16·99	18·83	29·21	16·04	16·12	12·02
Potassa	0·15	.....	.....	.....	.....	.....
Soda.....................	1·33	.....	.....	.....	.....	.....
Water	0·20	.....	.....	.....	.....	.....
Color.....................	.....	.....	.....	Clear Green.	.....	Yellow.
Hardness.................	8·00	.....	.....	.....	.....	.....
Specific Gravity...........	3·51	.....	.....	.....	3·514	3·45

[1] F. W. Clarke, U. S. Geol. Sur. Bull. No. 27.
[2] F. A. Genth, Proc. Am. Phil. Soc., Oct. 2, 1885.
[3] C. M. Bradbury, Chem. News, Sept. 7, 1883.
[4] W. F. Hillebrand, and Whitman Cross, U. S. Geol. Sur. Bull. No. 20.
[5] Rammelsberg, J. für pr. Ch. 96, 7.
[6] Silliman and Hitchcock, Am. J. Sci. (1) 11, 112.

the sides of the cavities in all positions, are often found doubly terminated. The crystals are from ⅛ to (rarely) 1 inch in length and ⅛ to ¼ inch across the prism. In color they are generally transparent and flawless, and are either colorless, pale-blue, or distinctly sherry-colored. A similar occurrence is noted by Mr. Cross, in the nevadite of Chalk Mountain, but the crystals are somewhat smaller. Chalk Mountain is situated at the juncture of Lake Eagle and Summit Counties in Colorado.[2] Many fine large topaz crystals have been found at Crystal Park, near Pike's Peak, El Paso County, Col. Three crystals from this lo-

[1] Report on the Development of the Resources of Colorado, p. 36, 1881–1882.
[2] Am. J. Sci. III., Vol. 27, p. 94, Feb., 1884.

cality, all of which are remarkable for their size and clearness, were very fully described by Whitman Cross and William F. Hillebrand, under title of "Minerals from the Neighborhood of Pike's Peak, Col."[1]   One of these, a fragment of a crystal, was found near Florissant with amazon-stone; it is remarkable on account of the probable size of the original crystal, which when complete must have been nearly a foot in diameter.   It was clear in parts and had a decided greenish tinge.   The specific gravity of a fragment was 3·578 and its chemical composition was entirely normal.[2] Another locality of importance in the vicinity is Devil's Head Mountain in the Colorado range, some thirty miles north of Pike's Peak. The pocket in which the topaz was found at this place is of irregular shape, being about 50 feet long, from 2 to 15 feet wide, and averaging 4 feet in depth. Owing to the disintegration of the rock at the surface, many of the crystals had been carried in the débris to a considerable distance down the mountain side, and were badly worn and broken. The topaz is found here in isolated and usually loose crystals, surrounded by distorted quartz crystals of smoky reddish shades, frequently the exact color of the topaz. The principal color of the latter was reddish, although wine-yellow, milky-blue, and colorless crystals were found.[3]   These Colorado localities have proved quite valuable.   Within a year after their discovery it was estimated that over 100 crystals had been sold for nearly $1,000, at prices varying from 50 cents to $100 each.[3]   A topaz crystal weighing 18½ ounces (587 grams) was found at Cheyenne Mountains, Col., during 1886; but, although very perfect, it had little gem value.   There is in the United States National Museum in Washington a cinnamon-tinted cut stone from Pike's Peak weighing 15 carats, that is superior in beauty to the brilliant white topazes from Brazil.   Several of the sherry-colored Colorado crystals have been cut in stones, two of the larger ones weighing 125 to 193 carats each.   (See Colored Plate No. 3.)   During 1882, crystals from Harndon Hill, in the vicinity of Stoneham, Me., were determined by the writer to be topazes, and further research

---

[1] Am. J. Sci. III., Vol. 24, p. 282.

[2] Contributions to the Mineralogy of the Rocky Mountains, p. 70, et seq., Bulletin No. 20 of the United States Geological Survey, Washington, 1885.

[3] Mineral Resources of the United States, 1886, p. 596.

resulted in the finding of large quantities of fragments. This locality furnished good, clear, and distinct crystals of topaz and has yielded the best crystals found in the East. The specimens are either colorless or faintly tinted with green or blue. The finest crystals were from $\frac{3}{8}$ inch to $2\frac{1}{2}$ inches (10 to 65 millimeters) across, perfect, and in part transparent. Several perfect gems have been cut from some of the fragments. They had the characteristic fluid cavities, and in hardness were the same as the Brazilian.[1] Some white opaque crystals, a foot in diameter, were blasted out by the writer. The finest crystal found at this locality is in the cabinet of Clarence S. Bement. (See illustration.) During 1888 nearly 100 crystals associated with phenacite were found on Bald Mountain, North Chatham, N. H., which is only a few miles from the Stoneham locality, both places being near the State line.[2] They were colorless, light-green, or cherry-colored on the outer sides and colorless in the center. The largest crystal measured $1\frac{1}{2}$ inches in height and the same in thickness. Almost all the crystals contained irregular hollow spaces from $\frac{1}{100}$ to $\frac{1}{10}$ inch (1 to 10 millimeters) across. In habit the crystals closely resemble those from Cheyenne Mountain, Col. Some of these crystals are equal in point of quality to any found in Colorado, although they are not as large. At Stoneham, Me., green and red damourite, altered from topaz, has been cut into different odd forms and charms by the local collectors.[3]

TOURMALINE

belongs to the rhombohedral system, and occurs in prisms, the sides of which are generally striated and channeled. The hardness of the transparent variety is 7·5, and its specific gravity ranges from 3·0 to 3·25. Its composition is very complex, as is shown in the table of analysis.

The question of color is an interesting one, particularly when the varying colors of the lithia tourmaline are concerned. The color of the iron and magnesian varieties depends on the amount of iron present, and passes from the colorless specimens

[1] See Topaz and Associated Minerals from Stoneham, Oxford County, Me. Am. J. Sci. III., Vol. 25, p. 161, Feb., 1883 ; and Vol. 27, p. 212, March, 1884.

[2] Am. J. Sci. III., Vol. 36, p. 222, Sept., 1888.

[3] Am. J. Sci. III., Vol. 29, p. 278, May, 1885.

from DeKalb through all the shades of brown to the black variety found in Pierrepont. On the other hand, the lithia tourmaline, containing more or less manganese, gives us the red, green, blue, and colorless varieties. The shades of color do not appear to depend on the absolute amount of manganese present, but rather on the ratios existing between that element and iron. When the ratio of manganese is to iron as one is to one, there is produced the colorless, pink, or very pale green tourmaline. An excess of manganese produces the red varieties, while if the iron be in excess, the result is various shades of green and blue. The finest green and red specimens are found in the province of Minas Geraes, Brazil, the deep red rubellite in Siberia, the yellow and brown in Ceylon, and Carinthia, Austria, and pink on the island of Elba. The hardness of the flawless variety is about 7·5, and the specific gravity varies from 3·0 to 3·25. It is very electric. The colorless variety is called achroite, the red, rubellite, the blue, indicolite, the green, Brazilian emerald, and the black, schorl.

Tourmaline is one of the most dichroitic of all gems. When a crystal is viewed through the side, it is transparent green, but when viewed through the end of the prism, it is either opaque or yellow-green. For instance, in tourmaline from Paris, Me., if two gems are taken from a green crystal, one with the top cut from the side of the prism and the other from the pyramid side, one will be bright green and the other yellow-green. It has frequently happened with specimens from Brazil that one would be green and the other opaque. Specimens that rival any found in the world have been obtained in Maine. The localities that have furnished fine ones are Mount Mica, near Paris, Auburn, Hebron, Norway, Mount Black, in Andover, Rumford, and Standish. In the two latter places, however, they do not count as gems. The famous tourmaline locality at Paris, Me., is situated on Mount Mica, a spur of Streaked Mountain, about one mile east of Paris Court House. It was discovered in 1820 by Elijah L. Hamlin and Ezekiel Holmes, while they were on a mineralogical and geological trip. Mr. Hamlin found a fragment of a transparent crystal lying loose upon some earth which still clung to the foot of a fallen tree, and procured about thirty beautiful crystals.

These were entrusted to Governor Lincoln of Maine to take to New Haven, and all but one were, at this time, lost. It is believed that these tourmalines are at present in the Imperial Mineralogical Cabinet at Vienna, since there were some fine specimens of tourmalines purchased with the collection of the well-known antiquarian, Vandervull, in 1830. These were recognized as being from Paris, Me., by Baron Lederer, the Austrian Consul in New York City, who was familiar with the crystals, having made collections in that locality. In 1825, Prof. Charles U. Shepard visited the locality, and after considerable work obtained some of the best crystals ever found, which are now in the Shepard Collection at Amherst College, having escaped the disastrous fire of 1882. Prof. John W. Webster, of Harvard College, found a large red crystal and some beautiful grass-green ones. In 1865 the locality was supposed to be exhausted, but excavations which have been made there since, from time to time, through the perseverance of Dr. Augustus C. Hamlin, have brought to light many fine crystals. In 1881 the Mount Mica Tin and Mica Company began operations, with Doctor Hamlin as president, and work has been carried on at intervals since. Some hundreds of tourmalines are the result of this mining, among them a blue indicolite crystal 9 inches long, somewhat shattered by blasting. (See Colored Plate No. 4.) It is light-blue at one end, shading gradually into dark-blue and deep blue-black. This would have been the finest crystal known, and would have furnished several hundred carats of fine stones, had it not been so broken. It is now in the State Museum at Albany, N. Y. The next summer's work brought to light material that cut into two of the finest gems, of a grass-green hue, weighing about 30 carats, which surpass in beauty anything hitherto found. (See Colored Plate No. 4.) The gems and crystals obtained by this company have been valued at over $5,000, and the value of all that have been taken from this locality, and sold at the highest rate asked for them as native gems, probably amounts to $50,000. The crystals of green tourmaline, inclosing red crystals of rubellite, found at Mount Mica, when properly cut across the prism form objects of great beauty. The centers have often furnished magnificent transparent gems, scarcely distinguishable by the eye from the true ruby.   It would be difficult to find a more wonderful mineral than that composing these crystals from Paris, which are white at the termination, then almost emerald-green, light green, pink, then colorless as water, and when broken are dark blue or red in the center, this center in turn being coated white, pink, and green.

The green tourmaline, which has been called Brazilian emerald, is used by the Brazilian clergy as their emblem.   Fine tourmalines have a greater brilliancy than the emerald when seen by artificial light, but have not the rich deep light of the latter.   Some of the finest cut rubellites and green tourmalines are in the possession of members of the family of Professor Shepard.   One of the finest known, which is 1 inch long, ¾ inch broad, and 1 inch thick, was described by Professor Shepard as of a chrysolite green, with a blue tinge, but less yellow and more green than chrysolite.   This, on comparison, he found to be finer than any of the gems in the Hope Collection, that was sold at auction in 1881.   It now belongs to his daughter, Mrs. James, wife of Judge James, of Washington. (See Plate No. 4.)   One fine achroite two-thirds this size, and one remarkable rubellite, the size of the largest tourmaline, are in the possession of L. E. DeForest of New Haven, Conn.   (See Colored Plate No. 4.)   The Hamlin cabinet,[1] the first crystal of which was found in 1820, contains many fine rubellites (red tourmalines), indicolites (blue tourmalines), and achroites (white tourmalines), as well as good examples of pink, yellow, green, and other colors, all from Paris, Me.   This is the best tourmaline collection in the world, and would furnish full suites for a dozen cabinets.   The crystals used by Dr. Hamlin to illustrate his treatise on the tourmaline are in this cabinet, as well as many other fine stones of nearly every known shade of the gem, including a wonderful dark gem of 28 carats (see Colored Plate No. 4), 1 inch in diameter, and an achroite of 23 carats.   One, the finest tourmaline of this collection, is shown as it now is.   (See Colored Plate No. 4.)   In the Peabody Museum at New Haven are some crystals collected by Dr. Sanborn Tenney, of Williams

[1] See The Tourmaline, by Dr. Augustus C. Hamlin, Boston, 1873.

College. A light-green crystal, about 2 inches long, has at one end a transparent, kernel-like nodule that would afford a gem of over 10 carats' weight. The center of a section of green and red tourmaline would cut one of the finest magenta-colored rubellites ever seen. The next important tourmaline locality in Maine is Mount Apatite, in Auburn, Androscoggin County. It was first worked in 1882, and since then fully 1,500 crystals have been found. They were colorless, light-pink, light-blue, bluish-pink, and light-golden, the sections showing the characteristic variety of color, such as blue and pink, green and pink, etc., when viewed through the end of the crystal. Some of the faintly-colored crystals afforded gems that were considerably darker after the cutting, but no gems over 6 or 8 carats were obtained here. Further working in 1883 or 1884 brought darker material to light, especially the green colors, some of which equal those found at Mount Mica. Rude black crystals 8 inches in diameter and 12 feet long (at times enclosing quartzite) were observed here. This, like the Mount Mica locality, gives promise of fine gems for some time to come. The collection in the United States National Museum contains a 1-carat blue indicolite, two lavender-colored stones of 1 carat each, a light emerald-green stone of $\frac{1}{4}$ carat, as handsome as an emerald viewed by artificial light, and also a suite of several dozen loose crystals of various colors. The tourmaline locality of Rumford is situated in the northeast part of the town, in Oxford County, Me., on the northwest slope of Mount Black, and is about 1,500 feet above sea-level. The vein, which has been covered for a length of about 250 feet, has been found to be quite irregular, varying from 30 to 100 feet in width, and dips northeast and southwest at an angle of about 60°. The rock is a coarse granite with mica schist overlying. The Mount Mica Company did some work here, and since they stopped E. M. Bailey has worked the solid ledge to a depth of from 3 to 10 feet. No gems have been found, though some interesting mineralogical specimens have been secured, among them specimens of lepidolite, which is found here of finer grain than that from any other Maine locality. One form is in scales not over $\frac{3}{100}$ inch (1 millimeter) across, quite compact, and in large masses of a beautiful lilac color, closely resembling the mineral from Altenberg, Saxony. A characteristic form is of a light lavender color, very compact, so that it could be used for ornamental purposes, and in this the scales are not more than $\frac{1}{100}$ to 1 inch (1 to 25 millimeters) in width. The mass is penetrated in every direction by crystals or rubellites, which are of the light or dark shade of red. This association is similar to that from Rozena in Moravia.

Rubellite, indicolite, and the green tourmalines are the common varieties at this locality. All exhibit a tendency to radiate, assuming this form when they occur side by side in one radiation. Crystals of green, red, and blue tourmaline have been found at Standish, which, although very good as crystals, are not of gem quality. Little work has been done in this locality, which may improve by development. The specimens at Bates College, Lewiston, Me., labelled "Baldwin," are supposed to have been found here. Bluish and brownish-green tourmaline is found in fine crystals, penetrating damourite and diaspore, in Newlin Township, but none of them transparent enough for gem purposes. A small, well-terminated, transparent green tourmaline was found by J. C. Mills, on Silver Creek, Burke County, N. C., also a black crystal 4 inches long, inclosed in a green beryl crystal. William Irelan, Jr., State Mineralogist, reports that fine crystals of translucent rubellite, but not of gem value, are found in California. Fine crystals of indicolite and green tourmaline are found with the cleavelandite feldspar at Chesterfield, Mass., but none transparent enough to furnish gems. They are interesting from the fact that the green crystals often inclose crystals of rubellite, and sometimes both red and white tourmaline. These afford very interesting specimens when cut in sections across the prism. St. Lawrence County, N. Y., has given to mineral cabinets the greatest number and the finest examples of doubly terminated crystals of black and brown tourmaline, both kinds occurring in a granular limestone, from which they can readily be broken out, or the limestone removed by acid. The result is that many thousands of specimens from Pierrepont, of the most highly polished black, doubly terminated crystals, although without value as gems, grace the mineral cabinets of the world. Nor have the products of any locality ever excelled in

TOURMALINE

Color.	Locality.	Silica.	Alumina.	Ferric Oxide.	Ferrous Oxide.	Manganous Oxide.	Lime.	Magnesia.	Soda.	Potassa.	Lithia.	Fluorine.	Boric Acid.	Water.	Specific Gravity.	Analyst.
Rose colored	Rumford, Me.	38.07	42.24	....	0.26	0.35	0.56	0.07	2.18	0.44	1.59	0.28	9.99	4.26	....	Riggs.[1]
Pale green	Auburn, Me.	38.14	39.60	0.30	1.38	1.38	0.43	....	2.36	0.27	1.34	0.62	10.25	4.16	3.07	" [2]
Light green	Auburn, Me.	37.85	37.73	0.42	3.89	0.51	0.49	0.04	2.16	0.62	1.34	0.62	10.55	4.18	3.00	" [3]
Dark green	Rumford, Me.	36.53	38.10	....	6.43	0.32	0.34	....	2.86	0.38	0.95	0.16	10.22	3.49	....	" [4]
Dark green	Auburn, Me.	36.26	36.68	0.15	7.07	0.72	0.17	0.16	2.88	0.44	1.05	0.71	9.94	4.05	....	" [5]
Black	Paris, Me.	35.03	34.44	1.13	12.10	0.08	0.24	1.81	2.03	0.25	0.07	....	9.02	3.60	....	" [6]
Black	Auburn, Me.	34.99	33.96	....	14.23	0.06	0.15	1.01	2.01	0.34	trace	....	9.63	3.62	....	" [7]
Black	Haddam, Conn.	34.95	31.11	0.50	11.87	0.09	0.81	4.45	2.22	0.24	trace	....	9.92	3.62	....	" [8]
Black	Stony Point, N.C.	35.54	33.38	....	8.49	0.04	0.53	5.54	2.16	0.24	trace	....	10.40	3.63	3.13	" [9]
Black	Pierrepont, N.Y.	35.61	35.29	0.44	8.19	trace	3.31	11.07	1.51	0.20	trace	0.27	10.15	3.34	3.08	" [10]
Dark brown	Monroe, Conn.	36.41	31.27	....	3.80	trace	0.98	9.47	2.68	0.21	trace	trace	9.65	3.79	....	" [11]
Dark brown	Orford, N.H.	36.66	32.84	0.10	2.50	trace	1.35	10.35	2.42	0.22	trace	trace	10.07	3.78	....	" [12]
Brown	Gouverneur, N.Y.	37.39	27.79	....	0.64	....	2.78	14.09	1.72	0.16	trace	0.50	10.73	3.83	....	" [13]
Light brown	DeKalb, N.Y.	36.88	28.87	....	0.86	....	5.09	14.53	1.39	0.18	trace	0.78	10.58	3.56	3.08	" [14]
Cinnamon brown	Hamburg, N.J.	35.25	28.49	....	1.14	....	1.60	14.58	0.94	0.18	trace	....	10.45	3.10	....	" [15]
Brown	Gouverneur, N.Y.	38.85	31.32	....	2.88	....	0.71	14.89	1.28	0.26	0.26	....	8.35	2.31	3.049	Rammelsberg[16]
Brownish black	Orford, N.H.	38.33	33.15	....	4.07	0.09	1.81	9.11	1.52	....	....	....	9.86	2.81	3.068	" [17]
Brownish black	Texas, Penn.	38.45	34.56	....	8.54	....	1.33	9.90	2.00	0.73	....	....	8.57	2.80	3.043	" [18]
Greenish black	Monroe, Conn.	39.01	31.18	....	13.23	....	1.02	8.60	1.82	0.44	....	....	8.95	2.82	3.068	" [19]
Black	Haddam, Conn.	37.50	30.87	....	12.55	0.51	....	6.33	1.60	0.73	....	0.31	9.02	1.81	3.136	" [20]
Black	Unity, N.H.	36.29	30.44	....	11.95	1.25	....	....	1.94	....	....	0.82	9.04	1.72	3.192	" [21]
Black	DeKalb, N.Y.	37.07	31.86	....	....	1.94	....	3.49	2.04	0.30	0.84	1.18	9.70	2.48	3.195	" [22]
Blue black	Goshen, Mass.	36.22	33.35	....	6.38	0.78	0.45	0.63	1.75	0.40	1.17	0.55	10.65	2.21	3.203	" [23]
Red	Paris, Me.	38.19	42.63	....	7.43	2.88	....	0.84	1.53	0.68	0.72	....	9.97	2.00	3.019	" [24]
Green	Chesterfield, Mass.	38.46	36.86	....	....	....	....	1.88	2.47	0.47	....	....	9.73	2.31	3.08	" [25]
Green	Chesterfield, Mass.	33.80	39.61	....	....	....	....	....	4.97	....	....	....	3.88	0.78	3.102	Gmelin.[26]
Black	New Hampshire.	37.65	33.46	....	....	....	....	10.98	2.55	....	....	....	....	....	....	" [27]

1 to 25, inclusive, Robert B. Riggs, Am. J. Sci. III., 35, 35.
26 to 27, inclusive, Rammelsberg, Mineral Chemie, p. 538 to 545.

26 Am. J. Sci. (1) 35, 389; Am. der Phys. Pogg., Nov., 189.
27 C. T. Jackson's Final Rep. Geol. N. Hamp. p. 31.

beauty or size the wonderful crystals of brown tourmaline from Gouverneur, N. Y. Many thousands of perfect crystals, measuring from 1 inch to 6 inches in length, doubly terminated, rich in modifications, but rarely affording a gem over 1 carat, have been taken from this locality. One of the finest is in the Root Collection, at Hamilton College; many fine ones are in the Peabody Museum, New Haven, Conn. ; but the best series of both varieties is in the cabinet of Clarence S. Bement, in Philadelphia. At Richville, near De Kalb, N. Y., achroites (white tourmalines) have been found in fine crystals. The choicest of these, in the cabinet of Mr. Bement, is over 1 inch in length, and would cut into a gem weighing over 10 carats. Crystals of brown tourmaline were obtained by Charles E. Beecher, at Newcomb, Essex County, N. Y. Portions of these crystals were very free from flaws, and material enough was found to cut hundreds of gems weighing from 1 to 10 carats. In the eupyrchroite locality, near Crawford, N. Y., William P. Blake obtained beautiful transparent brown and light-brown tourmaline in crystals large enough to cut gems of several carats' weight. A large number of green tourmalines, some quite thick and several inches in length, have been obtained at Franklin Furnace, Essex County, N. J., but, although they are an important addition to our mineralogical collections, and the outer parts of some of the crystals are of a rich, almost chrome green, not a single crystal has been found that would cut a transparent gem of even 1 carat. Professor Genth mentions beautiful light-yellow, brownish-yellow, and at times white crystals of tourmaline, at Bailey's limestone quarry, East Marlborough, Pa. ; yellow crystals at Logan's limestone quarry, West Marlborough ; brown and light-yellow, which are at times transparent, at John Niven's limestone quarry, New Garden Township; and green tourmaline in talc near Rock Spring, Centre County. Specimens of black tourmaline as fine as are ever obtained, are found near Leiperville, Delaware County, in well terminated crystals, 5 inches in length and 1½ inches thick ; also at Marple Township, terminated with two low rhombohedra.

CHAPTER V.

Garnet Group—Essonite, Spessartite, Almandite, Pyrope, Ouvarovite, Schorlomite.

THE garnet represents a group of minerals which, although chemically quite different, crystallize in the isometric system. The following are the varieties that have been used as gems.

Essonite, which has been confused with zircon, the only true hyacinth, is still called hyacinth by the jeweler. It has a hardness of 7 and its specific gravity is 3·68. Grossularite is the pale-green or yellowish variety of essonite.

Almandine garnets vary in color from violet or purple through brownish-red to deep red. The scarlet and the crimson varieties, when cut en cabochon, are called carbuncles. The finest almandines are from Siriam, India. Their hardness is 7·5 and specific gravity, 4·1 to 4·3.

Pyrope, or blood-red garnet, is commonly known from its use in cheap Bohemian jewelry and is found extensively at a number of places in Bohemia, and also of fine quality at the Kimberley Mines in South Africa. Its hardness is 7·5 and its specific gravity 3·7 to 3·8.

Ouvarovite is of a brilliant emerald-green color and is found at Bissersk, in Siberia, but is rarely large enough to furnish gem stones. Its hardness is nearly 8 and specific gravity 3·45.

Demantoid is a variety of green garnet called Bobrowsaka garnet or Uralian emerald, and is found near Poldnewaja, district of Syssersk, in the Ural Mountains, in nodules varying from $\frac{1}{10}$ to $\frac{1}{4}$ an inch across. The color ranges from yellowish-green or brownish-green to almost yellow emerald-green. The refractive power of the garnet on light is so great that it shows a remarkable amount of "fire" by artificial light. Its hardness is only about 5 and specific gravity is 3·85. It has not been found in the United States.

Essonite, cinnamon garnet, cinnamon-stone, or the hyacinth of the jeweler, has been found of good quality in Oxford County, Me. Very fine essonites, red and yellow, were formerly found at Phippsburgh, Me., and at Warren, N. H. Beautiful essonite crystals, $\frac{1}{4}$ inch in diameter, entirely transparent and quite flat, have been found between plates of mica at Avondale Quarry, Pa., and near Bakersville, N. C. Some of these would cut into fine gems over a carat in weight. In 1882 grossularite was found in perfect, yellow-green, opaque crystals, nearly 1 inch across, in the Gila Cañon, Ariz. The finest in the United States are the rich, dark, oily-green dodecahedral crystals, $\frac{1}{4}$ inch in diameter, from the Tilly Foster Mine, Brewster, N. Y. William P. Blake mentions a green grossularite found in copper ore near Petaluma, Sonoma County, Cal. In the cabinet of Dr. Isaac Lea are transparent crystals of a dark oily-green grossularite, from 1 to 5 millimeters long, that were found at the Good Hope Mine, California. Some fair crystals of a rich, green color, from 1 to 5 millimeters in diameter, were found at Hebron and West Minot, Me. At none of these localities, however, was the mineral of gem value. At Amelia Court House, Va., a large quantity of spessartite garnet, which is a variety of essonite in which part of the alumina is replaced by manganous oxide, has been found in masses several inches across, and of a dark brown, dark red, or honey-yellow color. These are the finest specimens of this variety of garnet ever found, and have been cut into gems from 1 carat to 100 carats in weight, almost rivaling the essonites from Ceylon. Some of the most beautiful natural gems are the microscopic yellow garnets, evidently spessartite, found in cleaning out a small cavity at this place. The beautiful little red spessartites found in the rhyolite cavities with topaz, at Chalk Mountain, Nathrop, Chaffee County, Col., and in Ruby Valley,

Elko County, Nev., are perfect gems, so splendent are they, but they are generally too small or too dark in color for jewelry.

| The finest pyrope garnets in the United States are found in New Mexico, Arizona, and southern Colorado, where they are often called rubies.   In New Mexico they are to be found, it is believed, only on the Navajo Reservation, where the Indians collect them in large quantities from ant-hills and scorpion-hills, in the sand, and also, it is believed, pound them out of the rock. They are found associated with olivine and chrome pyroxene, and in northeastern Arizona they are found in loose sand, having probably been brought by the action of water from a point fifty miles to the north, where they are supposed to occur in a peridotite rock, from which it is said the Indians pound them out with stones.   In the western part of Arizona, on the same parallel with Fort Defiance, on both sides of the Colorado River, garnets have been observed associated with grains of peridot, a chrome pyroxene, and a hyaline chalcedony.   They are also found on the ant-hills and near the excavations made by scorpions, having been taken therefrom by the busy occupants as obstructions to the erection of their galleries and chambers.   They are collected by soldiers and Indians, and sold to the Indian traders, who send them to the large cities in lots of from an ounce upward.   The garnets have never been found in place by any of the geologists or any surveyor of the United States Geological Survey, and it is suggested that they are derived from some lower cretaceous sandstone; but it is very evident, from the associated minerals, that they have weathered out of a peridotitic rock.   They are from $\frac{1}{8}$ to $\frac{1}{4}$ inch in diameter, rarely over $\frac{1}{3}$, and but a few have been seen that measure $\frac{1}{2}$ inch across.   In form they are generally quite round and pitted, often, however, with fractured edges, as if they had been rolled. They average well for quality; one-half are worth cutting, and one-quarter will furnish good stones, but fine ones are quite rare. An interesting fact in connection with these garnets is that a large proportion of them contain a network of fine acicular crystals, evidently rutile from their arrangement, as has been suggested by Babinet and Dr. Isaac Lea.[1]   Occasionally these grains or pebbles of garnet break in two with a conchoidal fracture, reveal-

[1] Proc. Acad. Nat. Sci., Phil., Vol. 21, p. 119, May, 1869.

ing in the center a small grain or kernel of transparent quartz. These garnets are found in a large variety of tints of red, claret, almandine, and even yellow essonite-colored stones. They are often believed by the finders to be spinels or rubies, and have been sold as Arizona or Colorado rubies.

Although the garnets found in washing and mining diamonds at the Cape of Good Hope, the so-called " Cape Rubies," are of larger size than those found in Arizona and New Mexico, and perhaps equal to them in color by daylight, the latter are much superior by artificial light, only the clear, blood-red hue being visible, while in the " Cape Rubies" the dark color remains unchanged. They are extensively used as gems, the annual sales amounting to about $5,000 worth of cut stones. A few remarkably fine ones have brought $50 each, though stones equally good have frequently sold for much less. Fine stones of 1 carat sell at from $1 to $3 each, the exceptional ones rarely for $5. They seldom exceed 3 carats in size. Pyrope garnet of good color, that has furnished gems, has been found in the sands of the gold-washings of Burke, McDowell, and Alexander Counties, N. C. In the peridote rock of Elliott County, Ky., are found deep ruby-red grains of pyrope garnet, locally regarded as rubies, having a specific gravity of 3·673, and varying in size from $\frac{1}{16}$ to $\frac{1}{4}$ inch in diameter. They are especially abundant along the line of the peridote trap-dykes in the soil resulting from the disintegration of the rock, and would cut into gems almost as beautiful as those from Arizona. Garnets are found in many localities in California ; at Roger's Mine, in the eastern part of El Dorado County, they are associated with specular iron, calcite, and iron and copper pyrites ; in the Coosa district, Inyo County, they are found in large, semi-crystalline masses, of a light-yellow color, some specimens of which were taken to San Francisco under the impression that they contained tin. Three miles from Pilot Hill, El Dorado County, garnet rock is found in blocks several feet thick. They also occur in Plumas, Mono, Fresno, Los Angeles, and San Diego Counties, Cal. In Burke, Caldwell, and Catawba Counties, N. C., are found large dodecahedral and trapezohedral almandite garnets, coated externally with a brown crust of limonite, the result of superficial alteration, but usually showing a bright and very compact interior when broken. They are sometimes as fine in color as the Bohemian garnets, and should find a ready use for watch jewels and other like purposes. Some of the crystals which have been found, weighing 20 pounds each, although not fine enough for gems, might be cut into dishes or cups measuring from 3 to 6 inches across. A very large quantity of these garnets has been found about eight miles southeast of Morgantown, and also near Warlick, in Burke County, N. C., and in Rabun County, Ga. Many of them are transparent, varying in color from the purple almandine to pyrope red. Tons of these have been crushed to make " emery " and the sand-paper called garnet paper. The peculiar play of color observed in the North Carolina garnets is often due to the inclusions. In those secured in Rabun County, Ga., at times nearly one-quarter of the entire specimen is taken up by fluid cavities containing acicular crystals of rutile. Quantities of fine purple almandine garnets, which are found in the gravel of the placer mines near Lewiston, Idaho, in rolled and pitted grains from $\frac{1}{16}$ to 1 inch across, would cut into good gems or jewels for watches. Hoffmann mentions good small crystals from Black Cañon, Colorado River, Nev. Fine small almandine garnets are also found in the trachyte of White Pine County, Nev. At Acworth, Grafton, and Hanover, N. H., garnets of gem value have often been found. In Essex County, N. Y., many tons of common garnets are mined annually to be ground into abrasive materials. Many small pieces would furnish clear garnets, and occasionally of fine color. The feldspar quarry at Avondale, Pa., has furnished some of the finest known crystals of common garnet ; one of them, perhaps the finest specimen of this mineral in crystal form, measuring $2\frac{1}{2}$ inches across, imbedded in a mass of quartzite, is of a rich purplish-red color, with high natural polish and remarkably sharp angles. It is in the cabinet of Mr. Bement. At Ruby Mountain, three miles from Salida, Chaffee County, Col., is a remarkable deposit of almandite-garnet crystals in a bed of green chlorite. These crystals vary in weight from 1 ounce to 3 or 4 pounds each, and occasionally 10 or 12 pounds. Two very perfect crystals, weighing respectively 14 and $14\frac{1}{2}$ pounds, were obtained from this locality. They were simple dodecahedrons in form, and were altered to chlorite superficially to the depth of $\frac{1}{10}$ of an inch. Inside they are very compact, and often show two or three distinct zones of color, but are not transparent, hence not of gem value. From the fact that they occur in so soft a matrix, the crystals literally fall out of it when it is broken, and hence are generally perfect. At least 5 tons of these crystals have been sold to collectors and tourists for cabinets, for use as paper-weights and ornaments. They are compact enough to make them valuable for watch jewelry or for ornamental dishes. At Russell, Mass., a vein of garnet, very dark in color, and called there black garnet (not melonite), was opened about 1885, and many fine crystals were obtained and exchanged for minerals, or sold as specimens, to the value of over $1,000. The colophonite from Millsborough, N. Y., although of a beautifully rich, iridescent color, has never been utilized, except as a substitute for emery, owing to the small size of the grains and the friability of the large masses. At Franklin, Sussex County, N. J., immense crystals of the different varieties, melonite, polyadelphite, colophonite, etc., have been found, but rarely in crystals transparent enough to afford a gem. The iron-alumina garnet is found in Concord Township, at Deshong's Quarry, Shaw & Ezra's Quarry, and at Upland, near Chester; also in Darby, Acton, Low Providence, Haverford, and Radnor Townships, Pa. A dark-red variety, similar to pyrope in color, is found in the bed of Darby Creek, near the Lazaretto, in Delaware County. Some peculiar garnets of a deep blood-red color have been mistaken for pyrope. Many garnets from both Chester and Delaware Counties have been cut, and some of them have proved of fine quality and rich color. The Alaska garnets, which are so well known for their remarkably perfect crystals, forming such a beautiful contrast to their dark-gray matrix, occur in great quantities near the mouth of the Stikeen River, in the vicinity of Fort Wrangel, Alaska. They are found about one mile from the river in a bed of mica schist, and after being quarried out, are transported on the backs of men to the river, and thence by boat to Fort Wrangel. As groups of crystals, they are the finest that have been found anywhere, and

SPESSARTITE (MANGANESE-ALUMINA GARNET)

	Color	Locality	Silica	Ferric Oxide	Alumina	Ferrous Oxide	Manganous Oxide	Lime	Magnesia	Soda	Potassa	Water	Hardness	Specific Gravity	Analyst
Theoretical Composition			36.29		20.76		42.95								
Typical Analysis	Red	Fairmount Park, Philadelphia, Pa.	36.34	4.57	12.63		44.20	1.49	0.47				6.5	4.20	C. M. Bradbury[1]
Analysis	D'k red'ish b'n	Fairmount Park, Philadelphia, Pa.	38.27	2.26	19.58	13.58	25.39	0.55					7.0	4.23	W. C. Robinson, Jr.[2]
"	D'k red'ish b'n	Fairmount Park, Philadelphia, Pa.	38.21	2.28	19.46	13.62	25.21	0.53					7.0	4.23	W. C. Robinson, Jr.[3]
"	Dark red	Nathrop, Col.	35.60	0.32	18.55	14.25	29.48	1.15						4.23	L. G. Eakin[4]
"	Reddish	Haddam, Conn.	35.83		18.00	14.93	30.96	0.58		0.21	0.27	0.44		4.128	Seybert[5]
"	Reddish	Haddam, Conn.	36.16		19.76	11.10	32.28	0.22	0.22			0.66		4.275	Rammelsberg[6]

ALMANDINE (IRON-ALUMINA GARNET)

	Color	Locality	Silica	Alumina	Ferric Oxide	Ferrous Oxide	Manganous Oxide	Lime	Magnesia	Specific Gravity	Analyst
Theoretical Composition			36.1	20.6		43.3					
Typical Analysis	Light red	Lake Superior	38.03	20.83		36.15	2.14	2.73	0.97	4.11	Penfield & Sperry[7]
Analysis	Dark red	Milford, Penn	35.92	19.18	4.92	34.47	4.86	2.38	3.70	4.03	E. F. Smith[8]
"	Red	Yonkers, N. Y.	38.32	21.44		30.23	2.46	1.38	6.29		Taylor[9]
"	Red	Delaware Co., Penn.	40.15	21.16		26.66	1.85	1.83	8.08		Kurllann[10]
"	Red brown	Salida, Col.	37.61	20.77		33.83	1.12	1.44	3.61	4.163	Penfield & Sperry[11]

GROSSULARITE (LIME-ALUMINA GARNET)

	Color	Locality	Titanic Oxide	Silica	Alumina	Ferric Oxide	Ferrous Oxide	Manganous Oxide	Lime	Manganese Oxide	Magnesia	Hardness	Specific Gravity	Analyst
Theoretical Composition				40.00	22.8				37.2					
Typical Analysis	White	Wakefield, Canada		38.80	22.66	1.75			35.00		0.68		3.598	Bullman[12]
Analysis	Brown yellow	Delaware Co., Penn.		39.80	21.16	3.14	0.72		34.00	0.30	trace	6.0	3.637	Koenig[13]
"	Green	Delaware Co., Penn.		39.08	23.26	0.80	0.86		28.90			6.0	3.238	R. B. Chipman[14]
"	Cinnamon	Santa Clara, Cal.	0.16	42.04	17.76	5.06			35.01		0.13		3.59	J. L. Smith[15]

PYROPE (MAGNESIA-ALUMINA GARNET)

	Color	Locality	Silica	Ferric Oxide	Alumina	Ferrous Oxide	Manganese Oxide	Lime	Magnesia	Soda	Water	Specific Gravity	Analyst
Theoretical Composition			44.67		25.96				29.77				
Analysis	Red	Elliott Co., Ky.	41.32	4.21	21.21	7.93	0.34	4.94	19.32	0.07	0.17	3.673	T. M. Chatard[16]
"	Blood red	Santa Fé, N. M.	42.11		19.35	14.87	0.36	5.23	14.01		0.07	3.738	F. A. Genth[17]

[1] C. M. Bradbury, Am. J. Sci. III., 25, 334.
[2],[3] Robinson, Jour. Anal. Ch. Vol. I., 251, 1887.
[4] Am. J. Sci. III., 31, 435. L. G. Eakin, Analyst.
[5] Seybert, Am. J. Sci. (1), 6, 155.
[6] Rammelsberg, J. fur pr. ch., 55, 461.
[7] Penfield and Sperry, Am. J. Sci. III., 32, 306.
[8] E. F. Smith, Am. Chem. Jour. Vol. V., p. 272.
[9] Taylor, Am. J. Sci. II., 39, 20.
[10] Kurllann, Am. J. Sci. III., 39, 20.
[11] Penfield and Sperry, Am. J. Sci. III., 32, 306.
[12] Bullman, Am. J. Sci. III., 27, 306.
[13] Koenig, Zeitschrift fur Cryst. and Min. 1, 301, 1878.
[14] Same as [11]. R. B. Chipman, Analyst.
[15] J. L. Smith, Am. J. Sci. III., 6, 434.
[16] T. M. Chatard, Bull. U. S. Geol. Survey, No. 42. Proc.
[17] F. A. Genth, Am. J. Sci. III., 33, 156.

ANDRADITE, COLOPHONITE, MELANITE, POLYADELPHITE (LIME-IRON GARNET)

	Variety.	Color.	Locality.	Silica.	Titanic Oxide.	Alumina.	Ferric Oxide.	Ferrous Oxide.	Manganous Oxide.	Lime.	Magnesia.	Specific Gravity.	Analyst.
Theoretical Composition.	....	....	....	35·43	....	....	31·50	....	....	33·07	....	....	....
Typical Analysis.	Andradite....	Yellow........	Lehigh Co., Penn.	35·25	....	....	32·17	0·92	....	30·86	trace	5·05-1·7	E. F. Smith.[1]
Analysis.	Andradite....	Brown........		31·25	3·19	....	31·80	....	....	33·80	0·46		Stromeyer.[2]
"	Colophonite.	Red........	Franconia, N. H.	38·85	....	....	....	28·15	....	32·00	....	....	Wm. Fisher.[3]
"	Melanite........	Bl'k to yellow brown	New Haven, Conn.	35·09	....	trace	29·15	2·49	0·36	32·86	0·24	3·74	E. S. Dana.[4]

OUVAROVITE (LIME-CHROME GARNET)

	Color.	Locality.	Silica.	Alumina.	Chromic Oxide.	Ferrous Oxide.	Lime.	Magnesia.	Water.	Analyst.
Theoretical Composition.	....	....	44·12	....	30·54	....	33·53	....	....	....
Analysis.	Green........	Orford, Can.	36·65	17·50	6·20	4·97	33·20	0·81	0·30	T. S. Hunt.[5]

SCHORLOMITE (TITANIUM GARNET)

	Color.	Locality.	Silica.	Titanic Oxide.	Alumina.	Ferrous Oxide.	Ferric Oxide.	Manganous Oxide.	Lime.	Magnesia.	Calcium Carbonate.	Analyst.
Analysis.	Colorado........		30·71	4·47	2·26	3·29	22·67	trace	32·41	0·30	3·36	G. A. Koenig.[6]
"		Magnet Cove, Ark.	25·80	12·46	1·00	4·44	23·20	0·46	31·40	1·22	....	" "[7]

ANDRADITE, COLOPHONITE, ETC.
[1] E. F. Smith, Am. Chem. J., Vol. V., p. 272.
[2] Stromeyer, Jahrb, 1864 (13), 23.
[3] Fisher, Am. J. Sci. II., 9, 84.
[4] E. S. Dana, Am. J. Sci. III., 14, 225.

OUVAROVITE.
[5] T. S. Hunt, Final Report Geol. Can., 1863, p. 407.

SCHORLOMITE.
[6],[7] Koenig, Proc. Acad. Nat. Sci., Phil., 1886, p. 355.

many thousands of specimens have been brought from Alaska in the past ten years. Some time ago the United States man-of-war "Corwin" visited the place, and brought away specimens, which are now in the United States National Museum.

The beautiful and rare species, known as ouvarovite or chrome garnet, was first described as occurring in the United States by Prof. Charles U. Shepard, who found it in minute, nearly transparent, emerald-green crystals, $\frac{1}{10}$ inch in diameter, at Wood's chrome mine, Lancaster County, Pa. Ouvarovite is found in large quantities at Orford, Canada, adjoining Newport, Vt., on Lake Memphremagog, sometimes in masses measuring over 1 foot across. The crystals, however, are very small, rarely over $\frac{1}{10}$ inch in diameter, though usually of good color. The white garnet of that locality, described by Dr. T. Sterry Hunt, although not in crystals, is identical with the fine crystals found at Wakefield, Canada, and has been cut into gems. The Wakefield ouvarovite is much finer than the Orford variety. It has been described by Waldemar Lindgren as occurring in small crystals associated with a chromiferous chlorite related to kotscheubeite, from Green Valley on the American River in California. The crystals are of very fine color, but not transparent enough for gems.[1] Schorlomite, which has recently been referred to the garnet group, is really a titaniferous garnet, and occurs at Magnet Cove, Ark. It is generally penetrated by white crystals of apatite, but at times it is free from all foreign matters, and very compact, breaking with a bright conchoidal fracture. On cutting it yields a dead black stone, having a lustre not quite as metallic as that of rutile, but rather between it and black onyx. As it occurs in sufficient quantity, it is suggested as a mineral that will afford a new and fine mourning gem. Stones can be cut of any size up to perhaps about 20 carats, as the mineral is found of sufficient size. The first stone cut was over 6 carats in weight. Prof. George A. Koenig, of the University of Pennsylvania, describes a titaniferous garnet from southwestern Colorado, and also gives an analysis of so-called schorlomite from Magnet Cove, Ark., which he finds to be titaniferous garnet.[2]

[1] See Proc. Cal. Acad. Sci. II., Vol. 1, Dec., 1887.
[2] Proc. Acad. Nat. Sci., Phil., 1886, p. 355.

CHAPTER VI.

Beryl (Emerald Aquamarine), Chrysoberyl, Phenacite, and Euclase.

THE emerald and aquamarine are mineralogically included in the species of beryl. Their difference in color is due to slight traces of other compounds. They crystallize in the rhombohedral system, almost always in six-sided prisms. The specific gravity of the transparent beryl is very nearly 2·7, the hardness of the aquamarine being 8 and the emerald variety about 7·8. The emeralds from Muso are less hard than the aquamarine from Siberia. They are also found in Takowaja, Siberia, and at Zabara, near the Red Sea, in upper Egypt, and in Habachthal, Tyrol. This latter locality evidently furnished some of the material used in ancient Rome. The finest emeralds are found in isolated crystals and in geodes with calcite quartz, iron pyrites, and parisite, and in a clay slate rock containing fossiliferous limestone concretions, at the Muso Mine, near Santa Fé de Bogota, New Grenada. Fine blue and green beryls are found in Brazil, Hindoostan, Ceylon, and in the mica schist of the right bank of the Takowaja River, Ekatharinenburg, Siberia. The emerald variety of beryl is among the most remarkable of American gem minerals. In Alexander County, N. C., emeralds, or beryls suggesting them, have been found at five different points, with quartz, rutile (some of the finest ever found), dolomite, muscovite, garnet, apatite, pyrite, etc., all in fine crystals. One of these localities, Stony Point, is about thirty-five miles southeast of the Blue Ridge Mountains, and sixteen miles northeast of Statesville, N. C.   The surface of the country is rolling, the altitude being about 1,000 feet above sea level.   The soil, which is not very productive, is generally a red, gravelly clay, resulting from the decomposition of the gneissoid rock, and under these circumstances it is easy to find the sources of minerals discovered on the surface.   Prof. Washington C. Kerr's theory of the "frost-drift" is strongly confirmed by the conditions that prevail throughout this region.   The unaltered rock was found at Stony Point at a depth of 26 feet and is unusually hard, especially the walls of the gem-bearing pockets.   A corporation called the Emerald and Hiddenite Mining Company was organized to work the property at Stony Point, and has prosecuted the search for gems irregularly, for periods varying from one week to eight months of each year.   The entire output, including specimens and gems, has amounted to about $15,000.   The history of the discovery of the deposit and its subsequent development is best told in the words of William E. Hidden, the Superintendent.   Recounting the discovery of the mine, he says:[1] "Sixteen years ago the site of the mine now being worked was covered with a dense primitive forest. Less than ten years ago (1871), this county was mineralogically a blank; nothing was known to exist here having any special value or interest.   Whatever we know of it to-day is due directly or indirectly to the earnest field work done here in the past seven years by J. A. D. Stephenson, a native of the county, now a well-to-do and respected merchant of Statesville, N. C. Under a promise of reward for success, he engaged the farmers for miles around to search carefully over the soil for minerals, Indian relics, etc., and for several years he enjoyed surprising success in thus gathering specimens. . . . The amount and variety of the material gathered in this way was simply astonishing, and his sanguine expectations were more than realized.   To be brief and to the point I will state that from a few localities in the County Mr. Stephenson would occasionally procure crystals of beryl of the ordinary kind, but now and then a semi-transparent prism of beryl, having a decided grass-green tint, would be brought

---

[1] The Discovery of Emeralds in North Carolina, by W. E. Hidden.   Privately printed, 8vo, 4 p., 1881, and also Trans. N. Y. Acad. Sci., 1882, p. 101–105.

to him. These the farmers named 'green rocks or bolts,' and became the principal object of the people's searchings. Mr. Stephenson had told them that a dark-green beryl would be valuable if clear and perfect, would in fact be the emerald, and for them to search more carefully than ever to find one. Surely, he had informed the people aright, and had given them a rara avis to look for. It is sufficient to say that within a period of about six years there was found on three plantations in this county, loose in the soil, a number, say ten, of veritable emeralds, none of which, however, were dark-colored or transparent enough for use as gems. All of these specimens went into Mr. Stephenson's collection, with the single exception of one very choice crystal obtained at that locality by the late John T. Humphreys, which crystal is now in the New York State Museum at Albany, after first being in the collection of the late Doctor Eddy of Providence." The original find consisted of nine crystals, one of which was 8¼ inches in length (see Colored Plate No. 5), and weighed 9 ounces; one was 5 inches; others were over 3 inches in length. For two months during the summer of 1885, mining was carried on with flattering success. In the soil overlying the rock, nine crystals of emerald were found, all doubly terminated and measuring from 1 inch to ⅜ inch (25 to 77 millimeters) in width. This latter crystal is very perfect as a specimen; it is of a fine light-green color, is doubly terminated, and weighs 8¼ ounces, or only ¼ ounce less than the famous Duke of Devonshire emerald crystal. Another crystal, doubly terminated, and measuring 2½ inches (63 millimeters) by 1⅛ inch (23 millimeters) is filled with large rhombohedral cavities, formerly containing dolomite. As mineral specimens, these crystals are quite unique. The only gem which has been cut from this find was from a crystal found in a pocket at a depth of over 43 feet. In color it is a pleasing light green and weighs 4¾ carats. In 1887, at the depth of about 70 feet, another crystal that was cut into a gem of 5 carats was found. Both are too light in color to rank as fine gems. The two largest emeralds, and a series of the smaller ones, are in the cabinet of Clarence S. Bement. Some fine ones are in the British Museum mineral cabinet. The fine emerald color characteristic of many of the crystals is confined to the border from $\frac{8}{100}$ to $\frac{8}{100}$ inches in thickness around the edge and near the termination of the crystals.[1]   If this edge were thicker, fine gems could be cut from it.   The finding of fine beryls and emeralds of pale color, collected by Mr. Stephenson, one mile southwest of the Stony Point deposit and a short distance from the place where the same mineral was found by Mr. Smeaton, of New York, shows that the deposit is evidently not accidental, and that there is encouragement for future working in this new locality.

Some beautiful beryls were found at Haddam, Conn., over fifty years ago, the largest of which was 2 inches in length and 1 inch in diameter.   They were remarkable from the fact that part of the crystal was of a transparent green color and free from flaws, while below a certain line of demarcation the whole was white and opaque, as if it were a flocculent precipitate.   Fine specimens from this locality are in the Peabody Museum of Yale University, in New Haven, Conn., the William S. Vaux Collection, at the Academy of Natural Sciences, in Philadelphia, Pa., and the Bement Collection in the same city.   The largest beryls of the world are found at Grafton and Acworth, N. H.   From the former locality a crystal $6\frac{1}{4}$ feet long was quarried and another weighing over $2\frac{1}{4}$ tons.   One obtained from the Acworth Quarries was 4 feet long and $2\frac{1}{4}$ feet in diameter.   One of the best known is on exhibition in the rooms of the Boston Society of Natural History.   (See Illustration.)   It is a hexagonal prism, $3\frac{1}{4}$ feet long by 3 feet wide, and weighs several tons.   There is also an immense beryl in the United States National Museum, that weighs over 600 pounds.   These large crystals are of a pale-green color.   Some very large crystals still remain in the quarries, where they can be seen, but their extraction is a matter of considerable expense, as it involves the moving of a great deal of rock, and, moreover, it is very difficult to get them out whole, since the material of which beryls are composed is very brittle and filled with rifts, and a slight jar is sufficient to break them when they are not well supported; large crystals, consequently, have always been securely hooped before any attempt was made to move them.   Such specimens rarely have transparent spots so large as to

[1] Am. J. Sci. III., Vol. 33, p. 505, June, 1887.

allow the cutting of even a small gem. The beryls from Monroe, Conn., often present interrupted curvatures as shown in Fig. 2. During the last twenty years many beryls, approaching those from New Hampshire in magnitude, have been found in other localites, chiefly in Oxford County, Me., in North Carolina, and in Amelia County, Va., all of which have furnished crystals from 2 to 4 feet in length and 1 foot or more in diameter. Only occasionally small spaces are clear enough to afford gems. Mr. Stephenson called the attention of the writer to a crystal of dark-green beryl, weighing 25·4 ounces, part of which would furnish gems of some size, that was found in January, 1888, near Russell Gap Road, Alexander County, N. C., by a farmer plowing. This locality is about ten miles from the Alexander County Emerald Mine, and is the largest beryl deposit affording gems that has been found in North Carolina. It is noteworthy that the highly modified beryls of this region occur rarely, and only when associated with spodumene or albite, and also that the white or pale-greenish beryls are found with the deepest green spodumene. It has before been noted that the quartz and beryl of Alexander County are more highly modified when implanted

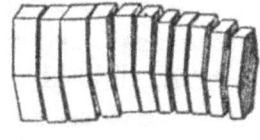

FIG. 2.
BERYL.

on the feldspathic layers of the walls of the pockets. Two emerald beryls, which were found in 1881, at a depth of 34 feet, were in a little pocket, the walls of which were almost covered with crystals of albite twinned parallel to the base. Only four emeralds were found, averaging about 1 centimeter in the three dimensions. The pocket was free from all decomposition whatever. The crystals were of good color, transparent, and had their commoner planes well polished, but they differed to some extent in habit.[1] Blue beryl in fine crystals that afforded fair gems was reported by William E. Hidden from Mitchell County, near the Yancey County line, N. C. In the State cabinet in Albany, N. Y., is a curious beryl found by S. C. Hatch at Auburn, Me. It is of imperfect structure and broken diagonally across, showing the structure to advantage. (See Fig. 2.) It is 8⅜ inches, 30 centimeters high, 8⅜ inches, 22

[1] Am. J. Sci. III., Vol. 33, p. 505, June, 1887.

centimeters wide, and has fifty different layers, twenty-five of beryl, the remaining twenty-five of albite, quartz, and muscovite. All the corners of the hexagonal prism are carried out in full, giving the beryl an asteriated appearance, and making it a striking and interesting specimen. Prof. Parker Cleaveland mentions' having seen several emeralds from Topsham, Me., of a lively green color, scarcely, if at all, inferior to the finest Peruvian emeralds; also two beautiful rose-colored beryls, over 1 inch across, have been found at Goshen, Mass., and are in the Gibbs Cabinet at Yale University. An emerald from Haddam, Conn., deep green in color, an inch in diameter and several inches in length, is mentioned in Bruce's "Mineralogical Journal"' as belonging to Col. George Gibbs' cabinet; but as no true emeralds from Haddam and Topsham are in existence, this may really be a dark-green beryl, as the species beryl is in that locality called emerald.

In the United States National Museum, at Washington, are three beryls, one 6 carats in weight, of a light-green color, another 1 carat, light-blue, from Royalston, Mass., and a third and perhaps the finest specimen ever found at the Portland, Conn., quarries, is 15 carats in weight, and of a rich sea-blue color, almost deep enough to rival in splendor the superb 3-carat Brazilian blue-stone that is in the same case. The writer obtained at Stoneham, Oxford County, Me., two beryls, exceptional for the United States. These were found in 1881, several miles apart, and several miles from the topaz region, by farmers who were traversing pastures in the township. The first was found in two pieces, as if it had been roughly used, and broken, and discarded as worthless, or else broken in taking from the rock and then rejected, its value not being known. This crystal measured 4$\frac{4}{5}$ inches (120 millimeters) long, and 2$\frac{1}{10}$ inches (54 millimeters) wide, and was originally about 5 inches (130 millimeters) long, and 3 inches (75 millimeters) wide. The color was rich sea-green viewed in the direction of the longer axis of the prism, and sea-blue of a very deep tint through the side of the crystal. In color and material, this is the finest specimen that has been found at any North

¹ Mineralogy and Geology, by Parker Cleaveland, p. 341, Boston, 1822.
² Vol. 5, p. 9, 1813.

American locality, and the crystals, unbroken, would equal the finest foreign crystals known. It furnished the finest aquamarine ever found in the United States, measuring 1⅜ inches (35 millimeters) by 1⅜ inches (35 millimeters), by ¾ inch (20 millimeters). It was cut as a brilliant and weighs 133¼ carats. The color is bluish-green, and, with the exception of a few hairlike internal striations, is perfect. (See Colored Plate No. 5.) In addition to this remarkable gem, the same crystal furnished over 300 carats of fine stones. The other crystal is doubly terminated, being 1⅝ inches (41 millimeters) long, and ⅝ inch (15 millimeters) in diameter. Half of it is transparent, with a faint green color, the remainder is of a milky green and only translucent. Where the two colors meet, the crystal, like the Haddam beryls, has the appearance of a solution in which a flocculent precipitate has almost completely settled, leaving the upper portion nearly clear.

Beryl, resembling the Siberian, is found in greenish-yellow and deep-green crystals, in the South Mountains, nine miles southwest of Morganton, Burke County; in the Sugar Mountains at Shoup's Ford, Dietz's, Huffman's, and Hildebrand's; and in smaller crystals in Jackson County, N. C. One fine blue-green crystal in quartz was found at Mill's Gold Mine, Burke County, and one fine transparent green crystal from that vicinity is now in the cabinet of M. T. Lynde, of Brooklyn, N. Y. Fine blue-green aquamarine occurs at Ray's Mine on Hurricane Mountain, Yancey County, N. C. Clear green beryls have been found at Balsam Gap, Buncombe County; Carter's Mine, Madison County; Thorn Mountain, Macon County, and at Wells, Gaston County. Some crystals 2 feet long and 7 feet in diameter, small pieces of which would cut into gems with small, clear spots, occur four miles south of Bakersville Creek, and still larger crystals, not of gem value, at Grassey Creek, N. C. Beautiful transparent beryls have been found at Streaked Mountains, Norway, Lovell, Bethell, and Franklin Plantation, Me., and very good ones also at Mount Mica and Grafton, Me. At Albany, Me., have been found beautiful transparent golden-yellow beryls that would cut into perfect gems of over 2 carats each. A fine sea-green aquamarine beryl, weighing about 7 carats, was found near Sumner, Me. Some very clear white stones are obtained at Pearl Hill, Fitchburg, Mass., and are there sold by the local dealers. A very fine golden-yellow beryl of 4 carats, from this locality, is in the collection of Doctor Hamlin. Fine crystals of beryl, of almost emerald-green color, also beautiful yellowish-green and bluish beryls, are found in Deshong's Quarry, near Leiperville, Pa. At Shaw & Ezra's Quarry, near Chester, in Upper Providence, and in Middletown, Concord, and Marple Townships, fine specimens have been found. Fine beryls also have been found at White Horse, three or four miles below Darby, Pa. Bluish-green and blue beryls occur in the vicinity of Unionville, Newlin Township, and on Brandywine battlefield, in Birmingham Township. One crystal, of a dark tourmaline green tint, over $\frac{1}{4}$ inch long, in the cabinet of Michael Bradley, of Chester, Pa., is from Middletown, Delaware County, and would afford a fine gem. Some of the stones here have much the appearance of bluish emeralds. The finest golden-yellow beryls are found at the Avondale Quarries, Delaware County, Pa. A 20-carat gem is in the cabinet of Mrs. M. J. Chase, of Philadelphia, and the material for another is in the cabinet of Clarence S. Bement. In 1882 B. B. Chamberlin found in Manhattanville, New York City, six fine yellow beryls that cut into stones of 1 to 2 carats each. At a mica mine in Litchfield County, Conn., between Litchfield and New Milford, were found during the past four years a quantity of deep-yellow, light-yellow, yellow-green, light-green, and white beryls, which were cut into gems and extensively sold as jewelry, the former under the name of golden beryl. Several thousand dollars' worth of beryls from this locality were annually sold. These beryls were at first placed on the New York market as an entirely new stone, said to be very nearly as hard as the sapphire, and to be from some South American locality. Prof. Eugene A. Smith, State Geologist of Alabama, obtained from Coosa County, Ala., some light, golden-yellow beryls of sufficient transparency to furnish small gems. Large masses weighing many pounds, of translucent, light sea-green beryl, were obtained at Branchville, Conn., in connection with other minerals described by Prof. George J. Brush and Prof. Edward

S. Dana. These were handsome enough to furnish ornamental objects, small balls 1 or 2 inches in diameter, charms, etc.

Aquamarine has been found in a number of localities in the United States, the principal among them being Royalston, Mass.; Acworth, N. H.; Grafton, Vt.; Burke County; Stony Point, N. C.; Paris, Me.; Fitchburg, Mass.; and Avondale, Pa. The richest colored gems from any known locality have been found at Royalston, Mass. Although small, they are almost as blue as sapphire. Large, clear gems of light-blue and sea-green tint have been found at Acworth, Grafton, and Stony Point; at the latter locality shading into beryl-emerald. The crystals of beryl found associated with phenacite on Mount Antero, Chaffee County, Col., are obtained at an altitude of from 12,000 to 14,000 feet, and vary in size from 1 to 4 inches in length and from $\frac{1}{16}$ to 1 inch in diameter. As crystals, they are re-markable for the fact that portions of them have been entirely dissolved or eaten away, which gives them a peculiar etched appear-ance. In a number of instances not only have the ends of the crystals entirely disap-peared, occasionally leaving long, needle-like projections, but holes have been eaten through the crystal. In color they vary from a very light-blue to quite a dark sky-blue, almost as rich as some of the finest Brazilian crystals. They would furnish gems up to 10 carats in weight, the largest one cut weighing 5 carats.

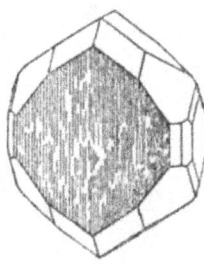

FIG. 3.
CHRYSOBERYL.

A variety of blue beryl, called goshenite, occurs at Goshen, Mass., in pieces transparent enough to afford gems. Chryso-beryl occurs in orthorhombic prisms, and frequently more or less modified as shown in Fig. 3. Its hardness is 8·5, next to that of sapphire, and its specific gravity is from 3·65 to 3·85. In color it varies from yellow or golden-yellow through brown and green, including a large series of sage-green and leaf-green, as well as rich brown. Alexandrite is the variety of chrysoberyl that is colored by chromium. It is, by natural light, of a deep leaf or olive-green color, but by candle-light appears a rasp-berry or columbine-red shade. The true cat's-eye is a variety of chrysoberyl that owes its chatoyancy to minute internal striations

BERYL

	COLOR.	LOCALITY.	Silica.	Alumina.	Ferric Oxide.	Manganous Oxide.	Ferrous Oxide.	Beryllia.	Caesia.	Potassa.	Soda.	Lithia.	Lime.	Water.	Specific Gravity.	Analyst.
Theoretical Composition.			66.80	19.10				14.10								
Typical Analysis	Leek green	Alexander Co., N. C.	66.28	18.60			0.22	13.61						0.83	2.703	F. A. Genth[1]
Analysis	Yellow	Amelia Co., Va.	65.24	17.05	2.20			12.64	2.92		0.68		0.57	0.70	2.702	R. W. Baker[2]
"	Colorless	Hebron, Me.	65.10	18.89			0.49	10.35	1.61	0.10	1.82	1.17	0.35	2.33		S. L. Penfield[3]
"	Yellow	Norway, Me.	64.12	17.89			0.16	12.23			1.21	0.75		2.24	2.747	F. L. Sperry[4]
"	Colorless	Litchfield Co., Conn.	65.02	17.86	0.37		0.18	13.50			0.54	0.10		2.34	2.716	Penfield & Sperry[5]
"		Willimantic, Conn.	65.72	18.40		0.12	0.26	13.08	0.03		0.75	0.28		2.06		Penfield & Sperry[6]
"	Bluish green	Stoneham, Me.	65.54	17.75	0.21		0.38	13.73		0.12	0.71	trace		2.01	2.725	Penfield & Harper[7]
"	Clear green	Green Co., Tenn	65.39	19.10				13.35						1.76	2.706	F. W. Clarke[8]
"	Green	Acworth, N. H.	68.33	17.60	trace		0.05	14.00								C. T. Jackson[9]
"		Royalston, Mass	67.52	17.42				14.35								T. Petersen[10]
"		Goshen, Mass	66.97	17.22	trace		2.03	12.99							2.65	J. W. Mallet[11]

[1] F. A. Genth, Trans. Am. Phil. Soc., Aug. 18, 1882, p. 402.

[2] R. W. Baker, Am. Chem. Jour., 7, 175.

[3] L. Penfield, Am. J. Sci. III., 28, 25.

[4],[5],[6] Penfield & Sperry, Am. J. Sci., Nov., 1888, p. 317.

[7] S. L. Penfield & D. N. Harper, Am. J. Sci. III., 32, 111.

[8] F. W. Clarke, U. S. Geol. Sur. Bull. No. 9.

[9] C. T. Jackson, Final Rep. Geol. and Min. of N. H., p. 182.

[10] T. Petersen, Jahrnb. 1866, p. 905.

[11] J. W. Mallet, Am. J. Sci. II., 17, 180.

of the composite crystals of which the latter is made up, or the twinning of the crystal; or when certain minerals have been deposited between the layers during crystallization, the stone, being cut en cabochon across these lines, exhibits the phenomenon. The stellate effect frequently produced by the twinning of chrysoberyl is shown in Fig. 4. Alexandrite was named after Alexander I., Czar of Russia, on whose birthday it was discovered. Large crystals that occasionally furnish gems are found in Takowaja, Siberia. Fine gems, up to 67 carats each, have been found during the last ten years in the kingdom of Kandy, Ceylon, associated with the true cat's-eye, and the yellow, brown, and green chrysoberyl. This is also found in the alexandrite variety, but it is extremely rare. Beautiful light-golden chrysoberyls (the chrysolite of the jeweler, valued at nearly as high a rate as the

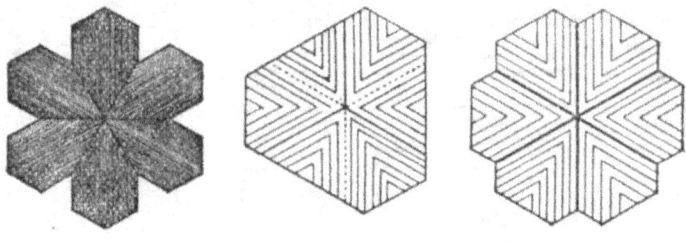

FIG. 4.

STELLATE EFFECTS FREQUENTLY PRODUCED BY THE TWINNING OF THE CHRYSOBERYL.

diamond in the time of Louis XIV.) have been found in Brazil, also fine light-yellow cat's-eyes.

Chrysoberyl, of sufficient transparency to be of gem value, is not found in North America. It has been found at Stoneham, also at Canton, Peru, Norway, and Stow, Me., but thus far not in fine specimens. Some of the small yellow crystals occurring in the fibrolite at Stoneham are, however, quite perfect in form. Small crystals occur at Canton and Stow, Me., together with large, coarse crystals. At Stow[1] it has been found in masses weighing about 5 pounds each. A single distorted crystal 3 by 5 by 1 inches, opaque, and of a dull yellow-gray color, has also been found, which may in part furnish very poor chrysoberyl cat's-eye. Nathaniel H. Perry found one small, very perfect crystal at Tubb's Ledge, Me., and it has also been observed at

[1] Trans. N. Y. Acad. Sci., Vol. 2, p. 64, Jan. 22, 1883.

Speckled Mountain, Norway, Me., by Prof. Addison E. Verrill. A crystal 3½ inches long by 1 inch wide, from Topsham, Me., and one 1½ inches long by 1 inch wide, from Buckfield, Me., are in the collection of Prof. George J. Brush, of New Haven, Conn. Rev. Frederick Merrick stated that he had collected, fifty years ago, some crystals that he believed would furnish gems, but perhaps not of the finest quality, at Haddam, Conn., the old and well-known locality now exhausted. At Greenfield, one mile north of Saratoga Springs, N. Y., now also exhausted, were found many beautiful crystals; also in New Hampshire in granite, at the deep cut of the Northern Railroad, at Orange Summit. None of these localities, however, has furnished a fine gem. The most promising localities are those in Maine, and gems, if found at all, will be likely to occur there. Haddam, Conn., has furnished

CHRYSOBERYL

Chemical Composition and Properties.	Theoretical Composition	Locality Haddam, Conn. Analyst, Damour.[1]	Locality North America. Analyst, Thomson.[2]	Locality Haddam, Conn. Analyst, Damour.[3]	Locality Haddam, Conn. Analyst, Damour.[4]
Ferrous Oxide......	....	....	4·49	....	....
Alumina	80·20	76·99	76·75	76·02	75·43
Beryllia	19·20	18·88	17·79	18·41	17·93
Ferric Oxide.......	....	4·13	....	4·51	4·06
Titanic Oxide......	....	....	....	....	....
Loss on Ignition ...	....	....	0·48	....	....

[1] Typical Analysis, Rammelsberg's Mineral Chemie, p. 128. [2] Phillip's Mineralogy, 1852, p. 268.
[3], [4] Damour, Ann. Chim. Phys. III., 7, 173.

many fine twin crystals. Among some rolled quartz pebbles sent from North Carolina for examination, a transparent yellow chrysoberyl was observed, which would afford a ¼-carat stone. The alexandrite variety of chrysoberyl has not been observed.

Phenacite crystallizes in the rhombohedral system. Its hardness is about 8 and its specific gravity about 3·0. It is a silicate of glucinum. The colorless, transparent variety is one of the most brilliant stones known, occasionally showing prismatic colors (or fire), by candle or artificial light. The finest large specimens known are found at Takowaja, fifty-six miles east of Ekaterinburg, in Siberia. Phenacite was first identified in the United States in 1882, when it was discovered in the Pike's Peak Region, Col.,[1] and more recently on Bald Mountain, North

[1] Am. J. Sci. III., Vol. 24, p. 282, Oct., 1882.

Chatham, N. H. Both localities have furnished crystals of sufficient size and quality to be cut into fair gems. The first occurrence of this mineral in the United States was mentioned by Whitman Cross and W. F. Hillebrand,[1] who published a short description and figure of a crystal occurring with mica and amazon-stone in El Paso County, Pike's Peak Range, Col., where some of the largest crystals in the United States have been found. The largest and finest phenacite crystal ever found in the United States is the one in the possession of Clarence S. Bement. It is from Crystal Park, Col., and weighs 59 penny-weights and 6 grains, and measures nearly 2 inches (46·5 milli-meters) in length, and 1¼ inches (32 millimeters) in thickness. Occasional transparent spots are noticeable in it. A second locality is at Topaz Butte, near Florissant, about sixteen miles

FIG. 5.
CRYSTAL OF PHENACITE FROM COLORADO.

from Pike's Peak. The crystals here are usually implanted on amazonstone and topaz. Many hundreds of them were found, varying in size from 1/25 inch to ⅘ inch (1 millimeter to 20 millimeters) in diameter, of which quite a number were transparent. They are often readily detached, or occur in a brown mass believed to be fayalite. The other Colorado locality is Mount Antero, where the crystals are found at an altitude

[1] Am. J. Sci., III., Vol. 24, p. 282, Oct., 1882. See also full description of phenacites from Crystal Park and Florissant, Col., by Whitman Cross and W. F. Hillebrand in Bulletin No. 20, of the United States Geological Survey, Washington, 1885. Phenacite from the Florissant locality was described later by William E. Hidden, Am. J. Sci., III., Vol. 29, p. 249, March, 1885. The crystals at Florissant were first found by J. G. Heistand, of Manitou, Col. See Samuel L. Penfield, Am. J. Sci., III., Vol. 36, p. 320, Nov. 1888.

of about 14,000 feet, in a region of almost perpetual snow, which is accessible for only a short period during the summer.   Hundreds of crystals have been found attached to and implanted on quartz crystals, transparent beryl, and Baveno twin crystals of orthoclase feldspar.   The largest crystal found measured over 1 inch across and was nearly 1 inch long.   The crystals are nearly all quartzoids or simple rhombohedrons.   (See Fig. 5.) Some have a faint wine-color and others a smoky, bluish tinge. Some smoky quartz crystals, with crystals of phenacite in the center, were observed.   In May, 1888, E. A. Andrews, of Stow, Me., discovered some crystals of phenacite on Bald Mountain, North Chatham, N. H., near the State line between Maine and New Hampshire, and in the neighborhood of Stoneham, Me. They were found in a vein of coarse albitic granite,[1] associated with crystals of smoky quartz, topaz, and muscovite, some implanted on smoky quartz, a few attached so loosely to the matrix by one of the rhombohedral faces that they could be removed without being broken.   They were about fifty in number, lenticular in shape, and measured from $\frac{1}{8}$ inch to $\frac{1}{2}$ inch (3 millimeters to 12 millimeters) across, and from $\frac{1}{25}$ inch to $\frac{1}{8}$ inch (1 millimeter to 3 millimeters) in thickness.   They were all white or colorless, with polished faces, and for the most part very simple in form. The series from Pike's Peak, Col., has been described by Prof. Samuel L. Penfield.[2]   Few of the phenacites found in the United States have been cut into gems, but several thousand dollars' worth have been sold as mineralogical specimens, and now adorn the cabinets of the world.

Mention is made of euclase in the United States as follows : Several crystals were reported as having been found at Mills's Spring, Polk County, N. C., by Gen. Thomas L. Clingman, in washing the gold sand at this locality, but Prof. Frederick A. Genth says[3] that they were not euclase.   It was also mentioned as having been found in connection with topaz at Trumbull, Conn., but this report proved incorrect.[4]   With the full series of glucinum found in this country, it is not unlikely that euclase

[1] Am. J. Sci. III., Vol. 27, p. 212, March, 1884.
[2] Am. J. Sci. III., Vol. 33, p. 131, Feb., 1887.
[3] Minerals and Mineral Localities of North Carolina, p. 54, 1881.
[4] Am. J. Sci. I., Vol. 43, p. 366, July, 1842.

will soon be reported. It has been found in magnificent crystals at Villa Rica, Brazil, S. A., but it is of extremely rare occurrence, is highly cleavable, and is scarcely known except to mineralogists.

The peridot of the jeweler, which is the chrysolite or olivine of the mineralogist, is found in abundance and of a good quality, in the form of small, olive-green, pitted grains or pebbles associated with garnet, in the sands of Arizona and New Mexico. Locally they are called Job's tears on account of their pitted appearance. This material affords smaller gems than those coming from the Levant, and as the demand seems to be for the large peridots of the richer olive-green color, which is not possessed by those from the United States, only a small number of the peri-

PHENACITE

CHEMICAL COMPOSITION AND PROPERTIES.	Theoretical Composition.	LOCALITY. Topaz Butte, Florissant, Col. Analyst, Sperry.[1]	LOCALITY. Topaz Butte, Florissant, Col. Analyst, Penfield.[2]
Silica...........................	54·25	54·46	54·42
Beryllia..........................	45·75	45·57	45·60
Soda............................	.....	0·21	.....
Lithia...........................	.....	Trace	.....
Water...........................	.....	0·26	.....
Color....	.....	Clear—Colorless.	Clear—Colorless.
Specific Gravity..................	.....	2·966–2·957	2·966–2·957

[1], [2] S. L. Penfield and E. S. Sperry, Am. J. Sci., Nov., 1888, p. 320.

dots found in the West have been cut into gems. Many of the so-called "emeralds" in European church treasuries, notably those of the "Three Magi" in the Cathedral of Cologne, are peridots and not emeralds, but the locality whence they were taken is now unknown. All the peridots that are sold in modern times are taken out of jewelry which is often two centuries old. The chrysolite of the French jewelers is chrysoberyl. From the meteoric iron that was found on Glorietta Mountain, Santa Fé County, N. M., in 1885, the writer[1] obtained some peridots of 1 carat in weight, that were transparent and yellowish-green in color. The meteorite that was found on Glorietta Mountain, Santa Fé County, N. M., and the one found at Eagle Station, Carroll County, Ky., is believed to be identical with the piece,

[1] Am. J. Sci. III., Vol. 32, p. 311, Oct., 1886.

CHRYSOLITE

Color.	Locality.	Silica.	Ferric Oxide.	Chromic Oxide.	Alumina.	Ferrous Oxide.	Nickel Oxide.	Cobalt Oxide.	Manganous Oxide.	Lime.	Magnesia.	Titanic Oxide.	Water.	Specific Gravity.	Analyst.
....	Meteorite from Carroll Co., Ky..	37.90	....	....	....	19.66	....	....	....	0.42	41.65	....	....	3.470	Mackintosh.[1]
....	Meteorite from Little Miami V'll'y.	40.02	....	....	....	14.06	....	....	{0.10	1.24	45.60	....	....	3.336	Kinnicutt.[2]
....	Waterville, N. H.	38.85	....	....	....	28.07	....	....	}	1.43	30.62	....	....	....	E. S. Dana.[3]
Pale grayish green..	Webster, Jackson Co., N. C.	41.89	{0.58		trace	7.39	0.35	trace	trace	0.06	49.13	....	....	....	F. A. Genth.[4]
Y'll'wrish olive gr'n..	Webster, N. C.	40.87	{1.27		trace	7.39	0.50	trace	trace	undet.	undet.	....	....	....	" " "
Y'll'wish olive gr'n.	Webster, N. C.	40.74	1.83		trace	7.26	0.39	trace	trace	0.02	49.18	....	....	....	Manice.[7]
....	Thetford, Vt	40.75	....	....	....	9.35	....	....	....	....	50.28	....	....	....	" " "
....	Webster, N. C.	41.17	....	....	....	7.35	0.41	....	....	0.04	49.16	....	....	....	F. A. Genth.[8]
....	Elliott, Ky.	40.05	2.36	0.24	0.39	7.14	trace	....	0.20	1.16	46.68	0.07	0.80	3.377	T. M. Chatard.[9]

[1] J. B. Mackintosh, Am. J. Sci. III., 33, 228.
[2] N. Jahrb. für Min., 1888. Vol. I., Part 2.
[3] E. S. Dana, Am. J. Sci. III., 3, 48.

[.] Am. J. Sci. III., 37, 225.

[4],[5] F. A. Genth, Am. J. Sci. III., 33, 200.
[7] Manice, Am. J. Sci. II., 31, p. 359.
[8] F. A. Genth, Am. J. Sci. II., 33, 199.

now in the meteorite collection at Harvard University, that was found on the altar of one of the Altar Mounds in the Little Miami Valley, Ohio, by Prof. Frederick W. Putnam. Both of these meteorites contained clear crystals of olivine that would cut into gems of over 1 carat each, so that these may be truly called " celestial precious stones."

The gem stone called zircon is sometimes known as jargon or jargoon, jacinth or the true hyacinth. Its hardness is about 7·5, and its specific gravity is generally 4·7, although it is variable, ranging from 4·1 to 4·9. It is a silicate of zirconium, containing zirconium oxide sixty-seven parts, and silica thirty-three parts. It has a large range of color owing to its high dispersive power, and exhibits more " fire " than any other known gem except the diamond. The finest gem stones come from Ceylon, Mudgee, and New South Wales. Those from Expailly in Auvergne, France, are exceedingly small and are of true hyacinth color.

Zircon has not often been found in the United States in pieces sufficiently clear to warrant cutting. Some very small crystals of good color have been found in Burke County, N. C., and the terminations of the zircons from St. Lawrence County, N. Y., might be cut into very small gems of 1 carat or less in weight. Near the Pike's Peak toll-road, almost due west from the Cheyenne Mountains, following a vein-like mass of white quartz in granite, is found a very interesting form of zircon. The crystals are either in the quartz or in a soft yellow material, and are generally a deep reddish-brown, pink, or pale honey-yellow, and T. Whitman Cross mentions a few small crystals of deep emerald-green color. The crystals are all pyramidal, having very little or no prism. The largest observed were about $\frac{1}{4}$ inch, but generally they are not more than $\frac{1}{16}$ to $\frac{1}{8}$ inch in length, and would only cut into minute gems. As crystals, however, these are perhaps the most beautiful known, owing to their transparency, brilliancy, and perfection.[1] Some white and colorless crystals sent from this locality were found to be the result of heating, which destroyed the natural color. Ceylon zircons treated in this way were formerly used for incrusting watches, which were then sold as diamond-incrusted, so greatly did this

[1] Am. J. Sci. III., Vol. 24, p. 285, Oct., 1882.

variety resemble the diamond. An opaque variety of zircon is found in several localities in the Pike's Peak District, in one instance associated with amazonstone, and in another with astrophyllite, also with flesh-colored microline. No material that would cut into gems has been found at any of these localities. In North Carolina zircon is abundant in the gold sands of Polk, Burke, McDowell, Rutherford, and Caldwell Counties, in nearly all the colors peculiar to Ceylon—yellowish-brown, brownish-white, amethystine, pink, and blue. The crystals are beautifully modified, but too minute to be of any value. Brown and brownish-yellow crystals, very perfect in form, occur abundantly in Henderson County, N. C., and in equal abundance in Anderson County, S. C. The latter are readily distinguished from the North Carolina crystals, as they are generally larger, often an inch across, and the prism is almost always very small, the crystal frequently being made up of the two pyramids only. Fine crystals of zircon have been found in Lower Saucon Township, Northampton County, Pa., three-fourths of a mile north of Bethlehem. The gravels of the Delaware and Schuylkill Rivers contain considerable quantities of very minute, nearly colorless, crystals of zircon. Some fine ones, over an inch in length, have been found at Litchfield, Me., and all through the cancrinite and sodalite rocks near that place. In the Canfield Cabinet at Dover, N. J., there are some of the finest known black zircon crystals, over an inch long, that were found near Franklin, N. J. Opaque green zircons in crystals an inch long and a half-inch across have been found by C. D. Nimms in the town of Fine, St. Lawrence County, N. Y. They were remarkable mineralogical specimens, but of no gem value. One found by Dr. Samuel L. Penfield, now in the United States National Museum, is nearly 4 inches long and doubly terminated. During 1886, the demand for minerals containing the rare earths, zirconia, thoria, glucina, etc., greatly increased, as they were then wanted to furnish the mantles or hoods of incandescent gas-burners. This demand led at once to active search by collectors and mineral-dealers in England, Germany, France, Russia, Norway, and Brazil, and especially in the United States. So thorough and successful has this search been that many minerals which were then considered rare are now so plentiful that they are quoted at one-tenth to one-hundredth of their former prices. The best zircon locality in North Carolina is on the old Meredith Freeman Estate, Green River, Henderson County. It was leased for twenty-five years by Gen. Thomas L. Clingman of that State, who, as early as 1869, mined 1,000 pounds of zircon, and during that whole period never lost faith in the incandescent properties of zirconia; but when these were proved and acknowledged, through some legal difficulties General Clingman had forfeited his leases, and hence failed to reap his reward. The Henderson County, N. C., and Anderson County, S. C., zircon

ZIRCON

Chemical Composition and Properties.	Theoretical Composition	Locality, Litchfield, Me. Analyst, W. Gibbs.[1]	Locality, Reading, Pa. Analyst, C. M. Wetherell.[2]	Locality, Buncombe Co., N. C. Analyst, C. F. Chandler.[3]	Locality, El Paso Co., Col. Analyst, G. A. Koenig.[4]
Silica	33·00	35·29	34·07	33·70	29·70
Zirconia...........	67·00	63·33	63·50	65·30	60·98
Ferric Oxide......	.....	0·79	2·02	0·67	9·20
Magnesia.........	.....	.....	.. .	.....	0·30
Water	.....	.....	0·50	0·41	.....
Color............	.....	.....	Chocolate.	.....	Iron Black.
Specific Gravity...	.....	4·7	4·595	4·607	4·538

[1] W. Gibbs, Ann. der Phys. Pogg., 71, 559.
[2] C. M. Wetherill, Trans. Am. Phil. Soc. 10, 346. Am. J. Sci. II., 15, 443.
[3] C. F. Chandler, Am. J. Sci. II., 24, 131.
[4] G. A. Koenig, Trans. Am. Phil. Soc. 16, 518, 1877.

is found in large quantities, loose in the soil, as the result of the decomposition of a feldspathic rock. The crystals are generally remarkable for their perfection, are distinct in each locality, and weigh occasionally several ounces. The recent demand has also brought to light the existence of enormous quantities of zircon in the Ural Mountains and in Norway. Though very large crystals, some weighing 15 pounds, have been found in Canada (Renfrew and the adjoining counties), they are so isolated that it would be impossible to obtain a supply there. The new demand has brought together more than 25 tons of zircon; and this mineral may prove of considerable value, for the earth it contains can be used as a refractory material for crucibles and furnaces. As new processes have cheapened and made available aluminum and magnesium, so zirconium may yet be called into use.

CHAPTER VII.

The Quartz Group:—Rock Crystal, Transparent Quartz, Amethyst, Smoky Quartz, Cairngorm Stone, Gold Quartz, Rose Quartz, Novaculite, Silicified Coral, Quartzite, Quartz Inclusions, Thetis Hairstone, Agate, Jasper, Silicified Wood, Opal, Hydrophane, etc.

THE quartz group consists of a large series of substances, which to the eye are very unlike each other, and pass under a great variety of names, but they are all chemically of one substance, namely, silica. The various colors are evidently due to the presence of metallic oxides, principally manganese or iron. Quartz may be divided into two groups, crystalline and cryptocrystalline. The former crystallizes in the rhombohedral system, generally as six-sided prisms, with a hardness of 7 and a specific gravity for the colorless of 2·65, the specific gravity of cairngorm and the amethyst being slightly higher.

Following is a list of the crystalline varieties :

AMETHYST.—Deep purple, bluish violet fading almost into pink.

ASTERIATED STAR-QUARTZ.—Containing between layers of the crystals a deposition of substances, so that when cut en cabochon across the prism it exhibits asterism.

AVENTURINE.—Transparent to opaque, either red or yellow, with iridescent spangles of mica distributed through it.

CAIRNGORM.—Transparent, smoky gray, yellow, yellowish brown, and brown.

CAT'S-EYE.—Translucent, gray or greenish, chatoyant when cut en cabochon, an effect due to fibres of asbestus or actinolite.

HYALINE.—Opalescent white, due to admixture of chalcedony.

MILK QUARTZ.—Opalescent, milky white, sometimes yellow by transmitted light.

MORION.—Deep black, almost opaque.

PRASE.—Translucent, leek-green, deep green.

ROCK CRYSTAL.—Transparent and colorless.

ROSE QUARTZ.—Rose red or pink, sometimes opalescent.

SAPPHIRINE QUARTZ or SIDERITE.—Translucent and grayish blue, indigo, and Berlin blue color.

SAGENITE.—Penetrated with acicular crystals of other minerals, generally rutile, tourmaline, göthite, stibnite, asbestus, actinolite, hornblende, epidote, etc.

SMOKY QUARTZ.—Transparent, and various shades of gray and brown.

FALSE TOPAZ, SCOTCH, SAXON, OR SPANISH TOPAZ.—Transparent yellow or light brown, generally the result of decolorization by heat.

The cryptocrystalline varieties are:

AGATE CHALCEDONY.—Jasper or rock crystal, mottled or in layers; when irregular, called fortification agate; when banded, banded agate.

AGATE JASPER.—A variety of agate containing jasper.

BASANITE, LYDIAN STONE, OR TOUCHSTONE.—A velvet-black siliceous stone, or flinty jasper, used by the jewelers for trying the purity of precious metals.

BEEKITE.—Silicified corals, shells, or limestones, resembling chalcedony.

BLOODSTONE.—Jasper, translucent to opaque, green with red spots.

CHALCEDONY.—Clouded or translucent, white, yellow, brown, or blue.

CHRYSOPRASE.—Translucent, pale bluish-green or yellow-green.

CARNELIAN.—Translucent like horn, yellow, brown, or red.

EGYPTIAN JASPER.—Opaque, concentric, with other layers of brown, yellow, or black.

HELIOTROPE.—With base of chalcedony colored with green delessite, red spots of iron oxide.

JASPER.—Impure, opaque-colored quartz, red, yellow, brown, or gray-blue, called ribbon jasper when striped.

ONYX.—Like agate, but consisting of distinct, even layers, so that it can be used in cutting cameos.

PLASMA.—Bright green, leaf-green, and almost emerald-green, very translucent.

PORCELAIN JASPER.—Different from true jasper in being visible gray, white, and pink.

SARD.—Translucent, red, brownish red, crimson, blue-red and blackish red, golden and amber.

SARD ONYX.—Like onyx, but having a stratum or several strata of sard.

A remarkable mass of rock crystal, weighing 51 pounds, was sent, in 1886, to Tiffany & Co., New York. It purported to be from Cave City, Va., but as it subsequently proved was found in the mountainous part of Ash County, N. C.[1] The original crystal, which must have weighed 300 pounds, was unfortunately broken in pieces by the ignorant mountain girl who found it, but the fragment sent to New York was sufficiently large to admit of being cut into slabs 8 inches square and from half an inch to an inch thick. The original crystal, if it had not been broken, would have furnished an almost perfect ball 4½ or 5 inches in diameter. A visit to the locality by the author showed that this specimen had been found near Long Shoal Creek, on a spur of Phœnix Mountain in Chestnut Hill Township. There have also been found at two places, 600 feet apart (about one mile from the former locality), two crystals, one weighing 285 pounds, that was 29 inches long, 18 inches wide, 13 inches thick, showing one pyramidal termination entirely perfect and the other partly so; also another specimen that weighed 188 pounds. These crystals were all found in decomposed crystalline rocks consisting of a

[1] Proc. Am. Ass'n Adv. Sci., Vol. 35, p. 229, 1886.

coarse feldspathic granite, and were obtained either by digging where one crystal had been found, or by driving a plough through the soil. Altogether, there have been found in this vicinity several dozen crystals, weighing from 20 to 300 pounds each, and future working will undoubtedly bring more to light. These large crystals are often very irregular and pitted, like many of the crystals of quartz from St. Gothard. Of those found, the most irregular was 20½ pounds in weight, with the entire surface rough and opaque like ground glass, and almost spherical in form but perfectly transparent. In a few instances, they had a coating of rich, green-colored chlorite that penetrated to the depth of an inch. This was left on the quartz, and it gave the cut object, after polishing, the effect of a pool of water with green moss growing on the bottom. A large piece weighing 11 pounds, brought from Alaska in 1884, originally formed a part of a mass that must have weighed 44 pounds. It afforded clear crystal slabs for hand-glasses, 3 by 5 inches. The superiority of this mineral over glass lies in the fact that it does not, like glass, detract from the rosiness of the complexion, as is well shown in the fine mirror of this substance in the green vaults at Dresden, Saxony. Transparent crystallized quartz is found in many places in the United States. At Lake George, in Herkimer County, and throughout the adjacent regions in New York State, the calciferous sandstone contains single crystals, and at times large cavities are found filled with doubly terminated crystals often of remarkable perfection and brilliancy. These are collected in numbers, and both natural and uncut specimens are mounted in jewelry and sold to tourists under the name of "Lake George Diamonds." Those sold in large cities under this name are, in nearly every instance, the so-called "paste," a lead glass which has more brilliancy and fire but does not have the same durability as the quartz. Of the Herkimer crystals, possibly $1,000 worth are sold yearly. On account of their remarkable brilliancy and perfect crystallization, rivaling even those found in the cavities of the Carrara marble, many collections of them have been made, notably one by Rev. Bogert Walker, formerly of Herkimer, N. Y. There are collections at Middleville, Little Falls, and Canajoharie, and very fine ones in the State Museum at

Albany, and Smith College at Northampton, Mass. These natural crystals are extensively sold along the railway, a two-ounce vial containing about 500 usually costing from fifty cents to $2. A specimen with a drop enclosed often commands from fifty cents to $30, and a single fine limpid crystal from ten cents to $25. Many of these crystals are whiter than any diamond and frequently as brilliant and transparent. They are often so small that in an ounce will be contained over 7,500 crystals, all perfect and doubly terminated. Curious groupings or inclusions, of great beauty, such as bitumen, pearl spar, and other substances, are eagerly sought for by collectors. Many fine specimens were obtained at Middleville, Newport, and Little Falls, N. Y., when the West Shore Railroad was opened. The old diggings at Little Falls have been worked so extensively from time to time that the roadway has been encroached upon, and to such a degree that further search has been rendered almost impossible. The mode of procedure was to tap the rock until a hollow sound indicated a cavity, and within these cavities the crystals were found, sometimes few in number, sometimes as many as a bushel. At Diamond Point and Diamond Island, Lake George, N. Y., crystals occur similar to those found in Herkimer County, and they have been extensively sold during the last forty years. At Crystal Mountain, Ark., and in the region around Hot Springs for about forty miles, large veins of quartz are frequently met with in a red sandstone. The exact geological horizon of the Arkansas quartz has not yet been accurately defined. The crystals are common in the millstone grit and in the underlying rock, occasionally the lower strata and also the millstone grit coming through the beds anywhere between the layer of carboniferous and the Cambrian. In some cases, detached crystals are found in beds of sandstone or quartzite, and again in quartz veins that traverse both the layers of the carboniferous and the underlying beds. They are often found in cavern-like openings, in one of which, a cavity 30 feet long and 6 feet high, were found several tons of crystals, the sides of the cavity being completely covered with them. (See Illustration.) Wagon-loads of these crystals are taken to Hot Springs and Little Rock by the farmers, who often do considerable blasting to secure them, and who search for them when their crops do not need attention. They are sold by the local dealers, principally as mementoes. Probably a hundred wagon-loads have been bought by visitors at these and other resorts.

Usually only half of the crystal is clear, and a clear space over two inches square is quite uncommon. The sale of uncut crystals from this region amounts to fully $10,000 per annum. At Hot Springs, Ark., clear rolled pebbles, that are found on the banks of the Washita River, are often sold and are more highly prized than the crystals, because of the mistaken belief that they will cut into clearer gems. The great demand for these pebbles, which are scarce, has so excited the cupidity of some of the inhabitants of the vicinity that they have learned to produce rolled pebbles by putting numbers of the crystals in a box, which is kept revolving for a few days by water-power. Any expert, however, can discern the difference, since the artificial ones are a little whiter on the surface. Many localities in Colorado yield fine specimens of quartz, and all along the Atlantic coast at Long Branch, Atlantic City, Cape May, and other places, transparent pebbles are found in the sand, and are much sought after by visitors, who often have them cut as souvenirs. At Narragansett Pier, R. I., some of the local lapidaries have been known to substitute for pebbles found on the beach, foreign cut quartz, cairngorm, topaz, crocidolite, moonstone from Ceylon, and even glass. At all of these resorts large quantities of the quartz pebbles are cut in gems and seals, and all manner of ornaments are sold as having been found in the vicinity. Sometimes even the stones that have been found by the visitors, and intrusted to lapidaries to be cut, are exchanged for cut stones, brought to this country from Bohemia, Oldenburg, and the Jura, where cutting is done on such a large scale and by labor so poorly paid that the cut stones can be delivered in this country at one-tenth of the price of cutting here, as the rock crystal in the articles themselves has but little value. The annual proceeds of the sale of cut stones and the money expended in cutting them at these different localities may amount to $20,000 or more a year, and the sale of specimens to as much more. The clear crystal used in the United States for optical purposes is almost entirely Brazilian, not on account of any defici-

ency in the quality of that found in this country, but because of
the cheapness of the Brazilian crystal. Cut spectacle-glasses can
be imported for less than the cutting costs here. Some of the
most magnificent groups of quartz ever found were formerly obtained at the Ellenville Lead Mines, Ulster County, N Y., and
some of the finest of these, by gift of Jackson Steward, are now
in the Museum of Natural History, New York City. Few, if
any, were cut into gems or used in the arts, although many were
sold in the vicinity as souvenirs. The Sterling Mine at Antwerp,
N. Y., furnishes small, fine, doubly terminated dodecahedral
crystals, and the same forms, with some slight differences, are
found in the specular iron at Fowler, Hermon, and Edwards, St.
Lawrence County. Diamond Hill, Lansingburgh, N. Y., is an
old but poor locality, and Diamond Island, Portland Harbor, Me.,
is well known for the small but bright crystals found there. The
highly modified crystals from Diamond Hill and Cumberland
Hill, R. I., also the fine ones from White Plains, in Surrey
County, N. C., and Stony Point, Alexander County, and from
Catawba and Burke Counties, N. C., are worthy of mention as
having formed the subject of the crystallographic memoirs by
Dr. Gerhard von Rath.[1] Prof. Frederick A. Genth mentions
the finding of fine specimens in Delaware and Chester Counties,
Pa., especially in East Bradford and Pocopson Townships.
Rock crystal seems to have been valued by the Indians of the
American continent. Dr. Daniel G. Britton, in a paper on the
folk lore of Yucatan, quoting Garcia, says that the natives practised witchcraft and sorcery, their wise men divining by means of
a rock crystal, which was believed to exert great influence over
the crops. The presence of crystals with abraded edges in the
mounds of Arkansas, North Carolina, and elsewhere, would lead
to the inference that they were not only collected to bury with
the dead, but were worn as charms and talismans, and having
been used for such purposes, were probably interred with the
dead as their property. Personal observation in Garland and
Montgomery Counties, Ark., forty miles from the Crystal Mountain locality, showed that these quartz crystals were found in
mounds, with a quantity of some of the smallest, finely-chipped

[1] See Naturwissenschaftlichen Verein, Westphalia.

arrow-points of chalcedony, yet not a single object made of chipped crystal was found.   In a number of the mounds leveled by the farmers in cultivating, and not examined systematically, single crystals of quartz were revealed, which may, however, have been kept for their beauty and symmetry by the Indians. The report of the finding at Bakersville, N. C., of transparent crystals of quartz, weighing 642 pounds and 340 pounds respectively, was premature, what was found proving to be, not crystals, but veins of translucent quartzite, with the crystalline markings of a group rather than of a single crystal.   The clear spaces, which were to be observed only on these crystalline sides, would hardly afford material for a crystal ball an inch in diameter, and with this exception they were almost an opaque white, with flaws. Specimens of rutilated quartz and of rock crystal, one mass of which weighed over 10 pounds, and was quite clear, though fractured by frosts, were found near Stuart, Va.   Near Trinidad, Col., there have been found large quantities of crystalline quartz, with small, doubly terminated crystals, resembling those from Herkimer County, N. Y.   Some of these crystals afford larger masses of clear rock crystal than have ever before been found in the United States, and suggest its use for art objects, such as the crystal balls, clock-cases, mirrors, etc., which are now to be seen in the Austrian Treasury at Vienna.   In Alexander and Burke Counties, N. C., crystals of white as well as of smoky quartz have been found, in which were spaces that would cut into clear crystal balls of from 2 to $2\frac{1}{4}$ inches.   One of these from Alexander County, measuring $2\frac{4}{16}$ inches, is in the State Museum of Natural History at Albany, N. Y.   A very interesting bead made of rock crystal, fluted and drilled from both ends, is in the collection of A. E. Douglas, in New York City. It is evidently native work, as it is improbable that foreign traders would use white rock crystal beads, when glass would answer the purpose as well.

Amethyst is found on Deer Hill, at Stow, Me., where there is a vein of amethystine quartz which has been traced fully one-quarter of a mile, and has furnished many thousands of crystals during the last twenty years, scarcely any of them, however, being of any gem value; but among some amethysts found during 1885 was one remarkable mass that yielded a gem weighing 25 carats, of the deep purple color of the Siberian amethyst. (See Colored Plate No. 6.) Some fine amethysts have been found at Mount Crawford, Surry, Waterville, and Westmoreland, N. H. At Burrillville, and at Bristol, on Mount Hope Bay, R. I., fine amethysts were found, and used as ornaments, over sixty years ago. J. Adams says[1] that some were taken from a quartz vein in a coarse granite, and others were found in the sand at the foot of the hill at low tide. An amethyst nearly equal in color to the finest Siberian, and that would afford a gem nearly ⅜ inch across, was found a mile and a half from Roaring Brook, near Cheshire, Conn. When the West Shore Railroad tunnel at Weehawken, N. J., was being blasted out, there were found a few very fair specimens of amethyst on the trap rock. The finest one of these is in the State Museum at Albany, N. Y. Professor Genth[2] mentions magnificent specimens from Delaware and Chester Counties, Pa. Some of the principal localities are the townships of East Bradford, Pocopson, Birmingham, Charlestown (where about a quart of loose crystals was obtained), and Newlin (where about 100 pounds have been found, but none of it of gem value). W. W. Jefferies[3] announced that amethysts of a rich purple color had been found in the northern part of Newlin Township. Crystals, of fine quality, though not affording gem material, one weighing 7 pounds, have been found in Upper Providence. Amethysts of large size, and with very perfect single crystals, well adapted for cutting, were found here in a vein of oxide of manganese and solid walls of sandstone, quartz, and quartzite, often extending to a depth of over 25 feet. The finest, perhaps, of the crystals of this locality, was found in November, 1887. (See Colored Plate No. 6.) In these gems the purple coloring is unevenly distributed in the crystals, as in the case of the Siberian amethysts, both of which, when properly cut, disseminate the color in an unevenly tinted amethyst, making a rich royal purple tint equal to that of any known gem. As a precious stone the large crystal has little value, but

[1] Am. J. Sci. I., Vol. 8, p. 199, Aug., 1824.
[2] Preliminary Report on the Mineralogy of Pennsylvania, p. 57.
[3] Proc. Acad. of Nat. Sci., Phil., Mineralogical Section, p. 44.

as a crystal it is quite unique.   Amethysts have also been found
in Astor, Concord, Marple, and Middletown Townships.   In
Birmingham, in one locality, they are found in clusters; in
another, in fine isolated crystals.   At Chester and Thornbury,
Delaware County, Pa., also, many fine gems have been found by
collectors.   Perhaps the most unique gem of the collection of the
United States National Museum at Washington is a piece of an
amethyst found at Webster, N. C., and deposited by Dr. H. S.
Lucas.   The present form is just such as would be made by a
lapidary in roughly shaping a stone, preliminary to cutting and
polishing it.   It was turtle-shaped when found, which shape was
unfortunately destroyed by chipping, and was said to have borne
marks of the handiwork of prehistoric man.   It now measures
3$\frac{3}{4}$ inches (7 centimetres) in length, 2$\frac{3}{4}$ inches (6 centimetres)
in width, 1$\frac{1}{2}$ inches (4 centimetres) in thickness, and weighs 4$\frac{3}{4}$
ounces (135·5 grams).   It is perfectly transparent, slightly smoky,
and pale at one end, and also has a smoky streak in the center.
This coloring is peculiar to the amethyst, however.   In Haywood
County, N. C., were found quite a number of crystals of amethyst which were cut into very fine gems.   Amethysts of a light
purple and sometimes of a pink color are found in abundance,
in crystals 3 inches long and over, at Clayton, Rabun County,
Ga.   At times these have large liquid cavities containing movable
bubbles of gas.   They are of little gem value, although fine as
specimens.   At the Lake Superior watering-places there are sold
many fine groups of amethyst from Prince Arthur's Landing,
Lake Superior.   These groups are generally composed of crystals from $\frac{1}{4}$ inch to 5 inches in size, the groups ranging from a
few inches to several feet.   Lake Superior crystals have one peculiarity : they are spotted with the red, moss-like markings so
well known, giving the moss amethyst effect if cut, though as a
rule the coating is so even as to cover the entire surface, nothing but a brick-red color being visible unless the crystals are
broken.   Notwithstanding its abundance, but few gems could be
cut from the mineral in this locality.   Hoffmann mentions the
finding of amethyst on the mesa near the mouth of the Rio Virgin, Nev.   In Llano and Burnet Counties, Tex., some very fair
amethysts have been found ; and also at Grand Rapids, Wood

County, Wis., in the amygdaloid on the Lake Superior shore, and in trap rock at Keweenaw Point and elsewhere in the upper peninsula. At Amethyst Mountain, in the Yellowstone National Park, and at Holbrook, Ariz., amethyst varying in color from light pink to dark purple lines the hollow trunks of agatized trees, and forms a beautiful contrast with the pale chalcedony and banded agate sides of the tree-trunks. It is also found in small crystals at Nevada and neighboring localities on Bear Creek, and on the summit of the range east of the Animas, Col.

Smoky quartz, also known as smoky topaz or cairngorm, and citrine are found in large quantities at and near Pike's Peak, Col.; also, to some extent, at Mount Anteros Summit, Col., Magnet Cove, Ark., Burke and Alexander Counties, N. C., and at other points. At Pike's Peak, Col., it occurs in pockets in a coarse, plegmatic granite, often associated with beautiful crystals of amazonstone and flesh-colored and other feldspars. The largest crystal that has as yet been found, measuring over 4 feet in length, is in the cabinet of the Marquis of Ailsa. A doubly terminated crystal 13 inches long and over 5 inches in diameter, which would furnish a 5-inch ball, is in the Kunz Collection in the State Museum at Albany, N. Y. The Pike's Peak material is sent abroad in large quantities to be cut, and the larger part is returned to be sold in tourists' jewelry, principally at Denver and Colorado Springs, Col., Hot Springs, Ark., and in other Western cities and summer resorts. The sum realized from the cut material amounts to about $7,500 annually, and that from the crystals sold to $2,500 more. Most of the cut articles of smoky quartz sold at the tourist resorts are of foreign material, or of material found in the United States and cut abroad. Smoky quartz pebbles are occasionally found along the coast of Long Branch, Cape May, and cut as souvenirs. Crystals of smoky quartz, a foot in length, are frequently found at Sterling, Mont. Of these, a remarkably fine specimen was presented by J. E. Davis to the cabinet of the California State Mining Bureau, in San Francisco. The quartz of Herkimer County, N. Y., and Diamond Island and Diamond Point, Lake George, are occasionally found in a variety of beautiful smoky tints which are exceptionally transparent. Fine smoky quartz has been found at Goshen, Mass. In 1884, a fine, clear mass, weighing over 6 pounds, with clear spaces several inches across, was found on Blueberry Hill, Stoneham, Me., and a broken crystal that weighed over 100 pounds and a crystal over 4 inches long and 2 across, very clear in parts, were found near Mount Pleasant, Oxford County, Me., and a fine crystal at Minot, Me. Professor Genth[1] mentioned the occurrence of smoky quartz near Philadelphia; on the Schuylkill, near Reading, Berks County; near Hummelstown, Dauphin County; in Upper Darby, near Garret's road toll-gate, and near the Kellyville school-house, all in Delaware County; at the tunnel near Phœnixville, in East Nottingham and Birmingham Townships, Chester County. In certain parts of Delaware and Chester Counties the amethyst and smoky quartz gradually shade into each other, a characteristic peculiar also to many specimens from North Carolina. Some fine crystals have been found at Iron Mountain, Mo., and Magnet Cove, Ark. Citrine is mentioned by Hoffmann[2] as occurring at Tuscarora, Gold Mountain, and in Palmetto Cañon, Nev. At Taylorsville and Stony Point, N. C., a number of clear pieces of this material were found that cut fair stones weighing over an ounce each. In Alexander, Burke, Catawba, and adjacent counties, N. C., smoky quartz crystals which would afford fine gems are frequently met with. They are generally from 1 to 5 inches in diameter, and often of a citron or light yellow color.

When clear, compact, white quartz contains veins, or streaks, or spots of fine gold, it is worked into jewelry and souvenirs on a considerable scale in San Francisco, and to a less extent in many of the large towns in the mining regions. Some of the mines in California, Oregon, Idaho, and Montana have furnished very fine specimens, especially when the quartz is clear and the gold penetrates in compact stringers. Gold miners, however, often have a prejudice against what are known as "specimen mines," that is, mines furnishing ore of this kind. The gold found in California quartz is worth about $16.50 an ounce, but jewelers willingly give from $20 to $30 for each ounce of gold contained in material that they can use. The price of specimens

---

[1] Preliminary Report on the Mineralogy of Pennsylvania, p. 58.
[2] Mineralogy of Nevada.

is governed by their beauty, varying from $3 to $40 per ounce of quartz. The specific gravity of the mineral is first taken, after which the gold value of the quartz is ascertained by Price's table. The amount of this material sold in the rough for jewelers' purposes is variously estimated at from $40,000 to $50,000 a year, $1,000 to $2,000 worth being often purchased at one time. One lapidary at Oakland, Cal., where most of the cutting of this material is done, bought nearly $10,000 worth within a year, and a large jewelry firm in San Francisco, during the same time, purchased nearly $15,000 worth. In the selection of the quartz, great care is necessary. The stone used must be large enough to bear the rough treatment of the diamond-saw and the lap-wheel of the polisher. All of the rock quartz is friable, and some of it crumbles to pieces while undergoing these processes. The saw, catching in the gold in the slitting, prevents the cutting of large pieces, as the wafer-like slabs are apt to be broken by this resistance while being detached from the mass. For this reason, all the pieces set in cabinet work are small. Pieces 4 by 2 inches are quite rare, although fine pieces 4 inches square are at times seen. Rarely more than half of the rough material purchased finds its way into the market, owing to breakage while being trimmed into shape. The white gold quartz of California is mainly supplied from the counties of Butte, Calaveras, El Dorado, Mariposa, Nevada, Placer, Sierra, Tuolumne, and Yuba. The black gold quartz, a quite recent novelty, is found at the Sheep Ranch Mine, Calaveras County, and at Sutter Creek, Amador County. The so-called rose gold quartz is made by backing a translucent quartz with the desired shade of carmine paste, and forms an effective contrast to the opaque white and black gold quartz with which it is usually mounted. Single specimens for scarf-pins, rings, and sets of pins and ear-rings sell from $2 to $10 each. Exceptionally fine or curious pieces bring higher prices. It is within a few years that gold quartz has been utilized to any great extent in jewelry. At first the designs were usually simple and the mountings modest, but the demand has created a supply of elaborate designs, and at present the quartz is used in every conceivable form of jewelry, and in articles of personal adornment and decoration of almost unlimited variety, such as canes, paper-weights, writing-cases, perfume-bottles, fan-sticks, bracelets, watch-chains, and lace-pins, the latter in such designs as shovels, picks, and other mining emblems. In certain new furniture, it has been used as paneling; and here, as in jewelry, the effect is better brought out by added colors, such as are afforded by agate, moss agate, native silver in a matrix, smoky quartz, iron and copper pyrite, cinnabar, malachite, turquoise in the matrix, and other bright minerals. By slitting and piecing, as is done with malachite, an entire table-top can be made from a few pounds of gold quartz. Much of the jewelry made of this material is sold to tourists from the Eastern States and elsewhere. Eleven hundred dollars worth was purchased, some years ago, by an Asiatic embassy, and scarcely any one visiting California fails to secure a specimen. The best taste is not often exercised in the designs for this material. Many are too large and ungainly for personal adornment, and others are not as well mounted as the jewelry sold with them. There is much room for improvement in these respects. One of the large designs made of gold quartz, representing the Cathedral of Notre Dame, at Paris, is valued at $20,000. It stands about a foot high, and is perhaps the finest piece of gold quartz work ever produced. A mass of gold quartz[1] weighing 160 pounds was taken out of the bank of the Nevada Hydraulic Company at Gibsonville, Cal. The boulder was smoothly washed and had the appearance of having been ground in a pothole. Its estimated value was $2,500, but its real worth was more than this, since it was valuable for lapidary purposes. The gold penetrating amethystine quartz from Hungary is very beautiful, but the California quartz is the finest known.

Some years ago a method was devised of fusing quartz, by throwing in lumps of heavily alloyed gold, and allowing the material to cool in molds of required shapes. It was said that the mingling of the metal and the quartz was complete, but the quartz had a milky, unnatural, glasslike appearance entirely unlike the gold quartz it was intended to represent. The firm of LeDuc, Connor & Laine, in San Francisco, applied for a patent for an imitation gold quartz produced by means of electricity,

[1] Jewelers' Circular, Vol. 14, p. 258, Sept., 1883.

but found that a similar patent had been issued nearly fifty years earlier to a resident of New York. Though thus unable to obtain the monopoly, they undertook the manufacture of jewelers' quartz, but the venture proved so unsatisfactory that they soon abandoned it.

Prase is found, always crystallized, at various limonite deposits on Staten Island, N. Y. As specimens the mineral is very good, occurring in groups of crystals often 8 to 10 inches across, although the crystals themselves are rarely over ⅓ inch long and ⅛ inch in diameter, and of no gem value. The color is generally a dark leek-green. William P. Blake mentions a greenish-tinged quartz, resembling datolite in color, from the French lode, Eureka District, Cal. Hoffmann, in the "Mineralogy of Nevada," mentions the occurrence of prase in crystals at Reese River, San Antonio, and occasionally on the mountains near Silver Peak. A translucent leek-green variety of chalcedony and quartz occurs in the syenitic range of the Lehigh, especially at the allanite locality, five miles east of Bethlehem, Pa. Prase is found at Blue Mill, Delaware County, in doubly-terminated crystals, and in curious crossings and rosettes, often several inches across ; also in inferior specimens near Dismal Run, Delaware County. Very fine specimens of massive green quartz occur in Bucks County ; in Delaware County at Radnor ; and in East Bradford Township, Chester County. At none of these localities is it of any value as a gem.

Rose quartz occurs in large masses at Albany and Paris, Me. ; Southbury, Conn. ; and at many other places in the United States, but as yet it has not been used in the arts or as a gem. At Stow, Albany, Paris, and a number of other localities in Maine, the veins of quartz shade from white—transparent and opalescent, resembling hyaline quartz, often without any imperfections—through faintly tinted pink and salmon into a rich rose color, thus forming a beautiful series of tints for gems or for ornamental stone-work. Specimens of this rose quartz, when cut into double cabochons, or sphere-shaped objects, distinctly show the asteria effect, similar to the star sapphire, if viewed by sunlight or artificial light, a peculiarity which has also been observed in specimens obtained from a number of other localities. Possibly as fine transparent opalescent rose quartz as has ever been found was obtained at Round Mountain, Albany, Me., in pieces free from all flaws and of a fine rose-red, with a beautiful, milky opalescence, measuring 4 by 5 inches in size. A sphere 2 inches in diameter, a small dish, and other objects have been cut from this material. A vinaigrette or scent bottle was made from the rock crystal found in Ashe County, N. C., and exhibited at the World's Fair, held in Paris in 1889. (See Fig. 6.) A beautiful opalescent quartz has been found in Stokes County, N. C. Rose quartz is found in many localities in the granites of Colo-

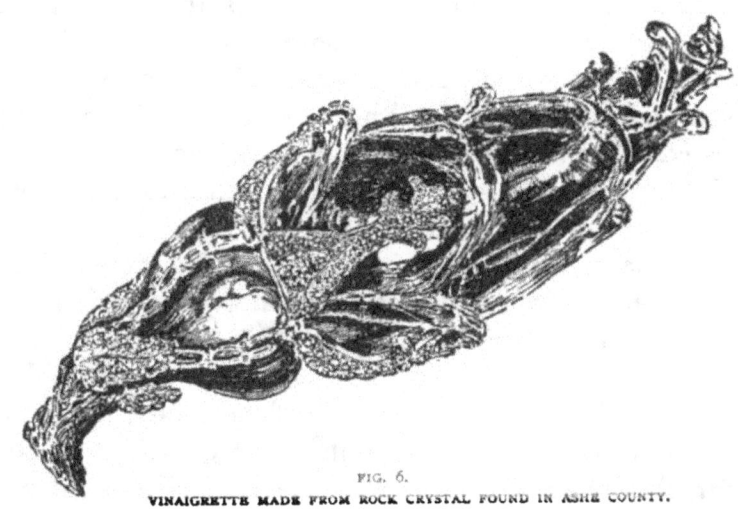

FIG. 6.

VINAIGRETTE MADE FROM ROCK CRYSTAL FOUND IN ASHE COUNTY.

rado, also in fine specimens at the head of Roaring Fork, near Clear Creek, and on Bear Creek. It is mentioned by Hoffmann as found at Tuscarora, Moray, Carlin, and Silver Peak, Nev.; and by Sweet as occurring in crystals from Grand Rapids, Wood County, Wis.

Aventurine quartz has not been observed in the United States in fine specimens, although mentioned by Dr. Frederic M. Endlich as occurring on Elk Creek, Col. Prof. John Collett, of Indianapolis, found a few small specimens of white aventurine quartz pebbles in the drift near Indianapolis, Ind., in 1885. From near Fairfax County, Va., James W. Beath obtained a quantity of quartz with many alternate green and white veinings, the green being produced by chloritic inclusions. When a crystal was cut between the lines of growth, it formed an interesting ornamental stone.

Novaculite is a fine, compact, sandstone-like substance, found in large pieces at Hot Springs, Ark., and employed to a limited extent for cutting into figures, such as birds, for jewelry. It is extensively used for whetstones, which have a world-wide reputation as "Washita whetstones." Its compactness and the purity of its white color make it a very pretty ornamental stone and it ought to be used for this purpose.

The true silicified corals found at Schoharie, N. Y., along the Catskills, and in many other localities in the United States, form very pretty gem stones. Some that are similar to the so-called fossil palm-wood from India have been observed in a few localities in New York State. One very interesting black silicious coral form with large white markings was found at Catskill, N. Y.; when cut across the large white columnar lines, the effect was very pleasing and ornamental.

The finest chrysoprase in the United States is found in a vein of serpentine in the nickel mines at Nickel Mount, near the town of Riddles, Douglas County, Ore. Here it occurs in veins over an inch thick in the nickel ore, and would furnish stones of a rich green color several inches square. In his treatise on quartz and opal, Traill mentions chrysoprase from Newfane, Vt., but Prof. James D. Dana identifies this mineral as green quartz and not chrysoprase, although it was so-called in the locality where it was found. A fine green-colored variety intermixed with black hornblende, that would afford gems an inch across, was found in Macon County, N. C. Thomas Taber mentions[1] in a letter to Dr. C. A. Lee the occurrence of chrysoprase in Chester County, Pa., without giving any description of its quality, though one would infer that it was of gem quality, since Mr. Taber was a jeweler. Dr. Frederic M. Endlich mentions chrysoprase as of rare occurrence in Middle Park, Col. William Irelan, Jr., reports from Tulare County, Cal., beautiful semi-transparent chrysoprase of fine color, and Beck, in his "Mineralogy of New York" (Albany, 1842), describes fine specimens of chalcedony and chryso-

[1] Am. J. Sci., Vol. 38, p. 61, Oct., 1839.

prase from Belmont's Lead Mine, in St. Lawrence County, N. Y.

The compact quartzite of Sioux Falls, So. Dak., has been quarried and polished for ornamental purposes. It is known and sold as "Sioux Falls Jasper," and is the stone referred to by Longfellow in his "Hiawatha" as being used for arrow-heads, when he says:

> "At the doorway of his wigwam
> Sat the ancient Arrow-maker;
> In the land of the Dacotahs,
> Making arrow-heads of jasper,
> Arrow-heads of chalcedony."

This stone is susceptible of a very high polish and is found in a variety of pleasing tints, such as chocolate, cinnamon, brownish-red, brick-red, peach-blow, and yellowish. Polishing works run by water-power have been erected at Sioux Falls, So. Dak., and so ingeniously are they contrived that pillars, pilasters, mantels, and table-tops are now made here as cheaply as abroad. Probably $30,000 worth of the polished material was sold during the year of 1887. The pilasters of the German American Bank and the columns in the doorway of the Chamber of Commerce building, in St. Paul, Minn., are of this beautiful jasper. It is likely to become one of our choicest ornamental stones, and is especially effective in combination with the Minnesota red granite. Its great tensile strength, its high, almost mirror-like polish, the facts that when polished, if used for tiling, the stone is not slippery, one of the properties that quartz possesses, and that large pieces can be quarried out, and its pleasing variety of colors, all combine to render it one of the most desirable of building stones. The mills are of sufficient capacity to polish $100,000 worth a year. In view of the unequaled facility with which it can be prepared for use, it could be employed to advantage for tablets, blocks, columns, tiles for fine interior and monumental work, and in the more artistic branches of stone-work. Some good results have been obtained with the sand-blast on polished surfaces. The material exists in almost unlimited quantities; the quarries already opened are 450 feet long, 100 feet wide, and 60 feet deep at the lowest point. More than 1,200 carloads were shipped from one quarry alone during the year of 1887, and the result has apparently justified the large expenditure of time and money necessary to prepare the stone for market.

The quartz inclusions in some varieties of minerals are of great beauty, and constitute an important part of the gem minerals of the United States. Some of the most interesting of these, and some that are quite rare and little known, are given here.

Sagenite, "rutile in quartz," "flèche d'amour" (love's arrow) or "Venus's hair stone," as it is variously termed, is found in many places in the United States, and is often cut into oval seals and charms for use as jewelry. The stone gives a very pleasing effect in either sunlight or gaslight. As much as $500 worth has been sold for gems and specimens in one year. The most magnificent specimens were found in boulders, from the vicinity of Hanover, N. H., during the years 1830 to 1850. None, however, were traced to their original locality. Three of these were remarkable specimens, equal in beauty and interest to anything known. One belonged to Dr. James R. Chilton of New York, and passed into the hands of William S. Vaux, of Philadelphia, and is now in the possession of his nephew, George Vaux. The rutile crystals in this specimen are of a rich red color, and are transparent by transmitted light, varying from the fineness of a needle to $\frac{1}{4}$ of an inch in diameter. In one part of the mass is a series of rutile crystals united into a single form $\frac{1}{4}$ of an inch wide and 5 inches long. The finest specimen found belongs' to Prof. Oliver P. Hubbard, of Dartmouth College. It is 6 inches long and 3 inches square, and of irregular shape. (See Colored Plate No. 7.) Both these pieces are evidently fragments of larger masses. The quartz itself is slightly smoky, almost clove-brown, and transparent, while slices cut from it are almost colorless, so that it is questionable whether the color is not due partly to the reflection from the rutile crystals, or perhaps to the presence of titanic acid in the quartz. The crystals of rutile in all these specimens vary in size from the fineness of a hair up to $\frac{1}{4}$ inch in diameter, are uniformly distributed through the quartz, cross and intersect each other in all directions, and are of a reddish-brown color with the lustre of polished copper. Of equal interest are the remarkable inclusions of ver-

[1] Proc. Am. Ass'n Adv. Sci., Vol. 4, p. 25, Washington, 1850.

micular chlorite which they contain.   Another piece, which was
cut from this specimen, is in the Silliman Collection at Cornell
University, Ithaca, N. Y.   Beautiful pieces of quartz 3 by 4
inches, and fine crystals penetrated by clove-brown and black rutile, were formerly found at Middlesex, Vt., and in 1848, when
the cut for the Central Vermont Railroad was made through a
perpendicular mass of talcose slate, at Waterbury, Vt., a vein or
pocket of quartz was found containing many fine crystals of rutilated quartz.[1]   Rutilated quartz of unexcelled beauty, the rutile
usually brown, red, golden, and black, has been found in many
places in Randolph, Catawba, Burke, Iredell, and Alexander
Counties, N. C., and in 1888, crystals of quartz, 3 inches in length,
and filled with rutile the thickness of a pin, were found at Stony
Point.   Beautiful series of these are in the collections of J. W.
Wilcox and Clarence S. Bement, both of Philadelphia.   Fine
pieces of quartz, 4 inches square, containing acicular rutile of
a rich red color, have been found near Amelia Court House, Va.
Some fine acicular crystals of rutile in limpid quartz, now in the
possession of Joseph Wharton, of Philadelphia, were found near
Kinger's, Lancaster County, Pa.   At Calumet Hill Quarry,
Cumberland, R. I., beautiful specimens of limpid milky quartz
from 2 to 6 inches square, and also quartz crystals, at times
$\frac{3}{4}$ inch to 2 inches long, are found penetrated by crystals of
black hornblende varying in thickness from a needle's diameter
to about $\frac{1}{16}$ inch, and these are at times 6 inches long, interlaced and penetrating the quartz in every direction, making
a very beautiful gem and ornamental stone.   Specimens of this
character are preserved by the quarrymen to sell to collectors.
Several hundred pounds of this material were sent abroad about
1883 to be cut into jewelry at Idar and Oberstein, but as work
has been suspended at Calumet Hill, the mineral is likely to
become somewhat scarce.   Cut specimens command prices ranging from twenty-five cents to $5 each.   The specimens found
here are quite equal to the variety found in Japan, and are even
better adapted for use in jewelry than the remarkable transparent masses, over a foot across, procured from Madagascar, in
which the crystals of hornblende are too large.

[1] Am. J. Sci. I., Vol. 10, p. 14, July, 1850.

Thetis' hair stone, found by Dr. Charles T. Jackson, near Sneatch Pond, Cumberland, R. I., is occasionally met with in fair pieces and is used to a very limited extent in jewelry. It is transparent quartz, so completely filled with acicular crystals of green actinolite as to make it quite opaque. Probably $100 worth was at one time sold annually to be cut into seals and charms. Acicular crystals of indicolite, somewhat resembling rutile in quartz, filling the quartz so completely as almost to render it opaque, were found in pieces over an inch square at the famous tourmaline locality, near Paris, Me. The mining operations at Stony Point, N. C., brought to light a number of crystals 4 by 3 inches, and masses of quartz 6 by 3 inches, some of the former filled with what appears to be asbestus or byssolite, forming interesting and pretty specimens susceptible of being cut into charms and other objects. The inclosures of what is seemingly göthite in red, fan-shaped crystals from North Carolina is also a beautiful and interesting gem stone. A fine limpid crystal of quartz, 1 inch long and $\frac{1}{2}$ inch in diameter, penetrated to the depth of half a millimeter by fine green crystals of actinolite, is reported from Virginia. The so-called Gibsonville emerald was a similar crystal of quartz, the crystals being 3 by 2 inches. It was plowed up in a field at Gibsonville, N. C., and when first found was believed to be an emerald. Some crystals of limpid quartz, containing particles of native gold, have been found in California. One of these was said to have been 1 inch long, and enclosed in the center was a scale of gold about the size of the lunula of a finger-nail. Two similar inclusions, though not so large, are in the possession of Rev. Horace C. Hovey, of Bridgeport, Conn. In Nevada County, Cal., in the Grass Valley Mines, quartz is occasionally found supporting gold between the crystals. Pellucid crystals of quartz, some 1 inch long and $\frac{1}{4}$ inch across, filled with a very brilliant stibnite projecting in all directions, and some of them curiously bent, were found at the Little Dora Mine, Animas Forks, San Juan, Col. This material is capable of being made into very beautiful gems. A fine crystal 2 inches long and 1 inch in diameter is in the Tiffany Collection. The crystals of quartz from the Herkimer, N. Y., North Carolina, and

Arkansas localities, containing fluid cavities with moving bubbles, are sometimes cut into ornaments which are not only interesting but pretty. One of these pure limpid crystals with a crescent-shaped cavity, from Little Falls, N. Y., was mounted in a pair of gold ice-tongs to represent a cake of ice. Such crystals are valued at from $1 to $25 each.

The fine amethyst from Rabun County, Ga., often contains fluid cavities nearly an inch long, and could be cut into interesting objects like those from Stow, Me. From a region twenty miles west of Hot Springs and extending westward for about sixty miles, the quartz crystals are generally all doubly terminated and detached, and are found loose in the sand between the breaks or veins in the sandstone, which in appearance strikingly resembles the calciferous sandstones of Herkimer, N. Y. At that part of the region called the "gem country" nearest Hot Springs, the crystals are quite white, but proceeding westward they gradually shade into the dark smoky color found at the other end of the district. The quartz is usually filled with fluid cavities. Some 400 crystals with liquid inclusions were collected by the writer as the result of three days' digging. The quartz pseudomorphs after calcite cleavages from the locality two or three miles northwest from Rutherfordton, Rutherford County, N. C., frequently contain irregularly shaped cavities filled with water, which, if broken out in good shape, can be utilized as curious ornaments. This variety of quartz was also found by J. A. D. Stephenson in Iredell County. Possibly the finest specimen is one that belonged to William B. Dinsmore, of New York City, and is believed to have been found in Georgia. It is coated with a beautiful, bluish-white chalcedony with a curious rough surface, is about an inch long, and is perfect on all sides, the bubble of air moving freely. Its walls are so thin that the liquid with which it is filled weighs fully twice as much as the quartz walls themselves. Among other inclusions that might be utilized for gems, the following may be mentioned: Crystals of transparent quartz filled with specular iron found at the Sterling Mine, Antwerp, N. Y.; quartz including scales of hematite from King's Mills, Iredell County, N. C.; rhomb-shaped crystals of dolomite in crystals of pellucid quartz from Herkimer County, N. Y.; crystals of quartz containing crystals of green spodumene (hiddenite) from Stony Point, N. C.; inclosures of muscovite mica, that are green when viewed through the side of the prism, and green chlorite from several localities in Alexander County, N. C.; and epidote in smoky quartz from Whitson's, near Sing Sing, N. Y.

Agates are not produced in sufficient quantity in the United States to admit of exportation. Indeed, $2,000 would cover the annual production and sale here. Nearly all the agate jewelry sold in this country, as elsewhere throughout the world, comes from Oberstein and Idar, on the river Nahe in the duchy of Oldenburg, where the manufacture of such articles has flourished for over three centuries. The supplies of agate material are obtained principally from Uruguay and Brazil, in South America,

QUARTZ

COLOR.	LOCALITY.	Silica.	Alumina.	Magnesia.	Soda.	Potassa.	Ferric Oxide.	Ferrous Oxide.	Titanic Oxide.	Specific Gravity.
White, waxy lustre	Hot Springs, Ark[1]	99·635	0·113	0·087	0·165	trace	....	trace	....	2·649
Pale blue to deep blue	Nelson Co., Va.[2]	99·392	....	....	....	....	0·539	....	0·069	....

[1] Analyst, C. E. Wait. D. D. Owen, 2d. Geol. Rep., State of Ark.; C. E. Wait, Chemical News, Nov. 29, 1873.
[2] Analyst, R. Robertson. F. P. Dunington, Chemical News, Oct. 31, 1884, p. 207.

and so extensive is this industry that it is not an uncommon thing to see in the tavern-yard of Idar great piles of from 10 to 100 tons of rough agate, varying in size from a few inches to several feet across, ready to be auctioned off in lots to suit purchasers. Prices usually range from five cents to several dollars a pound, the average probably not exceeding twenty-five cents. Agate, chalcedony, carnelian, sard, and other varieties of the agate group are found in great abundance at many places in the United States. At Agate Bay, Lake Superior, large numbers of small banded agates, often of a rich red color, are found. These are quite extensively cut. Often the natural pebbles are polished all over, then drilled at one end, and sold to tourists as charms, or they are placed in bottles of water, to show the markings to the best advantage, neatly arranged according to color and size, and sold as mementoes. Many fine agates, some of great beauty, are found in Colorado and through the Rocky Mountains, but only a small portion are polished, as the agates from Brazil and Uruguay can be cut in Germany, and sold at much lower rates, with the result that nearly all the polished agate specimens sold in America are from the German market. The trap rocks along the Connecticut River, especially at Amherst and Conway, Mass., and Farmington, East Haven, Woodbury, and Guilford, Conn., occasionally afford agates of considerable beauty, though rarely over 3 inches across. These were the so-called chalcedonic balls of Torringford and are very handsome when polished; the rich carnelian shades with milky translucency afford a very pleasing contrast. Many of these were cut into the forms of sealstones as early as 1837, and in the delicate arrangement of the layers and the richness of the colors were fully equal to any from abroad. At Natural Bridge, Jefferson County, N. Y., fine agates have been found. The Belmont Lead Mine, in St. Lawrence County, has afforded some very good chalcedony. Dr. W. H. Horton has described white, yellow, and blue chalcedony that was found in masses of good size near Bellvale, Orange County, N. Y.[1] Chalcedony is found in Delaware County, Pa., principally at Middletown and Marple. Brown botryoidal masses occur at the Hopewell Mine; also at Willistown, West Nottingham, West Goshen, and London Grove Townships, in Chester County; a pale variety at Cornwall, Lebanon County; near Rock Spring and Wood's Mine, in Lancaster County; between Clay and Hamburg; also, at Flint Mill, Berks County; in Cherry Valley, Monroe County; at Conshohocken, Montgomery County, and in other places in Pennsylvania. In many of these localities, especially in Delaware and Chester Counties, the residents wear ringstones, sealstones, and other ornaments, which they have had cut from local material. Dr. Lewis C. Beck in his " Mineralogy of New York " mentions agate nodules over 2 inches in diameter obtained from the trap rock near Paterson, N. J. J. C. and J. B. Anthony say: " Agate is found in great abundance at Diamond Hill and its vicinity, and is a mixture composed of quartz, chalcedony, and hornstone variously arranged in strips, spots, or irregular figures, and is susceptible of a fine

[1] Geological Survey of New York (1840), Report on Orange County Minerals.

polish and frequently combines a beautiful blending of colors."
Maryland chalcedony of a skyblue color, translucent and
beautiful, is found half a mile east of where the Western
Run crosses the York turnpike; agate and carnelian, in
thin coatings upon chalcedony, near the Jones Falls turnpike;
also at a point four miles from Baltimore, and again on the York
turnpike thirteen and a half miles from the same city.   A rich
fawn and salmon colored chalcedony has been found near Lin-
ville, in Burke County, N. C., and fine agates and chalcedony
at Caldwell's, Mecklenburgh County, near Harrisburg and Concord, Cabarrus County, and Granville, Orange County, and in
other localities in North Carolina.   Agate pebbles are found
all along the Mississippi River, especially in Minnesota,
and fine pebbles of chalcedony occur plentifully five miles
north of Grand Rapids, Wis.   Agate and chalcedony are both
found along Fox River, Ill., and agate, chalcedony, and carnelian near Van Horn's Well, Tex., and near Hot Springs,
Ark. In Pinal County, Ariz., are found large quantities of amygdules of beautifully banded agate, often coated with opal.   They
vary from 1 to 8 inches in diameter, and when broken are generally light bluish-gray or light gray in color.   They would be
extremely beautiful if cut and polished.   Seven miles south of
Cisco, Utah, are extensive beds of flesh-red, pink, and salmon-
colored agate, which received a great deal of notice by the
press a few years ago, under the name of blood-agate.

In Colorado, chalcedony is found eight miles south of
Cheyenne Mountain at the Los Pinos Agency at Chalk Hills; on
the bluffs near Wagon-Wheel Gap and along the upper Rio
Grande Valley ; in Middle South Parks, Buffalo Park, Fair Play,
Frying Pan, Trout Creek, Gunnison River, and frequently in
drift accumulations.   Agate is found in fine specimens lined with
amethyst on the summit of the range of the Animas; clouded
white and gray in the lower trachytic formations of the Uncompahgre; and in a variety of forms, clouded, banded, laminated,
and variegated, at the Los Pinos Agency; also in the drift in the
South Park, in the Lower Arkansas Valley, on the Frying Pan,
and throughout the Middle Park, in the form of onyx and sardonyx, on the lower Gunnison and adjacent regions.   William P.

Blake mentions the occurrence of large masses of white chalcedony, delicately veined and in mammillary sheets, near the Panoche, in Fresno County and in Monterey County, Cal.; on Walker River, Nev.; of a fine pink color near Aurora, Esmeralda County, Nev.; and in pear-shaped nodules in the eruptive rocks between Williamson's Park and Johnson's River, Los Angeles County, Cal. A very interesting form of chalcedony is found in the vicinity of Crawford, Darres County, Neb., where nearly all the narrow cavities in the large fossil bones found are entirely filled with cores of gray chalcedony, which are left scattered in great numbers over the ground when the bones are broken or have become weathered. It also fills all the seams in this formation, which after weathering leaves walls of chalcedony, varying in thickness from a few inches to that of paper, projecting from the ground often to the height of several feet, and sometimes extending across the country for miles. At Washougal, Wash., there has been found quite a variety of fine agates and moss agates in the form of pebbles from 1 to 4 inches in diameter. The corals and sponges of Tampa Bay, Fla., which are so often altered to chalcedony by the silicious waters, are at times filled with a fluid which was imprisoned while ' the regular deposition of the silica was closing the apertures that admitted the water. They are always lined with drusy quartz, as are those found in Uraguay, the so-called hydrolites, or water-stones, and, if not as beautiful as the latter, they are even more interesting, and have been sold from $2 to $20 each. Beautiful pebbles of agate and chalcedony are found in abundance along the beach of Crescent City, Cal., and are often cut as souvenirs. They are usually of a light color, but delicately veined and marked. Beautiful little agates from Pescadero Beach in California are sold in large quantities, and in different forms, polished and unpolished, loose or in vials of water. Occasionally some of these are found enclosing, like the hydrolites from Uraguay and the chalcedony from Tampa Bay, Fla., a pebble moving in liquid. These pebbles, which may well be called sealed flasks, vary from $\frac{1}{16}$ to $\frac{1}{4}$ inch, and rarely are 1 inch in diameter. They are also found at Yaquina Bay, Ore. In the pebbly drift of the Colorado River the agates are more highly colored, more abundant, and of larger size. Many of the surf-worn pebbles of Pescadero Beach, Cal., are agate and quartz, of very fine, bright colors, and are occasionally utilized as gem stones. Fine agates have been found with the jaspers on the Willamette, Columbia, and other rivers in Oregon. At Tampa Bay, Fla., red and yellow carnelian and sardonyx result from the silicification of the corals and sponges, and occur in rolled pebbles on the beach, and although the pieces are not large, the colors are very beautiful.

The silicified bones of Atlantasaurus, a great extinct saurian, found at Morrison, Col., have at times a coarse cellular structure, which has been infiltrated with carnelian, giving a very pleasing effect of brilliant red stripes and spots. Chalcedony coats and incloses the crystallized cinnabar of the Redington and other mines of California; and these crusts, if cut with the cinnabar, form some of the prettiest and most interesting gem stones ever found. The chalcedony coatings on the blue and green chrysocolla found in the cavities of the Copper Queen Mine, Ariz., are very beautiful if cut in the same manner. (See Agatized Wood.) No stone, used in jewelry, that is found in the United States is cheaper, more beautiful, or more plentiful than the moss agate. Those found in brooks and streams, called "river agates," are the most desirable. Nearly all are sent abroad for cutting, and returned for home use. When this stone was fashionable, fine stones were worth $10 each and upwards, and as much as $20,000 worth was sold in a year, but at present they are only sold to tourists or used in the cheapest jewelry. The principal sources of the supply are Utah, Colorado, Montana, and Wyoming. Large quantities of moss agates were found in the excavations formed in constructing the Omaha and Council Bluffs Bridge over the Missouri River, and near Cheyenne in Wyoming they are found by the ton. A so-called moss agate is found at Rock Springs, Lancaster County, and near Reading, Berks County, Pa. Moss agate was formerly found near Hillsborough, Orange County, N. C. The agatized trees from Holbrook and Specimen Mount show mosslike marking, more like that of the fine tree-stones from Brazil or the Mocha stones from India than of the common moss agate. One curious candle-shaped stalactite of chalcedony, about 3 inches long, had a black core of oxide of manganese, a secondary deposition in a chalcedony stalactite, running through its entire length, at first sight scarcely distinguishable from a half-burned candle; unfortunately it was cut into a number of matched stones for cuff-buttons, which were rendered quite unique by the black central dot. In the southeastern part of Humboldt County, Nev., are large quantities of moss agate of the dendritic and "fortification" forms. A beautiful moss agate is found in Trego County, Kan. (See Jasper and Moss Opal.) Moss agate has been little used since 1882, the sales not exceeding $1,000 a year. Since the introduction into cheap jewelry of the Chinese natural green and artificially-colored red and yellow moss agate, the sale of native stones has almost entirely fallen off.

Jasper is found in many places in the United States, and in a great variety of colors, though, for so common a stone, it is very little used in the arts, the entire annual sales not amounting to $500. Fine red jasper is found on Sugar Loaf Mountain, Me., and a yellow variety with chalcedony has been found at Chester, Mass., and red and yellow by Dr. Horton, at Bellvale, Orange County, N. Y. Pebbles of a fine red color occur along the Hudson River from Troy to New York, especially at Hoboken, Fort Lee, and Troy, where so-called jasperoid rock crops out. Jasper agate is found in considerable quantity at Diamond Hill, Cumberland, R. I., in all shades of white, yellow, red, and green, and with these colors intermixed in one specimen, usually mottled, and at times beautifully banded in irregular seams of white, creamy brown, greenish, and brecciated. It is found in large quantities. Fully 1,000 pounds are taken away yearly by visitors and collectors, but not over $100 worth is sold in a year. Large pieces of fine yellow jasper are found at Tyringham, and elsewhere in the Berkshire Hills, Mass. In Pennsylvania jaspers more or less impure are abundant in the drifts of the Delaware and Schuylkill Rivers; also in Berks County, near Reading; a yellowish brown variety is found at West Goshen, Chester County, a reddish-brown variety near Texas, Lancaster County, and a brown-banded variety near Bethlehem. The arrow-heads found in this vicinity and near Easton are mostly made of this jasper. The jaspery sandstone found near Mauch Chunk might be utilized with advantage for large ornamental works. In North Carolina fine jasper, banded red and black, is found in Granville, Person County; bright brick-red and yellow at Knapp's, Reed's Creek, Madison County; at Warm Springs; at Shut-in-Creek in Moore County; also in Wake County, and elsewhere. In Texas fine jasper has been found near Fort Davis, Jeff. Davis County, and at Barela Springs, where are obtained the jasper agates called Texas agates. The finest jasper is found in great quantity near Collyer, Trego County, Kan., where there is a remarkable bed of the banded variety; the colors are the various shades of red and yellow, with bands of white, so remarkably even that the stone would furnish an excellent material for cameo work, and should this style of jewelry come into vogue again this deposit may prove of considerable value; as it is, the beautiful red and yellow are so strikingly relieved by the white that it makes a fine ornamental stone. It affords blocks over a foot in length and 6 to 8 inches in width, and really merits the attention of workers in ornamental stone, as no banded jasper in the world can rival it, and it exists in unlimited quantities. A beautiful moss jasper, equal to any known, is found in this same locality, in pieces nearly a foot long and 5 or 6 inches in diameter. When polished, it is exceedingly beautiful. Dr. John T. Plummer mentions the occurrence, in Richmond, of "masses of beautiful breccia having a whitish base set with hornstone and bright red and other colored jasper,"[1] as well as of common jasper. Fine yellow, brown, and red jasper is found at the Los Pinos Agency; throughout the Middle and South Parks; along the Gunnison, in the Dakota group; on the Arkansas, Grand, White, Animas, and other rivers of Colorado; in the drift, and in some of the trachytes, mostly red, green, and brown. A very fine specimen was found at the junction of the Lost Trail Creek and the Rio Grande. Small but smoothly worn pebbles of jasper and agate are quite plentiful on the shores of Lake Tahoe, Cal. Red and green jasper are abundant in the neighborhood of San Francisco, where an impure variety of this stone has been used for buildings and sidewalks. Red, yellow, and brown jasper is found

[1] Suburban Geology of Richmond, Ind., Am. J. Sci. I., Vol. 44, p. 281, Jan., 1843.

at Murphy's, Calaveras County, Cal., in great variety and of superior quality.   Red jasper is also found on the Little Colorado River, in New Mexico, and on the Willamette in Oregon.   The latter region evidently furnished the material for the arrow-points of the Oregon Indians.

Blood-stone or heliotrope in beautiful specimens, with very fine red markings, is found in Chatham County, Ga.   Heliotropes from this vicinity are in the cabinet of W. W. Jefferis, of Philadelphia.   Heliotrope was formerly found in the veins in slate at Blooming Grove, Orange County, N. Y.   Good specimens have been found near the Willamette River, Oregon, near the South Park, Col., and below the Uncompahgre, near Grand River.   The so-called green jasper, which is really a chert of Norman's Kill, from the Hudson River slates at Albany County, N. Y., was used by the Indians for arrow-points.   A fine specimen of heliotrope or blood-stone is reported to have been found here, on the same authority that a similar and entirely unreliable occurrence was reported in Texas, and the stones from both are evidently of foreign origin.

Basanite (the Lydian stone, the touch-stone or test-stone of the jeweler) was found by Dr. Horton at Canterbury and Cornwall, N. Y.   It is also sparingly found in nearly all the drift north of New York City, and in that part of the Delaware River from Easton, Pa., down to the State line, also in many other parts of the United States.   A beautiful spear-point, 5 inches long, and a number of arrow-points, made from this material, have been found near Statesville, N. C.

Silicified wood, which is variously known as wood agate and wood opal, is found in great abundance in Colorado, California, and other Western States and Territories.   Of its mode of formation, Prof. Joseph Le Conte' says: " In a good specimen of petrified wood not only the external form of the trunk, not only the general structure of the stem—pith, wood, and bark—not only the radiating silver-grain and the concentric rings of growth are discernible, but even the microscopic cellular structure of the wood and the exquisite sculpturings of the cell-walls themselves are perfectly preserved, so that the kind of wood may often be

---

[1] Elements of Geology, p. 192.

determined by the microscope with the utmost certainty, yet not one particle of the organic matter of the wood remains.   It has been entirely replaced by mineral matter, usually some form of silica."   The general theory of petrifaction is derived as follows: When wood is soaked in a strong solution of iron sulphate (copperas), then dried, and the same process repeated until the wood is highly charged with this solution and then burned, the structure of the wood will be preserved in the peroxide of iron that remains; also it is well known that the smallest fissures and cavities in rocks are speedily filled by infiltrating waters with mineral matters; hence wood buried in soil soaked with some petrifying material becomes highly charged with the same and the cells filled with the infiltrating material, so that when the wood decays the petrifying material is left, retaining the structure of the wood. Furthermore, as each particle of organic matter passes away by decay, a particle of mineral matter takes its place, until finally all of the organic matter is replaced.   The process of petrifaction is therefore one of substitution as well as of interstitial filling. From the different nature of the process in the two cases, it happens that the interstitial filling always differs, either in chemical composition or in color, from the substituting material.   Thus the structure remains visible, although the mass is solid.   Prof. James D. Dana offers the following explanation of the phenomenon. "The wood or often trunks of trees, and sometimes standing forests, which have been petrified in the Rocky Mountain region, have in general been buried under volcanic débris, which constitutes beds of great extent in many regions.   This volcanic material, called tufa, undergoes partial alteration through the action of the waters or moisture it may contain, or that may filtrate through it.   In this alteration or partial decomposition much silica is set free, and makes the waters or moisture silicious.   The silicious solution then made penetrates the wood that is buried in the tufa.   Very slowly the silica is deposited in all the cells of the wood; and as the wood decomposes, silica takes the place of the particles of the fibres until finally the wood becomes wholly silica or quartz."   Concerning the color, he adds that the brownish-yellow is limonite, which if heated will turn red.   Among the great

American wonders is the silicified forest, known as Chalcedony Park, situated about eight miles south of Corrizo, a station on the Atlantic and Pacific Railroad, in Apache County, Ariz. The country formation is sandstone on volcanic ash, and the trees are exposed in gulches and basins where the water has worn the sandstone away, or are buried beneath the sandstone, their ends protruding from between the formations. (See Illustration.) The locality was noticed in 1853 by the Pacific Railroad Exploring Survey. The jasper and agate generally replaced the cell-walls and fibres, and the transparent quartz filled the cells and interstices, especially where the structure was broken down by decay. These cell-centers and cavities produced conditions favorable not only for the deposition of silica as quartz, but also for the formation of the drusy crystalline cavities of quartz and amethyst that so increase the beauty of the material. There is every evidence to show that the trees grew beside some inland sea. After falling they became water-logged, and during decomposition the cell structure of the wood was entirely replaced by silica from sandstone in the walls surrounding this great inland sea. Major John W. Powell, who has visited all these regions, says: "The wood consisted of logs water-rolled before burial, and are now gradually weathering out of their matrix. The enclosing rock is sandstone and cretaceous shale of the series known as Jura-trias and lying immediately above the Chinarump. Agatized wood containing much semi-opal has been formed in California (and possibly in Arizona) under volcanic deposits, but the wood in question is not associated with volcanic material; its matrix is sedimentary."

The red and yellow coloring matter is derived from the oxide of iron in the sandstone, which is red, and the black may be due to partial carbonization or to oxide of manganese. The bark in nearly every case has been decayed before silicification, and even part of the other layers of the tree is often gone; but the difference between the oxidation on the surface and inside is that the surface, to the depth of half an inch, is so altered and changed that it has the appearance of bark, and it is generally supposed to be such.

There is every indication that the deposit is of considerable depth.   Over the entire area, trees lie scattered in all conceivable positions and in fragments of all sizes, the broken sections sometimes resembling a pile of cartwheels.   A tree 150 feet in length is often found split into as many sections, of almost uniform length, presenting the appearance of having been sawn asunder for shingle-blocks by some prehistoric forester; or broken into countless fragments, ranging from the size of a small pebble to that of a fair-sized boulder, also fractured into perfect-shaped cubes, as if cut by a lapidary.   These multiplied fractures are the result of alternate heat and cold, produced by atmospheric changes, acting on the water collected in fissures of the tree.   A phenomenon perhaps unparalleled, and the most remarkable feature of the park, is a natural bridge formed by a tree of agatized wood spanning a cañon 45 feet in width. (See Illustration.)   In addition to the span, fully 50 feet of the tree rest on one side, making it visible for a length over 100 feet.   Both ends of the tree are imbedded in the sandstone.   It averages 3½ feet in diameter,—4 feet at the thickest part and 3 at the smallest.   Where the bark does not adhere the characteristic colors of jasper and agate are seen.   Although the wood is beautiful to the naked eye, a microscope is needed to reveal its greatest charms; not only does the glass enhance the brilliancy of the colors, but it renders visible the structure, which has been perfectly preserved even to the forms of minute cells, and is more beautiful now than before the transformation.   Dr. P. H. Dudley examined microscopically some sections of this wood, and found that part of it at least belongs to the genus Araucaria, one species of which, Araucaria excelsa, the Norfolk Island pine of the South Pacific Ocean, according to the same authority, grows to a height of from 100 to 200 feet.   Other portions were found to resemble our red cedar, Juniperus Virginiana, when grown in the extreme south.   The cell-structure of some of the wood indicates growth in a mild, uniform climate, the annual rings being marked only by one, two, three or more slightly smaller hexagonal or rounded, not tabular, cells as is usually the case.   The name "chinarump" has been suggested for this substance by Major John W. Powell, this being the Indian name for the material.   These trees, according to one of the Indian myths, were believed to be the bolts of the arrows used by their god. It has been extensively used by them in making spear and arrow-points.

William H. Holmes, of the United States Geological Survey, thus describes the locality in Utah known as Amethyst Mountain, opposite the valley of Soda Butte Creek: " Riding up the trail, a multitude of bleached trunks of the ancient forests are discerned. . . . In the steeper middle portion of the mountain face, rows of upright trunks stand out on the ledges like the columns of a ruined temple, on the more gentle slopes, farther down; but where it is still too steep to support vegetation, save a few pines, the petrified trunks fairly cover the surface and were at first taken to be the shattered remains of a recent forest. The exposures of strata in the first 300 or 400 feet at the base are not good, and but few of the silicified trunks appear above the covering of vegetation. At the height of 500 feet the occurrences become very numerous, and the great size and fine preservation of many of the trunks was a matter of much surprise. Prostrate trunks 40 and 50 feet in length are of frequent occurrence, and not a few of these are 5 or 6 feet in diameter. The standing trunks are generally rather short, the degradation of the compact inclosing strata being so slow that the brittle trunks break down almost as fast as they are exposed, and in many cases the roots are exposed and may be seen penetrating the now solid rock with all their original ramifications. One upright trunk of gigantic proportions rises from the inclosing strata to the height of 12 feet. (See Illustration.) By careful measurement it was found to be 10 feet in diameter, and as there is nothing to indicate to what part of the tree the exposed section belonged, the roots may be far below the surface, and we are free to imagine that there is buried there a worthy predecessor of the giant Sequoias of California. Although the trunk was hollow, and partly broken down on one side, the woody structure was perfectly preserved; the grain was straight and the circles of growth distinctly marked. The bark, which still remains on the firmer parts, was 4 inches thick and retained very perfectly the original deeply-lined outer surface. It was clear, however, that the tree is not a conifer. The strata inclosing the trunk consisted chiefly of fine-grained sandstones, indurated clays, and moderately coarse conglomerate, and contained many vegetable remains, such as branches, rootlets, fruits, and leaves.   In the stratum of sandstone occupying the horizon, nearly on a level with the present top of the giant tree, there was a large variety of most perfectly preserved leaves, specimens of which were determined by Leo Lesquereux to belong to the lower Pliocene or upper Miocene, and similar to the Chalk Bluff, Cal., specimens of Prof. Josiah D. Whitney.   At a point about a mile further east, trunks and fragments of trunks were found in great numbers and in all conceivable positions.   In most cases the woody structure was well preserved, but the trunks had a tendency to break in sections, and on the exposed ends the lines of growth from center to circumference could be counted with ease.   In many cases the wood was completely opalized or agatized, and cavities existing in the decayed trunks were filled with crystals of quartz and calcite.   Nearly all of the crystals found in the West have been formed in the hollow of silicified trees, notably in the case of the smoky quartz found in the Pike's Peak Region in Colorado.   Gen. William T. Sherman, while visiting Fort Wingate, N. M., during his trip across the continent, in the autumn of 1878, suggested to the officer in charge of that post the desirability of securing several large trunks of these fossil trees, found in that vicinity, for the United States National Museum.   In the following spring an expedition was sent out for this purpose, under the direction of Lieut. J. T. C. Hegewald, who states that in the locality of Lithodendron Valley, where they were procured, the soil was composed chiefly of clay and sand, and the petrified wood, broken into millions of pieces, lay scattered around the slopes of the valley.   Some of the large fossil trees were well preserved, though the alternate action of heat and cold had broken most of them in sections from 2 to 10 feet long, and certain of these he regarded as having been immense trees.   On measuring the exposed parts of several, it was found that they varied from 150 to 200 feet in length and from 2 to 4½ feet in diameter, and their centers often contained beautiful quartz crystals.   A microscopic examination shows the internal structure of all to have been tolerably well preserved, the cells having suffered but little from the pressure to which the trunks had been subjected. They all belong to the genus Araucarioxylon, and probably are of the same species. The two from Lithodendron Valley are absolutely identical in structure, and that from Fort Wingate is referred provisionally to the same species, although it lacks some of the essential characteristics. Some eight specimens were collected near Estherville, Ia., consisting of fragments, completely chalcedonized and stained a yellowish-brown color, of which the largest were only 6 inches in length and 4 in diameter. They were regarded by Prof. W J McGee as belonging to the Cretaceous age. Although found in the drift, the Cretaceous strata, from which it was originally derived, formerly extended over contiguous parts of Minnesota and were largely removed by glacial erosion during the Quaternary period. Specimens from Martin County, Minn., could not be distinguished from those obtained in Emmet County, Ia. Near Barrel Springs, in the Green River basin of Wyoming, Samuel F. Emmons, of the United States Geological Survey, found a silicified tree, the structure of which was admirably preserved, being filled in, wherever the wood had decomposed, with crystals of quartz. It was from 3 to 4 feet in diameter, and was exposed for 18 feet; both ends were imbedded in the soft earth of the Bridger beds of the Eocene formation.

Agatized wood in large quantities, consisting of trees from 12 to 35 feet in length and from 18 inches to 2 feet in diameter, has been found near Calistoga in Napa County, Cal. Specimens of agatized and opalized wood from the vicinity of Gallatin, Mont., were collected by Dr. Albert C. Peale and George P. Merrill, and later by Frank H. Knowlton, of the United States Geological Survey, who described it as white, banded and streaked with black and yellowish-brown. Although badly decomposed, it appeared to be dicotyledonous. From several specimens, camera-drawings were secured that resembled known forms of Betulinium and Quercinium, or representations of our modern beech and oak. Of specimens from the Yellowstone Park, examined similarly, some were found to be dicotyledonous and some coniferous, the latter mostly Cupressinoxylon, or fossil Sequoia.

The amount of silicified wood found in Apache County,

Ariz., is estimated as high as a million tons, but the material suitable for decorative purposes is comparatively small in quantity.

This material was selected to form the base of a beautiful silver testimonial made by Tiffany & Company for presentation to the French sculptor, Bartholdi. It was chosen on account of its superior hardness, and the warmth and pleasing combination of its colors; also, as the designer remarked, it was eminently fitting that the testimonial should rest "on a solid American base." The problem of polishing this exceedingly hard material having been solved,[1] its application for decorative purposes naturally follows. The combinations of color offer a great field for interior designs. In tiling floors, for mantels, and similar purposes, it is most valuable; for clock-cases and table-tops it also promises to take an important place, defying imitation, by reason of its marvelous colorings, close texture, and remarkable polish; and in the future the material may be worked into decorative columns for the interior of fine houses. The lustre of its finish cannot be marred or impaired by metal or acid, except hydrofluoric acid, with which it may be etched in the same way as glass. A column 1 foot in diameter and 2 feet long, bored out of the section of a tree across the grain of the wood, so as to display the heart in the center, was exhibited in New York City and was considered the most beautiful of all the polished specimens thus far shown. Smaller articles of jewelry, mosaic work, paper-weights, paper-cutters, toilet articles, handles for canes and umbrellas, and similar objects made from this material may find a ready sale. A number of pieces of this material was placed on exhibition during the early part of 1889, and attracted considerable notice from those interested in American minerals.

Opal showing a brilliant play of rainbow colors, either of the noble or of the fire opal variety, has been observed in the United States only, near John Davis River, in Crook County, Ore. The specimen found there is transparent, grayish-white in color, with red, green, and yellow flames. The play of colors equals in beauty that of any Mexican material, and it is the first opal found in the United States that exhibits color. It strikingly

[1] See Lapidary Work.

resembles and has the absorptive properties of tabasheer, the variety of opal which is formed in the joints of the bamboo, and which is used in India for medicinal purposes. Undoubtedly, better material of the kind exists where this was found. The opals sold so extensively at tourists' resorts are generally of Mexican origin. A beautiful fire opal without any opalescence occurs in a small vein about ¼ inch thick and 2 inches square, from Washington County, Ga.; this locality was first described by Prof. George J. Brush of the Sheffield Scientific School, and he has the finest piece of this opal in his cabinet. Common opal in small masses of a greenish and yellowish-white color, with vitreous lustre, is found at Cornwall, Lebanon County, Pa., also at Aguas Calientes, Gilson Gulch, Idaho Springs, Col., of a

OPAL

COLOR.	LOCALITY.	Silica.	Alumina.	Magnesia.	Water.	Analyst.
Prismatic (Fire Opal).	Washington Co., Ga.	91·89	1·40	0·02	5·84	G. J. Brush.[1]

[1] G. J. Brush, Dana, Mineralogy, 5th Ed. p. 200.

brownish color in narrow seams in the granite. J. W. Beath of Philadelphia, Pa., states that he had seen fine opal specimens showing play of colors, reported to have come from the latter place. William P. Blake [1] writes that a rich white variety of opal is found at Mokelumne Hill, Calaveras County, Cal.; and on the elevation near that place known as Stockton Hill, on the west side of Chile Gulch, a shaft had been sunk 345 feet, and opals were found there in a thin stratum of red gravel varying from the size of a kernel of corn to that of a walnut, and many of them containing dendritic infiltrations of oxide of manganese resembling moss. These stones were erroneously supposed to have considerable market value, and in 1866 about a bushel of them were raised to the surface in a day. A milky variety, similar to the above and without fire, is found with magnesite on Mount Diablo, Cal., thirty miles south of the mountain; also in the foothills of the Sierra at the Four Creeks. Yellow fire opals in small nodules not over an inch in diameter, from Mount Pleasant, Bergen

[1] Catalogue of California Minerals (1866) p. 18.

Hill, N. J.,' were described by the writer.   Common opal has been found at Sheffield, Mass.   Semi-opal is found together with chalcedony at the Los Pinos Agency and in trachyte north of Saguache Creek, Col., and coating agate, in Pima County, Ariz., and also at other localities in the United States.   Nodules from 1 to 4 inches in diameter, consisting of dead-white fire-opal filled with dendritic moss-like markings of beautiful moss-opal are found in South Park, Col.   A fine moss-opal, in pieces 3 to 4 inches across, is also found in Trego County, Kan.   A white opaque variety of hydrophane, in rounded lumps, from 5 millimeters to 25 millimeters ($\frac{1}{5}$ to 1 inch) in diameter, with a white, chalky or glazed coating, somewhat resembling the cacholong from Washington County, Ga., has recently been brought from Colorado.   It is quite remarkable for its power of absorbing liquid.   When water is allowed to drop slowly on it, it first becomes very white and chalky, and then, gradually, perfectly transparent.   This property is developed so strikingly that the finder has proposed for it the name " Magic Stone," and has suggested its use in rings, lockets, charms, etc., to conceal photographs, hair, or other objects which the wearer wishes to reveal only when his caprice dictates.   Specimens were examined by Prof. Arthur H. Church of Kew, England, and he proved that the volumes of the dry and the wet mineral were identical by weighing the bulk of mercury displaced in both instances.   His experiments were made in a small, flat-bottomed glass cup having a polished edge and accurately covered with a smooth glass plate.   The mean weight of mercury displaced in five concordant experiments was 7·415 grams, which figure, corrected for temperature, showed the specific gravity of the hydrophane was, when dry, 1·056, when wet, 1·545. The increase in specific gravity is due to the replacement of the original interstitial air of the mineral by water. From the above calculations it is further determined that the specific gravity of the opal, free from air, is 2·14.   The wet opal contained 47·75 per cent. by weight of water, and 52·25 per cent. by weight of silicic hydrate; hence the mineral absorbed rather less than half its bulk of water.   The specific gravity of several specimens furnished the writer[2] the fol-

---

[1] Trans. N. Y. Acad. Sci., Nov., 1888.
[2] Am. J. Sci., III., Vol. 34, p. 479, Dec., 1887.

lowing results: Nos. 1–3 were slabs 2 millimeters thick ($\frac{1}{13}$ inch), No. 4 was a natural lump with glazed coating. This stone is identical with one brought from China, several centuries ago, and described by De Boot, DeLaet, Boyle, and others, as the Oculus Mundi, or World's Eye, and as the Lapis Mutabilis. When wet, it became entirely transparent, except a central nucleus, possibly a core of chalcedony, that remained white. If the central core was black, evidently oxide of manganese, the stone was called Oculus Beli. Hoffmann mentions opalized wood in magnificent colors at San Antonio, Nye County, Nev., and states that on breaking some of the large trunks fine specimens were obtained. Fine large sections of trees altered to wood opal are found at Buena Vista, Col. The color varies from white to brown, and the structure of the wood is preserved. In the hydraulic mines of California,

	DRY GRAMS	WET GRAMS	WATER ABS.	WEIGHT (IN WATER).	SPECIFIC GRAVITY.
1	·880	1·342	·462	·463	2·110
2	·644	·934	·290	·3385	2·091
3	·730	1·109	·379	·382	2·097
4	1·8745		1·0595	·864	2·191

and at Murphey's in the same State, large and very beautiful masses of opalized wood, of fine brown, yellow, and black colors, have frequently been found.

Hyalite, or Muller's Glass, as it is called, occurs on the traprock at Weehawken and Orange, N. J.; with chalcedony at several localities in Yavapai County, Ariz.; at the Philips ore bed, Putnam County; with cachalong at Bellvale, Orange County, N. Y.; in Burke County, N. C., and Screven County, Ga.; in yellow fluorescent coating upon gneiss at Frankford, Pa.; at Avondale, Delaware County, Pa., in bluish-green; on the Wissahickon River in Pennsylvania; at Concord, Cabarrus County, and at the Culsagee Mine, Macon County, N. C. Associated with semi-opal, it is mentioned as occurring in the Mount Diablo Range about thirty miles south of Mount Diablo. It has also been found at Volcano Pass, Larimer County, Col. At none of these places, however, is it found in masses thick enough to afford even a mineralogical gem, and commercially it has no value.

During the survey of the Yellowstone National Park, in 1872, by Dr. Ferdinand V. Hayden, United States Geologist, a large number of geyserites and kieselsinter were collected by different members of the party.   Some of these specimens resemble the kieselsinter from Iceland, but in their general character they differ from anything heretofore found.   The analyses show but little variation from those of other localities.   Near the cone of the Giant Geyser, on the upper Geyser Basin of Fire Hole River, Dr. Albert C. Peale, assistant geologist to Doctor Hayden's survey, found near the base of the crater, apparently running through in nearly horizontal layers, a peculiar variety of geyserite, similar in some respects to the opal, which was named " pealite " after the discoverer.   It occurs in irregular tablets, sometimes with scalloped surfaces.   It is claimed that the position taken by this mineral, between quartz and opal, according to its chemical and physical characters, and the variance it shows from other geyserites and kieselsinter, justifies its distinction from them by a specific name. Although some specimens closely resemble semi-opal, it seems improbable that opal could be formed in the same way. They might well be used for small, odd ornaments, mounted as they are found, without any cutting.

CHAPTER VIII.

Spodumene (Hiddenite), Smaragdite, Diopside, Rhodonite, Enstatite and Bronzite,
Wollastonite, Crocidolite, Willemite, Vesuvianite, Allanite, Gadolinite, Epidote,
Zoisite, Axinite, Danburite, Iolite, Lepidolite, Scapolite, Cancrinite,
Sodalite, Elæolite, Lapis Lazuli.

TRANSPARENT spodumene has been found in two localities in the United States, the variety hiddenite or
"lithia emerald," at Stony Point, Alexander County,
N. C., and an amethystine colored variety at Branchville, Conn. The only variety that has gem value is that from
North Carolina. (For a history of the locality, see Emerald.)
About 1879, some crystals of yellow and yellowish-green mineral,
supposed to be diopside, were found at Stony Point, Alexander
County, N. C., associated with beryl, quartz, rutile, garnet, dolomite, etc. Shortly after their discovery, these crystals came into
the hands of J. A. D. Stephenson of Statesville, N. C., who sent
the best of them to Norman Spang of Pittsburgh, Pa. About
two years later Mr. Stephenson called the attention of William
E. Hidden to this mineral, and to the locality ; Mr. Hidden then
sent specimens for examination to Dr. J. Lawrence Smith, who
immediately discovered that the mineral was not diopside but a
transparent variety of spodumene. The crystals were first found
loose in the soil with emeralds, but systematic mining revealed
them in attached veins of the walls of the rock. The spodumene is generally more or less altered, hence its pitted or
eaten-out appearance ; but when found in the rock, the crystals
are quite perfect and unchanged. The mineral, which is always

transparent, ranges from colorless (rare), to a light yellow, into yellowish-green, then into deep yellow emerald-green. Sometimes an entire crystal has a uniform green color, but generally one end is yellow and the other green. Its hardness is on the prism faces, 6·5, and across them, according to Doctor Smith, nearly that of the emerald; but a series of experiments proved it to be somewhat less. At first considerable difficulty was experienced in cutting it, owing to its remarkably perfect prismatic cleavage, which is very lustrous. Gems have, however, been cut up to 2½ carats in weight. Specific gravity, 3·18 to 3·194.

The yellow color exhibited by the mineral in even the darkest green gems will prevent it from competing with the emerald, since it is this very quality that has kept down the prices of the Siberian demantoids, or Uralian emeralds, as the green garnets are variously termed. The finest crystal of lithia-emerald ever found is in the Bement Collection. (See Colored Plate, No. 5.) It measures 2⅜ inches (68 millimeters) by ½ inch (14 millimeters) by ⅓ inch (8 millimeters). One end is of very fine color, and would afford the largest gem yet cut from this mineral, weighing perhaps 5½ carats. In Dr. Augustus C. Hamlin's cabinet is a fine gem weighing about 2 carats, and a cut stone of fine color and a good crystal are in the collection of Col. W. A. Roebling. Dr. J. Lawrence Smith[1] says that the crystals, when cut and polished, resemble the emerald in lustre, though the color is not so intense as in the finer variety of the latter gem. Prof. Edward S. Dana says that, owing to its dichroism, it has a peculiar brilliancy which is wanting in the true emerald. Thomas T. Bouvé, of Boston says: "One might infer from the statement made of the greater brilliancy of both the hiddenite and garnet, when compared with the emerald, that this should decide their relative beauty; but it is not the case, for the emerald has a beauty of its own, in its deep and rich shade of color, that will ever make it rank at least an equal in loveliness with the newer aspirants for favor."[2] When the gem was first introduced, it had a considerable sale because of its novelty as an American gem and because of the newspaper notoriety it gained through the controversy

[1] Am. J. Sci. III., Vol. 21, p. 128, Feb., 1881.
[2] Proc. Boston Soc. Nat. His.,Vol. 23, p. 2, Jan. 2, 1884.

SPODUMENE

Color	Locality	Silica	Alumina	Ferric Oxide	Chromic Oxide	Ferrous Oxide	Manganous Oxide	Lime	Mag. nesit.	Lithia	Soda	Potassa	Water	Hardness	Specific Gravity	Analyst
Theoretical Composition	......	64.34	27.61	...	...	...	...	...	...	8.05	...	...	...	...	...	...
Emerald green	Alexander Co., N. C.	63.95	26.58	...	0.18	1.11	...	...	...	6.82	1.34	0.07	...	...	...	Genth.[1]
"	Alexander Co., N. C.	64.35	28.10	0.25	...	...	...	...	...	7.05	0.30	...	0.15	6.5-7	3.15-3.189	Smith.[2]
Gray green	Norwich, Mass.	63.79	27.03	...	...	0.39	...	0.73	0.21	7.04	1.10	0.12	...	...	...	Doelter.[3]
"	Goshen, Mass.	63.27	23.73	1.77	...	...	...	0.11	2.02	6.89	0.99	1.45	0.36	...	3.185	Julien.[4]
"	Chesterfield Hollow, Mass.	61.86	23.43	2.73	...	...	0.64	0.79	1.55	6.99	0.90	1.33	0.46	...	3.201	[5]
"	Norwich, Mass.	64.04	27.84	0.64	...	...	1.04	0.34	...	5.20	0.66	0.16	0.50	...	...	Smith & Brush.[6]
"	"	63.65	28.97	...	...	...	...	0.31	...	5.05	0.82	...	0.50	...	...	" [7]
"	"	63.90	28.70	2.55	...	...	...	0.26	0.06	4.99	0.80	...	0.60	...	...	" [8]
"	Sterling, Mass.	64.90	25.30	...	...	...	...	0.43	...	5.65	1.10	...	0.30 Ignition.	...	...	" [9]
Typical Analysis Red	Branchville, Conn.	64.25	27.30	0.20	...	...	...	...	...	7.62	0.39	trace	0.24	...	3.093	Penfield.[10]

[1] F. A. Genth, Am. J. Sci. (3), 23, 68 (Hiddenite).
[2] J. Lawrence Smith, Am. J. Sci. (3), 21, 128 (Hiddenite).
[3] C. Doelter, Min. u. petrog. Mittheil, 1876, l., 577.   Ann. N. Y. Acad. Sci. 1879, p. 318.
[4],[5] A. A. Julien, Ann. N. Y. Acad. Sci. l., 1879, p. 318.
[6],[7],[8],[9] Smith & Brush, Am. J. Sci. (2), 16, 57.
[10] Brush & Dana, Am. J. Sci. (3), 20, 257.   S. L. Penfield, Analyst.

that arose as to its discovery.   Hence for a time the demand exceeded the supply, which, from the desultory working of the mine, was limited.   Thus a 2¼ carat stone was sold for $500, and a number of stones brought from $40 to over $100 a carat. The total sale of all the gems found, from the beginning of operations in August, 1880, to the close of 1888, amounted to about $7,500, the yield in 1882, during which the preparatory work was done, being about $2,000.   At the time of the discovery, this was supposed to be the first occurrence of transparent spodumene: but Pisani, in the Comptes Rendus for 1877, announced a transparent yellow spodumene that had been found at Minas Geraes, Brazil, where it exists in large quantities and has been extensively sold as chrysoberyl.   The writer saw nearly a ton of broken crystals of this mineral at Idar, Germany, in 1881, whither it had been sent for cutting.   A stone from Brazil weighing 1 carat is in the United States National Museum, as also a series of crystals and cut stones from North Carolina.   At Branchville, Conn., spodumene is found in crystals 4 or 5 feet long and a foot in diameter, almost entirely altered to other minerals.   In spots, however, it is transparent enough to furnish small gems of an amethystine color.   The alterations which have taken place have entirely changed it to what might almost be called a defunct gem; otherwise, this material would have afforded gems over an inch in thickness and several inches in length.   The color before the alteration was probably much richer pink.[1]   It is of mineralogical value only.

Smaragdite is believed to be a variety of hornblende, and occurs plentifully at Cullakenee Mine, Clay County, N. C. In color it is bright emerald, grass-green, also grayish and greenish

SMARAGDITE

COLOR.	LOCALITY.	Silica.	Alumina.	Chromic Oxide.	Ferrous Oxide.	Nickel Oxide.	Magnesia.	Lime.	Soda.	Potassa.	Water.
Grass green..	Cullakenee Mine, N. C.[1]......	45·14	17·59	0·79	3·45	0·21	16·69	12·51	2·25	0·36	1·34
Light green..	North Carolina.[2]....	45·70	24·01	0·52	4·56	....	8·03	13·44	2·91	....	0·60

[1] Hardness, 5·5; Specific Gravity, 3·12. Analyst, T. M. Chatard. F. A. Genth, Proc. Am. Phil. Soc., 1873, Vol. 13, 361 to 407.
[2] Analyst, J. L. Smith. J. L. Smith, Am. J. Sci. (3), 6, 184. (Zoizite—Smith. Smaragdite—Genth.)

[1] See On Spodumene and its Alterations, by Alexis A. Julien, Ann., N. Y. Acad. Sci., Vol. I., p. 318, 1879.

gray. Masses through which the pink and ruby corundum occur disseminated, are exceedingly beautiful. The mineral is hard enough to admit of a fine polish and is worthy of attention as an ornamental or decorative stone.

Diopside, a variety of pyroxene, is found in the township of De Kalb, St. Lawrence County, N. Y., as short, stout, oily-green crystals, in color resembling the crystals from Piedmont, Ala. Parts in these have been found sufficiently large and clear to be cut into gems weighing from 6 to 8 carats each, and recently crystals have been obtained which in size and perfection rival any found in the world; some of these will furnish gems of 12 to 15 carats each. This is the only known locality in the United

DIOPSIDE

COLOR.	LOCALITY.	Silica.	Alumina.	Ferrous Oxide.	Manganous Oxide.	Lime.	Magnesia.	Soda.	Potassa.	Water.
....	St. Lawrence Co., N. Y.[1].....	55·12	0·40	1·12	....	25·04	18·15	0·45	0·02	0·17
Blue, large crystals.	Edenville, Orange Co., N. Y.[2]	55·01	....	4·95	....	22·80	16·95	....	....	0·36
Grayish green......	New Haven, Conn.[3]..........	53·124	1·062	6·008	0·598	23·62	14·50	0·468	....	0·468

[1] Analyst, E. S. Sperry. This analysis was made by Mr. E. S. Sperry under the direction of Dr. S. L. Penfield, in connection with some work for the U. S. Geological Survey, and is published with the consent of the Director of the U. S. Geological Survey.
[2] Specific Gravity, 3·294. Analyst, Rammelsberg. Rammelsberg, Mineralchemie, p. 366.
[3] Specific Gravity, 3·127-3·294. Analyst, Geo. T. Bowen. Geo. T. Bowen, Am. J. Sci. (1), 5, 344.—Var. SAHLITE.

States where this gem is found. At Richville, in the township of De Kalb, some very large crystals were found in 1884, several of which were over 3 inches long and 1 inch thick, with clear spots of gem material giving promise of cut gems weighing 20 to 30 carats each. The crystals generally averaged an inch in length. Associated with the garnets of Fort Defiance, Ariz., Gallup, N. M., and other localities in that part of the country, and in the detritus resulting from the decomposition of the peridotite in Elliott County, Ky., small pieces of almost emerald-green diopside are found, evidently a chromium diopside similar to that found with South African diamonds. They are generally too small to afford gems of any value, but a few pieces have been found that are of sufficient size for very small gems.

Rhodonite, a silicate of manganese, has been found in an extensive bed at Blue Hill Bay, Me., on Osgood's Farm, Mass., and in the neighboring towns; in Warwick, Mass.; in Irasburgh and Coventry, Vt.; near Winchester and Hinsdale, N. H.; and at Cumberland, R. I. The Alice Mine at Butte City, Mont., has produced a large quantity of rhodonite associated with rhodocrosite, which has been used to some extent as a gem stone. It has recently been described by Prof. William N. Rice as occurring at White Rocks, Middletown, Conn., but only in a limited quantity.

The variety of rhodonite known as fowlerite has been found at Franklin, N. J., in groups of rich, flesh-colored crystals finer than

RHODONITE

CHEMICAL COMPOSITION AND PROPERTIES.	Theoretical Composition.	Locality, Sterling, N. J.[1]	Locality, Jackson Co., N. C.[2]	Locality, Cumberland, R. I.[3]	Locality, Cumberland, R. I.[4]	Locality, Cummington, Mass.[5]	Locality, Cummington, Mass.[6]	Locality, Sterling, N. J.[7]
Silica	45·90	46·48	26·22	43·54	48·75	48·91	51·21	46·70
Manganous Oxide	54·10	31·52	15·54	15·49	30·66	46·74	42·65	31·20
Ferrous Oxide	.....	7·23	6·44	8·44	10·85	.....	4·34	8·35
Manganic Oxide	.....	.....	30·07	74·74	.....	.....	.....	.....
Peroxides of Iron and Alumina	.....	.....	7·95	4·65	.....	.....	.....	.....
Zinc Oxide	.....	5·85	.....	.....	.....	.....	.....	5·10
Lime	.....	4·50	1·65	4·02	6·21	2·35	2·93	6·30
Magnesia	.....	3·09	4·24	16·63	0·91	2·0	.....	2·81
Water	.....	1·0	7·94	2·25	0·80	.....	.....	0·28
Carbonic Acid	.....	.....	.....	.....	1·70	.....	.....	.....
Color	.....	.....	Gray Bl'ck to Black.	Blue Black.	Light Brown Red	.....	.....	.....
Hardness	.....	.....	3·5	5·0	.....	.....	.....	.....
Specific Gravity	.....	.....	3·66	3·65	.....	.....	.....	.....

[1] Typical Analysis. Dana Mineralogy, 5th Ed., p. 226. Hermann.
[2,3,4] A. H. Chester, Jahrb. für Min. 1888, Bd. 1, p. 187.
[5] Hermann, Jour. für pr. Ch., 42, 6.
[6] A. Schliefer, Dana Mineralogy, 5th Ed., p. 226.
[7] Rammelsberg, Mineral Chemie (1875), p. 394.

ever before known, some of them being 6 or 7 inches thick, forming groups a foot across. Although of value for gem material, it possesses higher mineralogical value. More than $1,000 worth was sold for specimens during the year of its discovery. The rhodonite of Cummington, Mass., of the richest flesh and light red color, was only found in boulders previous to 1887, when it was traced to the ledge by W. W. Chapman. Blocks were taken out weighing some hundreds of pounds each, having a rich pink and red color, and with large surfaces entirely free from streaks of black oxide and in other places beautifully mottled; they were equal in quality and beauty to the Russian rhodonite, which is made into vases and also table-tops and mantels. This material has recently been used very effectively in combination with unpolished or stone-finished silver, as handles for very fine ornaments, the rose-color streaked with black affording a pleasing contrast. Its hardness is only 6·5, but it is nearly as tough as jade. In Russia it is largely used as a gem and ornamental stone for jewelry, jewel caskets, tables, mantels, and altars and pillars of churches.

Enstatite and bronzite occur in many localities in the United States. The best varieties are found half a mile west of Texas, Pa., in beautiful massive foliated varieties. Bronzite was observed by Prof. Frederick A. Genth, in Pennsylvania, near Crump's serpentine quarry ; near Media, in Middletown Township ; in Marple Township, forming the mass of country rock ; in Newtown Township ; and near Radnor, Delaware County. Bronzite and enstatite are also found in large quantities at Bare Hills, near Baltimore, Md. If cut across the fibres, it shows a cat's-eye effect, but it is not fine enough to furnish gems for commerce.

A very interesting form of wollastonite, found by C. D. Nimms near Bonaparte Lake, Lewis County, N. Y., is described by Dr. Samuel L. Penfield. It occurs in distinct crystals and in all gradations to the fibrous form, and varies in color from a white to a faint yellowish pink. It has nearly the toughness and hardness of jadeite, and might be mistaken for Chinese jade.

Crocidolite was observed by Joseph Wilcox in long, delicate fibres of a blue color, in one of the western counties of North Carolina. Theodore D. Rand found a dark-bluish fibrous mineral at the Falls of the Schuylkill, and T. W. Roepper found it at Coopersburgh, Lehigh County, Pa., associated with white and brownish-white garnet and bluish-white crystalline fibrous coatings, which may belong here. It also occurs at Eland Fountain, Orange County, N. Y. Prof. Albert H. Chester, of Hamilton College, published analyses of the crocidolite from Beacon Hill Cumberland, R. I., a very interesting variety of this mineral though not in gem form.[1] It has not been found in gem form in

[1] See Am. J. Sci. III., Vol. 34, p. 108, Aug., 1887.

the United States. The altered tiger-eye variety from the Orange River, South Africa, has been sold in polished specimens, charms, umbrella handles, etc., at Pike's Peak and other places as domestic crocidolite; it is also extensively sold as a variety of petrified wood.

Willemite (anhydrous silicate of zinc) has been found at Franklin, N. J., sufficiently transparent to make a very fair gem. The color is a rich honey-yellow, in shade between the topaz and the chrysoberyl from Brazil, having, however, the vitreous lustre of the Tavetsch titanite. One crystal furnished a number of gems over 8 carats in weight, which are in the collection of Frederick A. Canfield, of Dover, N. J. This mineral is gener-

WILLEMITE

CHEMICAL COMPOSITION AND PROPERTIES.	Theoretical Composition.	Locality, Sterling, N. J.[1]	Locality, Sterling, N. J.[2]	Locality, Sterling, N. J.[3]	Locality, Sterling, N. J.[4]	Locality, Sterling, N. J.[5]	Locality, Sterling, N. J.[6]	Locality, Franklin, N. J.[7]
Silica	27·02	25·0	25·44	26·80	27·91	27·40	27·92	27·40
Zinc Oxide	72·98	71·33	68·06	60·07	59·93	66·83	57·83	68·83
Ferrous Oxide	.....	.....	.....	.....	5·35	0·06	0·62	0·87
Ferric Oxide	.....	0·67	} 6·50	.....	.....	.....	.....	.....
Manganic Peroxide	.....	2·66		.....	.....	.....	.....	.....
Manganous Oxide	.....	.....	.....	9·22	3·73	5·73	12·59	2·90
Magnesia	.....	.....	.....	2·91	1·66	trace	1·14	.....
Lime	.....	.....	.....	.....	1·60	.....	.....	.....
Water	.....	.....	.....	1·0	.....	.....	.....	.....
Color	.....	.....	.....	.....	.....	Apple Green.	Honey Yellow.	Yellow.
Specific Gravity	.....	.....	.....	.....	.....	4·16	4·11	4·155

[1], [2] Typical Analysis, Vanuxem and Keating. Dana, Mineralogy, 5th Ed., p. 262.
[3] Hermann, Jour. für pr. Ch., 47, 11.
[4] Wurtz, Proc. Am. Assn. Adv. Sci. 4, 147.
[5], [6] W. G. Mixter, Am. J. Sci. (2), 46, 230.
[7] Delesse, Ann. Min. (4), 10, 214.

ally opaque and of rich-brown or apple-green color, and it is not unlikely that fine transparent material of these shades that will cut into gems may yet be discovered.

At the Franklin (Sussex County, N. J.) zinc mines, zincite, yellow and yellowish-green willemite, and black franklinite occur mingled together in granular crystals not over ⅛ inch in diameter. This mixture, as well as the brown zinciferous serpentine from Franklin, described by Prof. Charles U. Shepard, is often ground into charms, paper-weights, and similar objects, the effect of the combination being very pleasing, although it does not admit of a high polish.

Vesuvianite or idocrase that would yield small gems has been found at Phippsburgh, Me. A beautiful wine-colored variety is mentioned as occurring near Hope, Bucks County, Pa. About a mile and a half from Sanford, Me., idocrase occurs in unlimited quantities, one ledge, fully 30 feet wide, being made up entirely of massive idocrase, associated with quartz and occasionally with calcite, which fills the cavities containing the crystals. Some of the crystals are 7 inches long, and occasionally the smaller ones would afford fair gems. Idocrase is mentioned by Dr. Frederic M.

VESUVIANITE

CHEMICAL COMPOSITION AND PROPERTIES.	Theoretical Composition	LOCALITY, Sandford, Me. Analyst, Rammelsberg.[1]	LOCALITY, Amity, N. Y. Analyst, Thomson.[2]	LOCALITY, Santa Clara, Cal. Analyst, Smith.[3]	LOCALITY, Newbury, Mass. Analyst, Greely.[4]
Silica	38·4	37·64	35·09	36·56	35·93
Alumina.	17·4	15·64	17·43	17·04	14·77
Ferric Oxide	. . . . .	6·07	6·37	5·93	. . . . .
Ferrous Oxide.	6·2	. . . . .	. . . . .	. . . . .	8·91
Manganous Oxide. .	. . . . .	. . . . .	2·80	0·18	trace
Lime.	38·0	35·86	33·08	35·94	39·46
Magnesia.	. . . . .	2·06	2·0	1·07	0·13
Potassa	. . . . .	. . . . .	. . . . .	0·51	0·44
Soda.	. . . . .	. . . . .	. . . . .	. . . . .	0·36
Water	. . . . .	. . . . .	1·68	2·0	. . . . .
Titanic Oxide.	. . . . .	2·40	. . . . .	. . . . .	. . . . .
Phosphoric Acid. . . .	. . . . .	. . . . .	. . . . .	. . . . .	trace
Color.	. . . . .	. . . . .	Yellowish Brown.	. . . . .	Brown.
Hardness.	. . . . .	. . . . .	. . . . .	. . . . .	6·0
Specific Gravity. . . .	. . . . .	3·43	. . . . .	3·45	3·55

[1] C. F. Rammelsberg, Pogg. Ann., 94, p. 92.   [2] T. Thomson, Mineralogy, vol. I., p. 143.
[3] J. Lawrence Smith, Am. J. Sci. (3), 8, 434. Ann. de Chim. III., 428, 1874. Comptes Rendus, 79, 813, 1874.
[4] James T. Greely, Technology Quarterly, May, 1888.

Endlich as occurring in large crystals on Mount Italia, Col., and north of the Arkansas River, in granite. This mineral, which was named vesuvianite by mineralogists, from the fact that it was first found in the lava at Vesuvius, splendidly crystallized, is sold by Neapolitan jewlers, and used to make the letters I and V in the manufacture of initial pieces of jewelry, in which some word or sentiment is spelled out, the initial of each letter being represented by a precious stone. Near Amity, Orange County, N. Y., is found a dark yellowish-brown variety, which, on the supposition of its being a new mineral, was named xanthite by Doctor Thompson; it has been found transparent enough to cut into small gems, that would serve as initial stones for the letter X in jewelry.

The allanite found in large masses and crystals in Amherst County, Va., is very compact and bright black in color. It would furnish a metallic black gem, which, however, would be of little or no value.

A large quantity of gadolinite has recently been found in Llano County, Tex. It is very compact, of deep velvet-black color, and furnishes a stone about the color of schorlomite.

Epidote is found in many places in the United States, and in very large crystals. It ranges from brown to green in color, and is generally translucent or semi-opaque, except in very small crystals. Fine crystals have been found at Haddam, Conn., which might yield small gems. The large crystals found in quartz at Warren, N. H., were all too opaque for gems, yet were fine as cabinet specimens. At Roseville, in Byram Township, Sussex County, N. J., epidote was formerly found in good crystals of deep green that would afford small gems of little value. The principal localities in Chester County, Pa., are West Bradford Township; East Bradford, where dark-green specimens occur; and West Goshen. In East Marlborough and Kennett Townships it occurs in yellowish-green crystals; in the limestone quarries of London Grove and Sadsbury Townships, in bottle-green crystals. Prof. Frederick A. Genth mentions [1] a crystal of epidote in the cabinet of the University of Pennsylvania, from the gold washings of Rutherford County, N. C. This crystal is strongly pleochroic, like the so-called puschkinite from the auriferous sands of Ekaterinburg, in the Ural Mountains, and would cut into a small gem. Some fine, highly complex forms have been observed at Hampton's, Yancey County, N. C., by William E. Hidden. These crystals might possibly afford cabinet gems, not so fine, however, as the Tyrolese epidote. In November, 1888, Dr. C. D. Smith sent the writer several dozen crystals of epidote from a place one mile from Rabun Gap, Rabun County, Ga., that are as fine in color, transparency, and habit as those from the famous Untersultzbachthal Tyrol locality. None were over an inch in length, but it is believed that proper working might

Minerals and Mineral Localities of North Carolina, Raleigh, p. 44, 1881.

develop as large crystals as those from the Tyrol, since they show the pleochrism beautifully, their color changing, as viewed in different directions through the prism, from dark grass-green to a rich yellow-green.

EPIDOTE

CHEMICAL COMPOSITION AND PROPERTIES.	LOCALITY Rowe, Mass. Analyst, A. G. Dana.[1] Typical Analysis.	LOCALITY Rowe, Mass. Analyst, A. G. Dana.[2]	LOCALITY Williamsburgh, Mass. Analyst, Thomson.[3]	LOCALITY Polk Co., Tenn. Analyst, F. A. Genth.[4]	LOCALITY Polk Co., Tenn. Analyst, Trippel[5]	LOCALITY Greenwood, Va. Analyst, T. P. Lippit.[6]
Silica	38·18	38·23	40·21	40·04	43·20	39·74
Alumina	24·57	24·66	25·59	30·63	29·60	21·55
Ferric Oxide	12·16	12·24	.....	2·28	2·88	15·29
Ferrous Oxide	.....	.....	7·68	.....	... ｡	.....
Manganous Oxide	0·57	0·57	trace	0·19	.....	.....
Copper Oxide	.....	.....	.....	0·24	.....	.....
Lime	21·64	21·54	23·28	25·11	22·72	22·75
Magnesia	0·12	0·14	1·71	trace	0·56	0·61
Potassa and Soda	0·37	0·37	.....	.....	.....	.....
Water	2·15	2·17	.....	0·71	0·26	.....
Color	Grayish Green.	Grayish Green.	.....	Gray to Blue.	Gray to Bluish White.	Green.
Specific Gravity	.....	.....	.....	3·34	.....	3·39

[1], [2] A. G. Dana, Am. J. Sci. (3), 29, 455.
[3] Thomson, Nicol's Man. of Min. (1849), p. 237.
[4], [5] Am. J. Sci. (2), 33, 197.
[6] J. W. Mallett, Chem. News (1881), 44, 189.

Zoisite is a silicate of alumina containing from 2 to 9 per cent. of oxide of iron. Its quality, as found in the United States, has not been such as to adapt it for use as a gem. Some beautiful specimens of yellowish-brown and greenish-gray crystals have been found at the Ducktown, Tenn., Copper Mines. The rose-red or thulite variety has been found at Deshong's Quarry, Delaware County, Pa., but this is not as handsome or as compact as the beautiful rose-red variety which occurs in considerable quantities at Trondhjem, Norway, some of which has been used for ornamental purposes.

No crystals of axinite have been found in this country of sufficient size to furnish gems. It has been observed near Bethlehem, Pa., at Cold Spring, N. Y., and associated with essonite and idocrase at Phippsburg and Wales, Me. The first-named locality, discovered by Prof. Frederick Prime, Jr., is in Northampton County, about three miles north of Bethlehem. Specimens from this place have been examined by Prof. Benjamin W.

Frazier,[1] who published his results. The crystals here are found in a rock containing crystalline hornblende, apparently mixed intimately with the axinite, and also traversed by numerous narrow veins of that mineral. He says: " Some of the crystals are colorless, others and the crystalline variety which fills the veins have a pale brown color." In some cases this is chiefly superficial from the presence of a thin brown incrustation which occurs at times in minute globular concretions and again in dendritic forms. The lustre of the crystals, which in size reach a length of

DANBURITE

CHEMICAL COMPOSITION AND PROPERTIES.	Theoretical Composition.	LOCALITY, Russell, N. Y.[1]	LOCALITY, Russell, N. Y.[3]	LOCALITY, Danbury, Conn.[2]	LOCALITY, Danbury, Conn.[4]	LOCALITY, Danbury, Conn.[5]
Silica.....................	48·90	48·23	49·70	56·00	48·10	48·20
Alumina..................	.....	} 0·47	} 1·02	1·70	0·30	1·02
Ferric Oxide..............	.....				}	.....
Manganous Oxide.........	.....	.....	.....	.....	0·56	.....
Lime....................	22·70	23·24	23·26	28·33	22·41	22·33
Boric Acid..............	28·40	26·93	25·80	.....	27·73	27·15
Ytthia...................	.....	.....	.....	0·85	.....	.....
Water	.....	.....	.....	8·00	.....	.....
Magnesia.................	.....	.....	.....	.....	0·40	undet.
Potassa..................	.....	.....	.....	5·12	.....	.....
Color....................	.....	Yellow.	Yellow.	Honey Y'llow	Yellow.	Yellow.
Hardness.................	.....	.....	.....	7·50	.....	.....
Specific Gravity..........	.....	.....	.....	2·83	.....	.....

[1] Typical Analysis. Analyst, Comstock. See G. J. Brush and E. S. Dana, Am. J. Sci. (3), 20, 111.
[3] Analyst, J. E. Whitfield. See Am. J. Sci. (3), 34, 225.
[2] Analyst, Chas. U. Shepard. See Nicol's Manual of Mineralogy (1849), p. 170.
[4, 5] Analysts, J. L. Smith and G. J. Brush. See Am. J. Sci. (2), 16, 365.

¼ of an inch, varies; some are dull, some highly brilliant. Specimens from Dauphin, France, and Scopi, Switzerland, are occasionally cut into beautiful stone-brown gems, but for gem collections only.

Danburite[2] has been found in the largest known crystals and in considerable abundance at Russell, N. Y., but only occasionally are the crystals clear enough to cut into gems. Its hardness is 7 to 7·25, its color usually either wine-colored, honey-yellow, or yellowish brown. Some of the crystals observed are 6 inches

[1] Am. J. Sci. III., Vol. 24, p. 439, Dec., 1882.
[2] Am. J. Sci. III., Vol. 20, p. 111, Aug., 1880.

long and 2 inches in diameter, but they are less beautiful than the small, colorless ones from Scopi, Switzerland.[1] The original locality, Danbury, Conn., never furnished any gems.

Iolite occurs at Haddam, Conn., in crystals occasionally 5 inches across, which are often dark blue and sufficiently clear for cutting as gems. Dr. John Torrey possessed a fine seal made of a cube of iolite from the albite granite of Haddam, Conn., which displayed to the greatest perfection its dichroitic properties, being blue when viewed in one direction, and white when viewed in the other, the blue being remarkably fine. This locality promised well, but the supply of gem material has been scant. An iolite-gneiss has recently been noticed by Edmund O. Hovey, at Guil-

IOLITE

COLOR.	LOCALITY.	Silica.	Alumina.	Magnesia.	Mangan-ous Oxide.	Manganic Oxide.	Ferrous Oxide.	Ferric Oxide.	Lime.	Water.
Blue...........	Unity, Maine.[1].......	48·11	32·50	10·14	0·28	....	7·92	....	....	0·50
....	Haddam, Conn.[2].....	49·62	28·72	8·64	1·51		11·58	....	0·23	....
Blue...........	Haddam, Conn.[3].....	48·35	32·50	10·00	0·10	....	6·00	....	....	3·10
Blue...........	Richmond, N. H.[4]....	48·00	35·00	10·00	1·00	....	6·00	....	....	....
Blue...........	Brimfield, Mass.[5].....	48·54	31·73	11·30	0·70	....	....	....	5·69	1·65

[1] Analyst, C. T. Jackson. C. T. Jackson, Final Rep. Geol. of N. H., p. 184; Dana, Min. 1844, 406.
[2] Analyst, Thomson. Thomson, Mineralogy, I., 278.
[3,4] Analyst, C. T. Jackson. C. T. Jackson, Final Rep. Geol. of N. H., p. 184.
[5] Analyst, L. C. Beck. Nat. Hist. of New York Mineralogy, by L. C. Beck, p. 451, 1842.

ford, Conn.[1] It was found near the Norwich and Worcester Railroad, between the Shetucket and Quinebaug Rivers, where the gneiss has been quarried for the road. At Brimfield, Mass., on the road leading to Warren, it occurs with andalusite in gneiss, and likewise near Norwich, Conn. It is also found at Richmond, N. H., with anthophyllite in a talcose rock. In the author's collection, there is a crystal of this mineral, found at Fort George, Manhattan Island, which is almost entirely altered to pinite, an alteration common to nearly all the crystals that were formerly found at Haddam, Conn.

Lepidolite is a mica containing lithia. Beautiful pink and lavender colored lepidolite has been found in large quantities at

[1] Danburite from Switzerland. Am. J. Sci. III., Vol. 24, p. 476, Dec., 1882 ; also, Vol. 25, p. 161, Feb., 1883.
[2] Am. J. Sci. III., Vol. 36, p. 57, Oct., 1888.

Mount Mica, Paris, Me., in masses of 50 to 200 pounds ; at Hebron and Norway, more recently at Auburn, and also at Mount Black, Rumford, Me., ranging from rose-pink through a variety of shades of pink-lavender to heliotrope color. As this mineral is used abroad to some extent for ornaments, such as dishes, vases, paper-weights, etc., the similar utilization of the American variety is suggested. This variety, like the lepidolite of Rozena, Moravia, contains crystals of rubellite. At Rumford the association is almost identical and the mixture can be as easily polished or worked as the former.

Pink, lavender, and purple scapolite, in compact masses 3 or 4 inches square, is found at Bolton, Mass., that will polish nicely and form a neat ornamental stone.

Cancrinite, sodalite, and elæolite are occasionally fine enough to be used as gems and ornamental stones. These minerals are found at Litchfield and South Litchfield, Me., in boulders varying in weight from a few pounds to many tons, that lie scattered over the ground for a distance of about four miles. One mile and a half west of this line, across a pond in West Gardner, these minerals are found associated with zircon, as in South Litchfield. In West Gardner are ledges of rocks which are believed to be the source of these boulders. The color of cancrinite varies from bright orange-yellow to pale yellow. There are three distinct types of this mineral, the bright orange-yellow, cleavable and transparent, in thin fragments; the pale yellow, not cleavable ; and the bright yellow, granular, which is the commonest form. These varieties all have been polished to some extent by collectors. Associated with cancrinite is found a bluish-colored mineral, which Prof. Frank W. Clarke has shown to be a mixture of cancrinite and elæolite. The sodalite found occurs in seams from ⅛ inch to 2 inches in thickness, and varies from violet to a deep azure-blue. This mineral when polished is almost as beautiful as lapis lazuli and it has been found in sufficient quantity to give it some gem importance. Hexagonal crystals of bright yellow cancrinite occasionally penetrate the deepest blue sodalite, forming an exceedingly beautiful stone when polished.

Lapis lazuli has not been found in North America, though it occurs extensively in the Andes Mountains of South America.

Lapis lazuli has been shown to be, not a definite mineral, but a mixture of a colorless and a blue substance, the latter (the mineral häuynite), predominating. The yellow spots of so-called gold are really iron pyrites. It was the sapphire of the ancients. Its hardness varies from 5 to 5·5 and its specific gravity is 2·4. The finest is brought over from Persia, but it is also found in Siberia, Tartary, China, and Thibet.

CHAPTER IX.

Feldspar Group.

THE greenish variety of orthoclase named lennilite by Dr. Isaac Lea[1] is found at Lenni Mills, Delaware County, Pa. The pearly variety found at Blue Hill, two miles north of Media, and called by Dr. Lea delawarite, is a bluish-green, sub-transparent cassinite, of an aventurine character, the bright particles being hexagonal hematite, and often fine enough in color to make a gem or ornamental stone. Elæolite has been found at Magnet Cove, Ark., in very compact nodules of rich flesh, cinnamon, and yellow-brown color, and in such abundance as to warrant its use for certain purposes in jewelry. That found at Gardiner and Litchfield, Me., admits of a very good polish, the color being greenish and of a good appearance, while part of that found at Salem, Mass., is also valuable. (See Cancrinite and Sodalite.) At Van Arsdale's Quarry[2] near Feasterville, Bucks County, Pa., orthoclase is found in crystals from ⅓ an inch to 2 inches in length, usually, however, in cleavage masses of a gray or grayish-black color, which show the blue chatoyancy finely and make a very fine variety of moonstone. The beautiful specimens of albite found at Mineral Hill, near Media, Delaware County, Pa., show

[1] Am. J. Sci. III., Vol. 36, p. 326, Nov., 1888.
[2] Preliminary Report on the Mineralogy of Pennsylvania (Harrisburg, 1875). p. 89.

the blue chatoyancy remarkably well. It is there called "moonstone," and may well be classed under this head, for it has the chatoyant effect and in appearance differs but slightly from orthoclase moonstone. The greenish-gray granular albite or oligoclase found in the serpentine at the magnesia quarries, West Nottingham Township, Chester County, Pa., shows a faint blue moonstone lustre. The beautiful feldspar found by W. W. Jefferis, with the sunstone at Pearce's paper mill, shows a blue chatoyancy as marked as that of any labrador spar. It may be the latter, or perhaps oligoclase. The finest examples of this material, very closely resembling that from Ceylon in quality, transparency, and color, and forming gems ½ inch across, have been found at the Allen Mica Mines, Amelia Court House, Va. It also occurs in opaque pieces 6 to 8 inches square and of good color, showing a delicate blue chatoyancy. Crystals an inch in length, of an opaque adularia feldspar, showing a beautiful blue chatoyancy, are found on Mount Beckwith, Col. Very good sunstone oligoclase, with fine reflections, has been found near Fairfield, Pennsbury Township, Pa., also at Mendenhall's lime quarries, Pennsbury, Chester County, and in Ashton Township, Pa., some of a grayish-white color with coppery reflections; and a curious variety in moonstone (albite), showing double reflections. The green and red sunstone found near Media, Pa., is very fine. In Middletown Township, Delaware County, Pa., in one locality moonstone and sunstone in small nodular lumps are scattered through the soil, and about a ton of the material has been removed since its existence was discovered; in another locality in the same township, moonstone is found in boulders. A very fine sunstone, the orthoclase of which is of a rich salmon color, quite transparent and streaked with white, showing the aventurine effect beautifully, is found at Glen Riddle, Delaware County, and another beautiful variety in the hornblende at Kennett Township, Chester County, Pa.; this Professor Genth thinks is probably an oligoclase. Greenish orthoclase, sometimes in bright green pieces, also pale green, and at times spotted with brownish tints, all showing a sunstone effect, is found at Mineral Hill, Middletown, and Upper Providence, Delaware County, Pa. The orthoclase of Frankford, Pa., with göthite disseminated through it, very closely approaches sunstone in appearance. Beautiful varieties of orthoclase sunstone were discovered near Crown Point, N. Y., by William P. Blake. On the Horace Greeley Farm, at Chappaqua, N. Y., small pieces of orthoclase sunstone were found, almost as fine as that from Swedestrand, Norway. It also occurs at Amelia Court House, Amelia County, Va. A very interesting variety of sunstone was found by J. A. D. Stephenson at the quarry in Statesville, N. C.; the reflections are as fine as those of the Norwegian, but the spots of color are very small. Several hundred dollars' worth from this locality have been sold as gems.

SUNSTONE—AVENTURINE ORTHOCLASE

	COLOR.	LOCALITY.	Silica.	Alumina.	Ferric Oxide.	Lime.	Magnesia.	Potassa.	Soda.	Water.	Hardness.	Specific Gravity.
Theoretical Composition	....	....	64·60	18·50	....	....	....	16·90	....	....	....	....
Typical Analysis	Delicate flesh red.	Ogden Mine, Sussex Co., N.J.[1]	64·80	19·02	0·23	1·29	0·61	15·22	....	0·26	....	....
Analysis.....	Flesh red.	Ogden Mine, Sussex Co., N.J.[2]	64·82	19·25		1·23	0·58	13·38	....	0·26	...	....
"	....	Delaware Co., Penn.[3]	67·70	19·98	1·47	1·47	0·11	1·36	8·86	....	6·5	2·59

[1, 2] Analyst, A. R. Leeds. A. R. Leeds, Am. J. Sci. (3), 4, 433.   [3] Analyst, A. R. Leeds. A. R. Leeds, Am. J. Sci. (3), 6, 25.

Labrador spar is found in large quantities in Lewis and Essex Counties, N. Y., and as boulders in the drift, all the way down to Long Island and New Jersey. In Lewis County the boulders are so plentiful in one of the rivers that it has been named Opalescent River. Large quantities of this labradorite rock are quarried at Keeseville, Essex County, N. Y., for monumental and building work. It is polished there for similar purposes at a cost of about one dollar a square foot, and finds a ready sale under the name of Au Sable granite. The Young Men's Christian Association Building at Burlington, Vt., and one of the public buildings in Minneapolis are built with it. Within a few miles of Amity, in Orange County, a boulder of fine material for specimens, weighing over 2 tons and showing the characteristic chatoyant play of colors, was found. In Pennsylvania labradorite occurs at Mineral Hill, Chester County, and opposite New Hope, Bucks County; and also in the Wichita Mountains, Ark. Mention is made by Professors Genth and Kerr [1] of a curi-

[1] Minerals and Mineral Localities of North Carolina, p. 48.

ous white variety occurring at the Cullakenee Mine, Clay County, also large crystals in the trap at Shiloh Church. On the road to Charlotte, Mecklenburg County, and near Bakersville, North Carolina, specimens showing a slight blue chatoyancy are also found. This domestic labradorite is scarcely used at all in the arts, as the mineral from Labrador is cheaper and of a much superior quality, and takes a finer polish.

At Pike's Peak, Col., amazonstone is found in cavities in a coarse pegmatite granite with smoky quartz crystals, often of huge size, flesh-colored and white feldspars. When associated with smoky quartz, it makes a most pleasing and effective mineralogical combination. The mineral here is finer than any found elsewhere. Many thousand amazonstone crystals of the most

LABRADORITE

	LOCALITY.	Silica.	Alumina.	Ferric Oxide.	Lime.	Ferrous Oxide.	Magnesia.	Soda.	Potassa.	Water.
Theoretical Composition Analysis......	....	52·90	30·30	....	12·30	....	....	4·50	....	....
"	Mt. Marcy, Essex Co., N. Y.[1]	54·47	26·45	1·30	10·80	0·66	0·69	4·37	0·92	0·53
"	Waterville, N. H[2]................	51·03	26·20	4·96	14·16	....	....	3·44	0·58	....

[1] Analyst, A. R. Leeds. A. R. Leeds, Am. Chem. J., March, 1877. Am. J. Sci. (3), 14, 240
[2] Analyst, E. S. Dana. E. S. Dana, Am. J. Sci. (3), 3, 48.

beautiful green color have been obtained, measuring from ¼ inch to over 12 inches in length and of different shades of green, from the lightest and most delicate to a deep apple-green. The crystals are often in groups, the bases of which are covered with white albite. The finest group of this character is in the New York State Museum in Albany, and the finest single crystals are in the collections of Clarence S. Bement of Philadelphia and Frederick A. Canfield of Dover, N. J. When this mineral was first exhibited at the World's Fair in Philadelphia, in 1876, it proved a great surprise to many, but especially to the Russians, who had brought over some small crystals valued at what would now be considered fabulously high prices. Some of it is cut into gems or ornamental stones, and large quantities are still sold annually to tourists. /Several localities in North Carolina also furnish this mineral. (Rockport, Mass., formerly afforded many finely colored pieces. Some fine green crystals have also been found at Paris, Me., and at Mount

Desert, Me., material that would cut into fair gems is occasionally met with. Several light-green crystals, over 6 inches long, and one over 10, were found in the Allen Mica Mines, Amelia Court House, Va. From the Pike's Peak locality one dealer sold over $8,000 worth as specimens, at prices as high as $200 for a single specimen. Over $1,000 worth from this place is annually cut into tourists' jewelry. In Middletown, Delaware County, Pa., many shades of green feldspar, passing into cassinite and delawarite, are found in the soil in loose boulders, up to 20 inches in diameter. In the Allen Mica Mines many hundreds of tons of rich green cleavages of amazonstone, some 6 or 8 inches across, were found in mining for mica, and this is so plentiful that it merits attention in the arts.

AMAZONSTONE

CHEMICAL COMPOSITION AND PROPERTIES.	Theoretical Composition	LOCALITY, Magnet Cove, Ark. Analyst, Pisani.[1]	LOCALITY, Mineral Hill, Penn. Analyst, Pisani.[2]	LOCALITY, Amelia Co., Va. Analyst, Page.[3]	LOCALITY Baltimore, Md. Analyst, Williams.[4]
Color	.....	Blue Green.	Blue Green.	Bluish Gray.	Blue Green.
Silica	64·60	64·30	64·90	64·12	65·41
Alumina	18·50	19·70	20·92	16·84	19·86
Ferric Oxide.......	.....	0·74	0·28	2·28	.....
Potassa	16·90	15·60	10·95	13·34	10·12
Soda...............	.....	0·48	3·95	1·88	4·61
Lime	.....	.....	.....	0·32	1·08
Magnesia...........	.....	.....	.....	0·26	.....
Water	.....	0·35	0·20	.....	.....
Specific Gravity....	.....	2·54	2·57	.....	2·56

[1], [2] Typical Analysis. Des Cloiseaux, Comptes Rendus, 1876, p. 885–91.
[3] F. P. Dunnington, Chem. News, Oct. 31, 1884, p. 208.
[4] G. W. Williams, Balt. Nat. Field Club, April, 1887. Neues Jahrb. für Min., 1888, 2 Band, 1 heft.

Perthite, abundant at Perth, Ontario, is found in the United States as boulders, and possibly in place. This mineral forms a very curious and rich-colored gem stone, with bright aventurine reflections.

Peristerite, associated with common orthoclase, has been found crystallized in great abundance in the town of Macomb, St. Lawrence County, N. Y. Many of the specimens show the beautiful light-blue chatoyant effect. It has also been observed as far north as Bythurst, Canada, nine miles north of Perth, in the townships of Pierrepont and Russell, and in at least a dozen other places in northern New York. Occasionally it makes a very fine gem stone, differing from labradorite and moonstone, the chatoyancy being an intermediate one between the white of moonstone and the dark blue of the former. It occurs in large masses at Cavendish, near Cavendish Falls, in the railway cutting, twenty-two miles northeast of Bellows Falls, Vt.

A compact variety of white or gray orthoclase, spotted black by hydrated manganese oxide, and called from its leopard-like appearance, leopardite, is abundant near Charlotte, Mecklenburgh County, and also in Gaston County, N. C. It is a variety of porphyry with disseminated crystals of quartz, and occurs in large masses as a rock, so that it would furnish a good ornamental stone, if polished. It might be also used for a gem stone.

In December, 1887, specimens of feldspar were sent to the writer[1] for examination by Daniel A. Bowman, who had found them at a depth of 380 feet in the Hawk Mica Mine, four miles east of Bakersville, N. C. They proved to be a variety of oligoclase, remarkable for its transparency. The clearest piece measured 1 by 2 by 3 inches. One of the two varieties is of a faint window-glass green color, and contains a series of cavities, surrounded and fringed by tufts of white, needle-shaped inclusions called microlites, which measure from $\frac{1}{50}$ to $\frac{1}{20}$ inch (0·5 to 1·5 millimeter) in diameter and are quite round, resembling those that are occasionally present in the Ceylonese moonstone. The wonderful transparency of the oligoclase and the whiteness of the inclusions give the whole mass a striking resemblance to the lumps of glass so commonly obtained from the bottom of a glass pot. It was mistaken for this until its highly perfect cleavage was noticed. Recently some material of a slightly different character has been obtained at the mine. Cleavage masses of a white, striated oligoclase, 3 inches long, were found, containing nodules about $\frac{3}{8}$ inch to $\frac{5}{8}$ inch (10 to 15 millimeters) square, which were as colorless and pellucid as the finest phenacite and entirely free from the inclusions found in the greenish variety. This transparent variety, like the other, shows no striæ.

The following analysis by Prof. Frank W. Clarke, made from a faint green variety, shows it to be a typical oligoclase. The specific gravity was determined to be 2·651. This has been

[1] See Mineralogical Notes, by George F. Kunz, Am. J. Sci. III., Vol. 36, p. 222, Sept., 1888.

cut into a transparent gem, and may be advantageously used for spectroscope, microscope, and other lenses.

Silica..	62·60
Alumina.......................................	23·52
Ferric Oxide..................................	08
Manganous Oxide.............................	trace
Lime..	4·47
Potassa.......................................	56
Soda..	8·62
Loss by ignition [1]...........................	10
	99·95

A very fine oligoclase occurs at Dixon's Quarry, New Castle County, Del.; and at West Chester, Delaware County, Pa., a striated variety which admits of a handsome polish.

Obsidian, a peculiar, glasslike stone of volcanic origin, is found along Pitt River, Cal., where handsome specimens of the streaked marekanite or "mountain mahogany" are found, also in Owen Valley, in the same State, where it occurs in red fragments, and also banded with alternate layers of black and brown. Near Sante Fé, N. M., it is found in rounded pebbles over an inch across, resembling moldavite, as the variety from Moravia is called, only not quite so green. A porphyrite and sperolite obsidian occurs under the trachyte on Gunnison River, and a heavy vein of porphyrite obsidian is found under the Grande pyramid, continuing from thence southward through the trachytic bed. Nodules are found in the lower members of the trachytic veins. There is a dyke of light-gray and clear obsidian, with concentric structure, near the Colorado Central lode, north of Saguache Creek, near Georgetown, Col. Obsidian in fine pieces is very abundant ten miles southeast of Silver Peak, Nev., and at Obsidian Cliff in the Yellowstone Park, Wyo. This locality is described by Joseph P. Iddings[2] who says: "The cliff presents the partial sections of a floor of obsidian, the dense glass constituting the lower portion, which is from 75 to 100 feet thick. One of its remarkable features is a prismatic column, forming its southern extremity, which rises 50 or 60 feet, and is only 2 to 4 feet in diameter. The

[1] E. L. Sperry's Analysis in Mineralogical Notes by S. L. Penfield and E. A. Sperry, Am. J. Sci. III., Vol. 36., p. 325, Nov., 1888.

[2] Seventh Annual Report of the United States Geological Survey, p. 254 et seq.

color of the material is for the most part jet-black, but some of it is mottled and streaked with brownish red and various shades of brown mountain mahogany, passing into dark or light yellow, purple, and yellowish green." Fine examples from this locality are in the United States National Museum Collection at Washington.

William H. Holmes, in an interesting paper in the "American Naturalist,"[1] states that while examining the locality it occurred to him that the various Indian tribes of the neighborhood had probably visited the place in order to procure material for arrowheads and similar implements, and after a short search he found a leaf-shaped instrument that was 4 inches in length, 3 in width, and $\frac{1}{4}$ inch in thickness, of very fine workmanship and made of the black opaque obsidian. Further search was rewarded by ten more or less perfect implements. The use of obsidian as points for arrows, spears, and cutting implements was noted by Squire and Davis, who found such articles, though mostly broken, in Indian altar mounds of the Scioto Valley in Ohio; and an object made of this material was found in Tennessee by Gerald Troost.[2]

John R. Bartlett,[3] commissioner of the United States from 1850 to 1853 for determining the boundary line between the United States and Mexico, found pieces of obsidian and fragments of painted pottery along the Gila River wherever there had been Indian villages. Specimens have been found along the ruins of the Casas Grandes in Chihuahua, Mex., as well as along the Gila and Salinas Rivers. Similar observations have been made by earlier and later travellers, among whom is Caleb Lyon, who in 1860 found the Shasta Indians of California making arrowheads from obsidian as well as from the glass of a broken bottle. In a letter, which was published by the American Ethnological Society, he describes the method of manufacture.[4] The beautiful

[1] Notes on an Extensive Deposit of Obsidian in the Yellowstone Park. Vol. 13, p. 247, April, 1879.

[2] Ancient Remains in Tennessee. Vol. 1., p. 361, New York, 1845.

[3] Personal Narrative of Explorations and Incidents in Texas, New Mexico, California, Sonora, and Chihuahua, during the years 1850–1853, Vol. 2, p. 50, New York, 1854. Humboldt's Essai Politique sur la Nouvelle Espagne, Vol. 2, p. 243, Paris, 1825. Clavirego's History of Mexico, Vol. 1, p. 157, Philadelphia, 1817.

[4] Bulletin of the American Ethnological Society, Vol. 1, p. 39, New York, 1861.

color of the different varieties recommended its use in the arts, and it exists in such immense quantities that it should receive some attention from jewelers and decorators.

Pitchstone is a variety having the lustre of pitch rather than glass. Pitchstones are often albite or oligoclase rather than or-

OBSIDIAN

COLOR.	LOCALITY.	Silica.	Alumina.	Ferric Oxide.	Lime.	Magnesia.	Soda.	Potassa.	Water.	Hardness.	Specific Gravity.
Greenish black....	Yellowstone Nat. Park[1].	77.00	13.40	trace	1.25	1.19	3.43	3.62	0.70	6.0	2.4
....	Mono Valley, Cal.[2].....	74.05	13.85	trace	0.90	0.07	4.60	4.31	2.20	....	....
Variety Pitch-stone.......	Isle Royale, Lake Superior[3]......	62.51	11.47	11.05	2.67	2.11	3.03	trace	7.14	....	....

[1] Analyst, Beam. Wm. Beam, U. S. Geol. Sur., Hayden, 1878, Part I., p. 453.
[2] Analyst, T. M. Chatard. T. M. Chatard, U. S. Geol. Sur., Bull. No. 9.
[3] Analyst, Foster & Whitney. Foster & Whitney, Rep. Geol., Lake Superior, Part II., p. 206. Am. J. Sci. (2), Vol. 17, p. 118.

thoclase; that is, they contain soda, or soda and lime, instead of potash as a base.

The best known crystals of chondrodite, and the finest known gems cut from this material, have been found at the Tilly Foster Mine, Brewster, N. Y. Recent working has brought to light transparent, garnet-colored crystals, measuring $\frac{1}{2}$ by $\frac{1}{4}$ inch, and a few over 4 inches across. One essonite-colored crystal is $\frac{1}{4}$ inch

CHONDRODITE

	COLOR.	LOCALITY.	Silica.	Alumina.	Magnesia.	Ferrous Oxide.	Ferric Oxide.	Fluorine.	Water.
Theoretical Composition	....	....	36.00	....	64.00	....	....	....	....
Typical Analysis	Brown......	Tilly Foster Mine, Putnam Co., N. Y.[1]	35.42	....	54.22	5.72	....	9.00	....
Analysis...	Garnet Red..	" " " "[2]	34.10	0.48	53.72	7.17	....	4.14	....
" ...	Deep Red...	" " " "[3]	34.05	0.41	53.72	7.28	....	3.88	....
" ...	Red..,.....	" " " "[4]	35.42	....	51.88	9.73	....	5.38	....
" ...	....	New Jersey............[5]	32.00	....	51.00	6.00	....	8.55	2.00
" ...	Yellow......	" "............[6]	33.06	....	55.46	3.65	....	7.60	....
" ...	Red..........	" "..............[7]	33.35	....	53.05	5.50	....	7.60	....
...	....	" "..............[8]	36.00	....	54.64	....	3.97	3.77	1.62

[1] Specific Gravity, 3.22. Analyst, Hawes. Am. J. Sci. (3), 6, 220 and 10, 96. E. S. Dana, Trans. Acad. Conn. III.—HUMITE.
[2] Typical Analysis. Specific Gravity, 3.2. Analyst, Breidenbaugh. Rammelsberg, Mineralchemie, p. 705.—HUMITE.
[3] Analyst, Hawes. E. S. Dana, Am. J. Sci. (3), 10, 96.—HUMITE.
[4] Analyst, Breidenbaugh. E. S. Breidenbaugh, Am. J. Sci. (3), 6, 212.—HUMITE.
[5] Analyst, Langstaff. Langstaff, Am. J. Sci. (1), 6, 172.
[6] Analyst, Rammelsberg. Rammelsberg, Pogg. Ann., 53, 130.
[7] Analyst, Fisher. Fisher, Am. J. Sci. (2), 9, 85. Phillips' Mineralogy, 1852, p. 353.
[8] Analyst, Thomson. Thomson, Ann. N. Y. Lyc. III., 54.

across. Others, still uncut, would furnish fine gems. The finest of these crystals are in the Allen Cabinet, now at the Johns Hopkins University, in Baltimore, Md., and in the Mineralogical Cabinet of the Peabody Museum, New Haven, Conn. The cabinets of Frederick A. Canfield and Clarence S. Bement also contain fine specimens.   The gems are so few in number as to be only mineralogical curiosities.

Andalusite is found in a number of places in the United States, but as yet no fine gem stone has been discovered. Among the localities most worthy of mention are Upper Providence, Delaware County, Pa.; Westford, Mass.; Mount Wiley, Standish, Cumberland County, Me.; and Gorham, near Sebago Lake, Me.   The first-named locality is remarkable for the crystals of unusual size it has produced. Prof. Edward S. Dana describes one crystal now in the cabinet at Yale University and also another weighing more than 7 pounds.[1] The crystals from Westford are not en-

ANDALUSITE

	LOCALITY.	Silica.	Alumina.	Ferric Oxide.	Manganic Oxide.	Lime.	Water.	Specific Gravity.	Analyst.
Theoretical Composition....	....	36·90	63·10	....	....	....	....	....	....
Analysis.........	Lancaster, Mass....	39·09	58·56	....	0·53	0·21	0·99	....	Bunsen.[1]
"	Lancaster, Mass....	33·0	61·00	4·00[a]	....	....	1·50	....	Jackson.[2]
"	Lancaster, Mass....	41·95	48·60	9·30	....	0·41	....	2·923	Petersen.[3]
"	Chester, Penn......	46·40	52·92	....	....	....	....	....	Thomson.[4]

[1] Dana, Mineralogy, 5th Ed., p. 372.—CHIASTOLITE.   [3] Bunsen, Rammelsberg's Mineralchemie, 1875, p. 578.
[2] Jackson, Jour. Nat. Hist. Soc., Boston, I., 55.—CHIASTOLITE.   [4] Thomson, Nicol. Man. of Min. p. 243, 1849.
[a] Protoxide of Iron.

tirely perfect, but are of a fair pink color, about 2 inches long and ¼ inch across, and of a quality to yield small gems.   Those from Mount Wiley are from ¼ to ⅜ of an inch in diameter, of good flesh-pink color, and would cut into very fair gems.   In this vicinity there are also to be found similar crystals, in a quartz ledge associated with pyrrhotite. This association, which is identical in three different places, six miles apart, suggests the probability of the existence of andalusite in some abundance, as the spots visited are only outcrops of the same rock. Further exploration would probably result in the discovery of fine specimens.   The crystals found at Gorham, as regards perfection, color, and size, are equal to those found at any locality where this mineral does not occur as a gem.   The color is generally a brownish-flesh color, although at times the pink color fades into a faint grayish-pink. The crys-

[1] Am. J. Sci. III., Vol. 4, p. 473, Dec., 1872.

tals occur in a quartzite vein in a brown mica schist, and scattered through it are small crystals of pyrrhotite.[1]

Among the many valuable ethnological additions to the United States National Museum, consequent upon the acquisition of Alaska, is that of a series of highly interesting objects, consisting of drills, adzes, and knife-sharpeners, collected at Point Barrow, Capes Nome and Prince of Wales, and at St. Michael's, Sledge and Diomede Islands. Prof. Frank W. Clarke found by analysis that these objects were true jadeite or nephrite.[2] A mineral was also found which was mistaken for jade, but was determined by analysis to be pectolite. (See Pectolite.) The jade was generally coarse in quality, but among the objects were some with a high finish, some translucency, and great beauty. In color they were yellowish green, olive-green, siskin-green, and blackish green. A critical analytical and microscopical examination at the laboratory of the United States Geological Survey gave the following results :[3]

CHEMICAL COMPOSITION AND PROPERTIES.	Cape Prince of Wales. Part of Adze. Mottled Yellowish Green.	Drill. St. Michael's. Siskin Green.	Small Knife. Diomede Island. Blackish Green, Mottled and Laminated.	Adze. Point Barrow. Nearly Black.
Specific Gravity............	29·89	3·006	3·010	2·922
Silica.....................	56·01	56·12	56·08	57·11
Alumina..................	1·98	0·63	1·01	2·57
Ferrous Oxide.............	6·34	7·45	7·67	5·15
Manganous Oxide..........	trace	trace	trace	trace
Lime.....................	12·54	12·72	13·95	11·54
Magnesia.................	21·54	20·92	19·96	21·38
Ignition, Loss by..........	1·91	1·42	2·03	2·06
	100·35	99·26	100·01	99·81

Proc. U. S. Nat. Mus., 1888, p. 117.

As regards origin, some early writers have attributed the Alaska nephrite to Siberian sources, but of late years it has been generally ascribed to a home locality. Native reports pointed to a source known as the Jade Mountains, north of the Kowak

[1] Andalusite from a New American Locality, by George F. Kunz, Proc. Am. Ass'n. Adv. Sci. Vol. 32, p. 270, Salem, 1883.

[2] Am. J. Sci. III., Vol. 28, p. 20, July, 1884.

[3] Proc. U. S. Nat. Mus. 1888, p. 115.

River, about 150 miles above its mouth; and after several attempts the spot was visited in 1882 by Lieut. G. M. Stoney,
U. S. N.   He collected specimens of jade in situ, and a number
of samples were examined.   They may be described as follows:
     A. Greenish gray, splintery, lamellar in structure; A, like
B, but more granular; C, paler, nearly white, closer grained;
D, brownish, highly foliated.   All four were analyzed with the
following results:

CHEMICAL COMPOSITION AND PROPERTIES	A.	B.	C.	D.
Ignition....................	1·78	1·38	1·76	1·73
Silica......................	58·11	55·87	56·85	57·38
Alumina....................	0·24	2·07	0·88	0·19
Ferric Oxide...............	5·44	5·79	4·33	4·43
Ferrous Oxide.............	0·38	0·38	1·15	1·25
Manganous Oxide..........	trace	trace	trace	trace
Lime	12·01	12·43	13·09	12·14
Magnesia..................	21·97	21·62	21·56	22·71
	99·93	99·54	99·92	99·83

     The foregoing evidence is sufficient to show the essential
identity of all the Alaskan jades, and to dispose of the theory
that their presence in Alaska is to be accounted for upon the
basis of trade with Siberia.   That theory is also negatived by the
discovery, announced by George M. Dawson, of small nephrite
boulders on the upper part of the Lewis River, not far from the
eastern boundary of Alaska.   But these nephrites are also strikingly like those from many other localities, and two of the latter
have been included in our comparisons.   First, a water-worn,
dark-green boulder from New Zealand, sent to the Museum by
Sir Julius Haast; and second, a small implement from Roben-
hausen, Lake Pfäffikon, Switzerland, out of the collection of
Thomas Wilson.   The latter specimen, also green, had a specific
gravity of 3·015, as determined by Dr. William Hallock, and a
more weighty distinction is based upon the presence of inclosures
of foreign matter in the Siberian nephrite, which are quite lacking in the specimens from Alaska.
     True jade or nephrite has not been observed in the United
States, although early mineralogists referred the bowenite of

Smithfield, R. I., to that mineral.   (See Bowenite.)   Near Easton, Pa., is found a mineral which James D. Dana says is a mixture and calls it pseudo-nephrite.   Of this there are two varieties, one pale green, almost white, the other darker green.   The former is found on the Delaware River about a mile north of Easton, and the darker green, about a mile west of this locality at Lerch's, the former on the south side and the latter on the north side of the Syenite ridge.

CHAPTER X.

Chiastolite, Cyanite, Datolite, Staurolite, Isopyre, Pectolite, Dioptase, Prehnite, Zono-chlorite, Chlorastrolite, Thomsonite, Lintonite, Natrolite, and Fluorite.

THE curious, cross-like markings of chiastolite (macle) have suggested its use for gem purposes. The illustration shows the many markings that may exist in different parts of one crystal, and the variety of ornamental effects that may be produced. It is used for a gem, and sold for that purpose abroad, but there is no demand for it in the United States. Chiastolites are found in Mariposa County, Cal., and at Lancaster and Westford, Mass. William P. Blake first observed this mineral in Mariposa, where, in the drifts of the Chowchilla River, near the old road to Fort Miller, he found crystals in great

FIG. 7.
DISSECTED CRYSTAL OF CHIASTOLITE.

abundance, showing the black crosses on the white ground in a remarkably perfect manner. They are also found in the stratum of conglomerate which caps the hills above the streams, and they were all doubtless originally in place in the slates a little higher up the river. Smaller and imperfect "macles" are found in the slates on the road to Bear River, at Hornitos, Cal. The Massachusetts localities have yielded many of the best specimens found.

Cyanites were found in the early part of the century at Chesterfield, Mass., where some of the finest mineralogical specimens were obtained. An example of these, a mass measuring 10 to 6 inches, and consisting of distinct crystals over 3 inches long, piled one upon the other, is in the British Museum at South Kensington, in London. The crystals are all distinct, of a fine dark-blue color, and would cut into small mineralogical gems. At Darby Creek, Moon's Ferry, Delaware County, Pa., have been found deep azure-blue blades 5 and 6 inches long, which might afford gems if the mineral were thicker. Blue, green, and gray specimens are found at East Bradford, Chester County, Pa. Fine crystals occur with lazulite at Chubb's and Crowder's Mountains, on the road to Cooper's Gap, in Gaston County, N. C. At Windham, Me., cyanite has been observed in crystals 6 inches long. The old

CYANITE

	COLOR.	LOCALITY.	Silica.	Alumina.	Ferric Oxide.	Analyst.
Theoretical Composition...	....	....	36·90	63·10	....	....
Analysis................	White....	Sinclair Co., N. C.....	37·60	60·40	1·60	J. L. Smith.[1]

[1] J. Lawrence Smith, Am. J. Sci. (2), 16, 371, 1853.

localities are Worthington, Blandford, Westfield, and Lancaster, Mass.; Litchfield and Washington, Conn.; Strafford, Salisbury, and Bellows Falls, Vt.; near Wilmington, Del.; at Willis Mountain, Buckingham County; also two miles north of Chancellorsville, in Spottsylvania County, Va. The finest cyanite is found at Bakersville, N. C.,[1] where it occurs in distinct isolated crystals that, for perfection, depth of color, and transparency, rival those from St. Gothard, Switzerland. They are found at an altitude of 5,500 feet, near the summit of Yellow Mountain, on the road to Marion, N. C., four miles southeast of Bakersville, in a vein of white massive quartz in a granitic bluff, associated with almandite garnet of a very light transparent pinkish-purple color. The vein has a dip of sixty degrees, bearing northeast and southwest. The color varies from almost colorless to deep azure-blue, as dark as the Ceylonese sapphire. Some of the crystals are

[1] Am. J. Sci., III., Vol. 36, p. 224, Sept., 1888.

2 inches long, while a few were observed ⅜ inch (15 millimeters) in width and ⅜ inch (10 millimeters) in thickness. Occurring in white quartz, they form beautiful specimens, and the loose crystals were extensively sold for sapphire at Roane Mountain, the summer resort. Some gems have been cut, and a fine example is in the United States National Museum. It is, however, too soft to admit of much wear.

Datolite, in compact, opaque, white, creamy, and flesh-colored varieties, found at the Minnesota, Quincy, Marquette, Ashbed, and other mines in the copper region of Lake Superior, admits of a very high polish, and makes an excellent opaque gem or ornamental stone. One especially fine nodule over 4 inches across, with a flesh-colored centre shading off into gray and creamy tints, found at the Delaware Mine, is in the cabinet of Clarence S. Bement. Some fine specimens of this mineral are also in the William S. Vaux Cabinet at the Academy of Natural Sciences in Philadelphia.

The staurolites of Fannin County, Ga., first described by Prof. Edward S. Dana,[1] are found twelve miles southeast of Ducktown, Tenn., a locality which has furnished some of the finest known twinnings of this mineral. From their resemblance to a cross, these staurolites have found sale abroad as ornaments and charms, and are as highly regarded as those that are found in Brittany, France, which, according to the legend, were supposed to have been dropped from heaven. The Fannin County staurolites occur twinned in single and double crosses, and are found in large quantities in a decomposed rock of mica schist. Of those taken out, perhaps one-tenth are perfect crystals. They all require a certain amount of scraping and cleaning. Brilliant crystals are found at Windham, Me., some of the twins forming fine crosses. Occasionally, transparent crystals are found here that if cut would afford mineralogical gems resembling poor garnets. Staurolite is also found at Franconia and Lisbon, N. H., in mica slate; on the shores of Mill Pond, loose in the soil; at Grantham, N. H.; at Cabot, Vt.; at Chesterfield, Mass.; at Bolton, Litchfield, Stafford, Tolland, and Vernon, Conn.; on the Wissahickon, eight miles from Philadel-

[1] Am. J. Sci., III., Vol. 11, p. 385, May, 1876.

phia, Pa., in reddish-brown crystals; and at the lead mine in Canton, Ga. At the Parker Mine, Cherokee County, N. C., it occurs in large, coarse, single crystals and twins; also along Persimmon, Hanging Dog, and Bear Creeks, Madison County, and Tusquitee Creek, Clay County. In the last-mentioned places staurolite is found in argillaceous and talcose slates. Some staurolite macles similar to a chiastolite, from Charlestown, N. H., are described by Dr. Charles T. Jackson. These pass by insensible shades or gradations into andalusite macles.

Isopyre is found in small veins from 1 to 3 inches in width in the magnetic iron at Dickinson Mine in Ferremonte, three miles from Dover, N. J. In color it very nearly resembles the darker green jasper, or, in other words, bloodstone without the red spots. It is used as a gem in the cabinets of collectors. Its hardness is 6·0 to 6·5.

Pectolite was found in quantity among the Esquimau implements collected by the United States Signal Service at Point Barrow, Alaska, and examined by Prof. Frank W. Clarke;[1] it was at first supposed to be jade, but on examination proved to be a new and interesting variety of compact pectolite, in two varieties, one pale apple-green, the other dark green. The specific gravity of the pale-green variety was 2·873, that of the dark-green 3·092. This forms an interesting and unexpected addition to the list of gem stones. During 1887 a massive white pectolite of unusually dense structure, and susceptible of a high polish, was announced by William P. Blake as occurring in Tehama County, Cal., in masses of considerable size. In a letter to the writer he gives the following description of it: "It occurs in a vein, and is broken out in rough tabular masses, from 2 to 3 or more inches in thickness, but it is reported that much larger masses can be obtained. It is exceedingly tough and hard to break. The fractured surfaces are irregular, without cleavage, but have a silky lustre, and a crypto-crystalline structure is exhibited in extremely fine inseparable fibres, which are radial, curved, and interlaced, and are, perhaps, imbedded in a silicious magma, but the fibres constitute the bulk of the mass. Color, white, with a delicate shade of sea-green; translucent. Exposed or weathered portions lose their porcelain-

[1] Am. J. Sci., III., Vol. 28, p. 21, July, 1884.

like translucency, and become white and somewhat earthy in appearance, and exhibit the crypto-fibrous structure with more distinctness. Specimens cut and polished across the end of a slab-like mass show on one side a narrow selvage of breccia made up of fragments of the pectolite and of a dark-colored rock, mixed and firmly cemented together. On the opposite side or border of the mass, there are distinctly-formed parallel planes of concentric layering, from the surfaces of which the fibres diverge. These layers and the breccilated border opposite show the vein-like

PECTOLITE

CHEMICAL COMPOSITION AND PROPERTIES.	Theoretical Composition.	Locality, Alaska.[1]	Locality, Isle Royale, Lake Superior.[2]	Locality, Isle Royale, Lake Superior.[3]	Locality, Bergen Hill, N.J.[4]	Locality, Point Barrow, Alaska.[5]	Locality, Bergen Hill, N.J.[6]	Locality, Isle Royale, Lake Superior.[7]	Locality, Lehigh Co., Penn.[8]	Locality, Alaska.[9]
Silica	54·20	53·20	53·45	55·66	54·62	53·94	54·00	55·00	55·17	53·94
Alumina	.....	....	4·94	1·45	.....	0·58	1·90	1·10	.....	0·58
Ferrous Oxide	.....	....	....	....	1·20	....	....	....	....	
Ferric Oxide	....	....	....	....		....	....	....	0·80	trace
Manganous Oxide	....	....	....	....		....	....	....	....	....
Lime	33·80	33·26	31·21	32·86	32·94	32·21	32·10	32·53	30·00	32·21
Magnesia	.....	0·64	....	....	....	....	....	....	....	1·43
Soda	9·30	9·40	7·37	7·31	8·96	8·57	8·89	9·72	9·02	8·57
Potassa	.....	....	trace	....	....	....	trace	....	0·37	....
Water	2·70	3·50	2·72	2·72	2·89	4·09	2·96	2·75	4·63	4·09
Color	.....	Light Green.	White.	White.	.....	Apple Green.	Silky White.	Silky White.	.....	Light Green.
Hardness	.....	7·0	....	....	.....	....	....	....	.....	
Specific Gravity	.....	2·85–2·86	....	....	.....	....	....	....	.....	2·873

[1] Analyst, Frenzel. Typical Analysis, Jahrb. d. Vereins f. Erd-Kunde, Dresden, 1884.
[2,3] Analyst, Whitney. J. D. Whitney, Jour. Boston Soc. Nat. His., 1849, p. 36. Am. J. Sci., II., 7, 434.
[4] Analyst, Whitney. J. D. Whitney, Am. J. Sci., II., 20, 205. Rammelsberg Mineralchemie, p. 380.
[5] Analyst, Clarke. F. W. Clarke, U. S. Geol. Survey, Bull. No. 9.
[6] Analyst, Kendall. Variety Stellite, same reference as 1 and 2.
[7] Analyst, Dickinson. Variety Stellite, same reference as 1 and 2.
[8] Analysts, Kneer & Smith. G. B. Kneer & E. F. Smith, Am. Chem. J., 6, p. 411.
[9] Analyst, Clarke. F. W. Clarke & T. M. Chatard, Am. J. Sci., III., 28, 20.

formation of the mass between walls. Its hardness is from 6 to 6·5. It may be found useful as an ornamental stone for making small objects, cups, plates, handles, or for carving figures, or inlaid work." This is identical with the pectolite from Alaska, described by Prof. Frank W. Clarke. (See Jade, Chapter on Mexico.)

Dioptase was first described by R. C. Hills as being found in the United States at the Bon Ton group of mines, about seventy miles from Clifton, Ariz., where it occurs in brilliant green crystals, $\frac{1}{16}$ to $\frac{1}{4}$ inch in length, lining cavities of what is called "mahogany ore," a dark-brown, compact mixture, consisting principally of limonite and oxide of copper in varying proportions.[1] It has since been found in larger and finer crystals, but notwithstanding its rich emerald-green color, its softness prevents its use as a gem. These crystals are not equal to those from the Khirghesse Steppes, Siberia. Single crystals from there are occasionally mounted entire without cutting.

Prehnite has been found in a number of localities in the United States, and gems have been cut from material found at Bergen Hill and Paterson, N. J. It is a stone of rich, oily-green

PREHNITE

CHEMICAL COMPOSITION AND PROPERTIES.	Theoretical Composition	LOCALITY, Cornwall, Penn. Analyst, Genth.[1]	LOCALITY. Lake Huron. Analyst, Thomson.[2]	LOCALITY, Isle Royale, Lake Superior, Analyst,[3]
Color...........................	.....	Bluish green.	Yellow green.	Light green.
Silica...........................	43·60	42·40	45·80	46·10
Alumina........................	24·90	20·88	33·92	25·90
Sulphur.........................	.....	.....	.....	0·90
Ferric Oxide....................	.....	5·54	.....	.....
Ferrous Oxide	.....	.....	4·32	.....
Lime........................	27·10	27·02	8·04	27·00
Magnesia.......................	.....	trace	1·72	.....
Water	4·40	4·01	4·16	.....
Hardness.......................	.....	.....	3·25	6·00
Specific Gravity................	.....	4·03	2·86	2·88

[1] Typical Analysis. F. A. Genth, Proc. Am. Phil. Soc., Aug. 18, 1881.
[2] Th. Thomson, Nicol's Man. of Min. (1849), p. 144.
[3] Albert Selle, Cours de Min. et de Geol. (1878), p. 326.

color, generally in botryoidal sheets or spheres. Quite a number of these, some an inch in diameter, were found at the Pennsylvania open cut, and in the cutting of the Morris and Essex Tunnel through Bergen Hill, N. J. When cut and polished, it resembles chrysoprase in color and lustre. Zonocholrite, chlorastrolite, and lintonite have been referred to this species, but Dr. George W. Hawes found by analysis that they were only impure varieties.

Zonochlorite was described by Prof. A. E. Foote in 1873, and was obtained by him in 1867 on a small island off Neepigon Bay, north shore of Lake Superior, where it occurs associated with quartz, amethyst, carnelian, etc. The largest pieces found are less than 2 inches across. Its hardness varies from 6·5 to 7. It received

[1] Am. J. Sci., III., Vol. 23, p. 325, April, 1882.

its name from its green color and banded appearance, from zona, a band, chloros, green, and lithos, stone. It takes a high polish, but it has been used only to a limited extent as a mineralogical gem.

Chlorastrolite is found only on the Isle Royale, Lake Superior. This island, which belongs to the State of Michigan, is forty miles long and five miles broad, and is about twenty miles from the mainland. The only inhabitant of the island is the lighthouse keeper, who, from time to time, entertains parties who come to fish or mineralogists who come for chlorastrolites. The underlying rock is an amygdaloid trap, in which the gem is found, but it is now collected in the form of rolled pebbles on the beach, having fallen or weathered out of the trap rock. It is entirely opaque, of green color, mottled and stellated, and admits of a high polish. Sometimes the stellations radiate from the centre,

CHLORASTROLITE

COLOR.	LOCALITY.	Silica.	Alumina.	Ferric Oxide.	Lime.	Soda.	Potassa.	Water.	Hardness.	Specific Gravity.	Analyst.
Bluish-green..	Isle Royale, Lake Superior.......	36·99	25·49	6·48	19·90	3·70	0·40	7·22	5·5	3·180	J. D. Whitney.[1]
....	Isle Royale, Lake Superior.......	37·41	24·25	6·26	21·68	4·88		5·77	6·0	....	J. D. Whitney.[2]

[1],[2] J. D. Whitney, Jour. Boston Nat. Hist. Soc., 5, 488. Am. J. Sci. (2), 6, 270. Rammelsberg, Mineralchemie, p. 639. Alber Selle Cours de Min. et de Geol., p. 306. Rep. Geol. of Lake Superior, 1851, II., p. 97. Dana, Mineralogy, 5th Ed., p. 412.

and show a beautiful chatoyancy, similar to the cat's-eye, crocidolite, and other fibrous minerals. Prof. A. E. Foote and a party camped for some months on this island in 1868, and chlorastrolite was first found by them in a vein-stone associated with native copper and epidote. Rounded pebbles of the rock containing the chlorastrolite are plentiful on the beach. One of the largest known perfect chlorastrolites is in the cabinet of M. T. Lynde, of Brooklyn, N. Y., and measures 1¼ by 1¼ inches (see Colored Plate No. 3); next in size is one belonging to Alfred Morrison, of London; and the third largest is owned by an American lady, now residing in London. A fine pair of oval chlorastrolites, over half an inch in length, are in the possession of Frederick A. Canfield, of Dover, N. J. About $1,000 worth are annually sold to tourists.

Thomsonite and lintonite, the latter first described by Peckham[1]

[1] Am. J. Sci. III., Vol. 19, p. 123, Feb., 1880.

and Hall, found at Good Harbor Bay, Grand Marias, on Lake Superior, Mich., in the basalt, and as rolled pebbles on the beach, result from the decomposition of the rock, the amygdules withstanding the action of the weather better than the rock. They vary from the size of a pinhead to over an inch in diameter. Many of the thomsonites are made up of series of concentric layers of various shades of color, in soft tones of flesh-red, creamy white, yellow, and green, and are excedingly pretty, especially when polished, when they resemble the eye-agate. Great numbers are annually sold to visitors at Lake Superior, especially at Duluth, Minn., and Grand Marias, Mich. The cutting of thomsonite consists almost entirely of a rounding off of the pebble, so as to show the concentric and other markings to the

THOMSONITE

COLOR.	LOCALITY.	Silica.	Alumina.	Ferric Oxide.	Lime.	Potassa.	Soda.	Ferrous Oxide.	Water.	Hardness.	Specific Gravity.
White...	Grand Marais, Minn. [1]	40·45	29·50	0·232	10.75	0·537	4·766	....	13·93	5·0–6·0	2·33–2·35
Flesh red	" " " [2]	46·02	26·717	0·813	9·40	0·390	3·756	....	12·80	5·0–6·0	2·33–2·35
Flesh red	" " " [3]	40·605	30·215	....	10·37	0·49	4·055	0·40	13·75	5·0–6·0	2·33–2·35
Green...	" " " [4]	40·45	29·37	0·88	10·43	0·42	4·28	....	13·93	5·0–6·0	2·33–2·35
White...	Colorado [5]	40·68	30·12	....	11·02	....	4·44	....	12·86	....	....
....	Minnesota [6]	41·23	29·00	....	11·60	....	4·86	....	14·06	6·0	2·316

[1,2,3,4] Analyst, Linton. Peckham & Hall, Am. J. Sci. (3), 19, 122, Feb., 1880.
[5] Analysts, Cross & Hillebrand. Cross & Hillebrand, Am. J. Sci. (3), 23, 452.
[6] Analyst, Koenig. G. A. Koenig, Naturalist's Leisure Hour, No. 8, Aug. 1, 1878.

best advantage. Some that have been polished are over an inch in diameter. The small ones are generally of the finest material. Lintonite is really a variety of prehnite, and takes a fine polish either alone or when associated with the flesh-colored forms of thomsonite.

Natrolite occurs in many localities in beautiful crystals, but too small to cut for gems. Many veins of it, and one large area containing over 300 square feet of the mineral, were met with in the sinking of Shaft No. 2 of the West Shore Railroad, at Weehawken, N. J. Scarcely any of the crystals were stout enough to afford gems. This beautiful, limpid white mineral occurs abundantly all along Bergen Hill where tunnelling has been carried on, and fine crystals have been found in the Lake Superior copper region. None have been sold for gems in the United States, although when suitable crystals are found, it is occasionally used for the letter N in initial jewelry.

Fluorite, in the colored transparent varieties, is designated as false ruby, emerald, sapphire, topaz, amethyst, etc. Thirty years ago many specimens of the green variety were found at Muscalonge Lake, St. Lawrence County, N. Y., where the mineral was taken out from a vein which ran under the lake; and in the autumn of 1888, an immense cavity lined with large cubic crystals of green fluorite was discovered at Macomb. This furnished groups measuring nearly two feet across and single composite crystals nearly a foot across, in all, several tons of fine crystals. The largest deposits in the United States are at Rosiclaire, Shawneetown, and Elizabethtown, Hardin County, Ill., and some thousands of tons are annually mined there, crystals of the richest purple, yellow, red, rose-colored, green, and other shades being very common. It differs from English

FLUORITE

	COLOR.	LOCALITY.	Calcium.	Fluorine.	Ferric Oxide.	Phosphate of Lime.	Specific Gravity.
Theoretical Composition..	....	....	51·30	48·70	....	trace	....
Analysis.................	Colorless..	Wheatley Mine, Penn.[1].	50·81	48·29	....	trace	3·15
"	Purple....	Lehigh Co., Penn.[2]....	50·87	49·20	trace	....	3·21
"	Green.....	Lehigh Co., Penn.[3]....	50·91	49·00	trace	....	3·21

[1] Analyst, J. L. Smith. J. Lawrence Smith, Am. J. Sci. (2), 20, 242, 253. Proc. Am. Ass'n, 1885. Erdm. J. fur pr. Ch. 66, 432-7, 1855.
[2,3] Analyst, E. F. Smith. E. F. Smith, Am. Chem. Jour., Vol. 5, p. 272.

fluorite in that the crystalline faces in nearly all the specimens are dull, and the colors show only by transmitted light. Crystals a foot across were observed here twenty years ago during the workings of the Rosiclaire lead mine. In the mounds of this region it has occasionally been found shaped into ornaments by the hand of prehistoric man. This is the only instance that is known of its being used as an ornament. The amount mined here for the arts amounts to over $15,000 a year. On the Cumberland River, Tenn., and at Pike's Peak, Col., some fine crystals of a blue-green fluorite have been found; also yellow crystals in the geodes of the limestone at St. Louis, Mo. One of the most remarkable varieties of this mineral is a chlorophane from the microlite localities at Amelia Court House, Va., which has been described by W. M. Fontaine,[1] who also noted the brilliancy of the phosphorescent light that it gives out at a low

[1] Am. J. Sci. III., Vol. 25, p. 330, May, 1883 Minerals in Amelia County, Va

temperature.   Pallas mentions a specimen from Siberia, of a pale violet color, which the heat of the hand caused to give out a white light merely ; the heat of boiling water, a green light ; and when on a live coal, it gave out a bright emerald-green light that might be discerned from a distance.   The writer found that while handling a few specimens of this mineral in the dark, phosphorescence resulted from the slightest attrition of the specimens, either one with another, or with a nail or any hard substance. In a dark room, at a temperature of about 80° Fahr., the Amelia County mineral shows a white, luminous glow, which is intensified by the warmth of the hand ; when placed in boiling water, it becomes green ; and on a heated iron plate, an intense emerald-green.   Most of the material is more or less flawed, so as to render it very friable under touch.   This variety is of a light-green or a yellowish-green color,' and seems to phosphoresce at even a lower temperature than the more compact form.   A stone cut from this material and placed in a vial of warm water fluoresced distinctly in a dark room, after being in the water a few minutes, thus giving a new form of gem, that is, a fluorescent gem stone, though not hard enough for continuous wear.

'Am. J. Sci. III., Vol. 28, p. 235, Sept., 1884.

CHAPTER XI.

Serpentine, Bowenite, Williamsite, Microlite, Meerschaum, Apatite, Beryllonite, Lazulite, Cassiterite, Hematite, Lodestone, Rutile, Octahedrite, Brookite, Arkansite, Titanic Iron, Titanite, Malachite, Chrysocolla, Azurite, Aragonite, Fossil Coral, Pyrite.

SERPENTINE is found in many localities in the United States, and of a quality to fit it for use for ornaments, although it is little used for that purpose, and finds its greatest demand for decorative and building purposes. The dark-green noble serpentine found at Newburyport, Mass., has been cut into oak and other leaf forms for ornaments. The golden and greenish-yellow serpentine of Montville, N. J., is of the precious variety, and takes an excellent polish. In this locality serpentine occurs associated with crystalline dolomite, and many fine specimens in different collections were obtained, during the process of quarrying this rock, for burning into quicklime or for flux in iron furnaces. It occurs in small seams or veins, or in isolated nodules from a few inches to several feet in diameter. George P. Merrill, of the United States National Museum, has written an exhaustive paper on this subject.[1] He has found that the white and gray nuclei which often exist in the centre of these nodules of serpentine are pyroxene, and by analysis and microscopic examination has proved that this serpentine is the result of an alteration from pyroxene. The beautiful series of polished specimens in the United States National Museum and in Yale University show all the changes from pyroxene to serpentine.

[1] See Proc. U. S. Nat. Mus., 1888, p. 105.

The serpentine of St. Lawrence County, N. Y., also that of Cornwall, Monroe, and Warwick, Orange County, N. Y., the ophiolite of New York City and vicinity, the serpentine of New Rochelle, N. Y., also some of the Hoboken, N. J., and the Staten Island varieties are useful for ornamental and decorative purposes.

On Deer Island, Me., serpentine of a very light-green color occurs. The serpentine from the neighborhood of Patterson, Caldwell County, N. C., is of a dark greenish-black color, and admits of a fine polish.[1] In several localities in Delaware County, Pa., it occurs in combination with calcite. Serpentine is quarried chiefly in three places, Roxbury, Vt.; Moriah, Essex County, N. Y.; and Dublin, Harford County, Md. The Vermont stone is deep green in color, traversed by white veins of calcite, and takes a beautiful polish. It compares very favorably with the Italian verde antique or verde di Prato, from the quarries in Tuscany. The Moriah stone, which is similar in color, but granular in texture, and spotted rather than veined, is found in the market in the form of mantels, table-tops, ornaments, and similar objects. The Maryland stone is more uniformly green in color than either of the others, and contains very little calcareous matter. It is within easy reach of Baltimore. According to Prof. Genth, who reported on this locality in 1875, it consists of a very large bed of green serpentine, about 500 feet in thickness, overlying a bed of black mottled serpentine about 800 feet in thickness; in the latter, masses of the green serpentine are frequently found imbedded. Beneath this immense bed of serpentine is a smaller bed of green serpentine, 180 feet in thickness, and beneath this, there is a third bed of green serpentine. Of its quality, he says: " Everywhere it shows exactly the same character, but, as should be expected, that which came from a greater depth showed a somewhat lighter color and greater compactness." He concludes that beyond doubt there is an inexhaustible quantity of this green serpentine in the most favorable position for mining on a large scale, and with an abundant water-power to manufacture it into marketable forms. A coarse serpentine, used for building purposes, but not suited for ornamental work, is quarried in considerable quantities in Chester County, Pa. The stone is dull green

[1] Minerals of North Carolina, p. 57.

in color, soft enough to work readily, and capable of producing most excellent effects, particularly in rock-faced work and rubble work. It has been used extensively in Philadelphia and vicinity, where it was employed in the construction of the buildings of the University of Pennsylvania and the Academy of Natural Sciences, and it has also been used to some extent in New York City and Washington.[1]

Bowenite is a variety of serpentine found in some quantity at Smithfield, R. I., varying in color from a pure white through light green to deep green. It is the "jade" and "nephrite" of the early American mineralogists, so-called on account of its remarkable toughness and its hardness. As yet, however, no archæological objects made from it have been found. Its rich color and peculiar toughness and hardness suggest it for use, to some extent, where jade has previously been employed. Prof. Genth[2] mentions as having been found at Easton, Pa., a bowenite of a greenish and reddish-white color and of great tenacity, frequently containing a small quantity of tremolite. The ease with which this material is worked, and the effective designs that can be made from it, suggest it for decorative purposes. Analyses of serpentine from Hartford Co., Wilmington, Del., by Professor Genth have been made with the following result:

SERPENTINE

Theoretical Composition and Properties.	Deep Green Translucent.	Black Mottled.
Silicic Acid	40·06	40·39
Alumina	1·37	1·01
Chromic Oxide	0·20	trace
Nickel Oxide	0·71	·23
Ferrous Oxide	3·43	·97
Manganous Oxide	0·09	trace
Magnesia	39·02	38·32
Water	12·10	12·86
Magnetic Iron	3·02	6·22
	100·00	100·00
Hardness	4·0	4·0
Specific Gravity	2·668	2·669

Williamsite is a variety of serpentine found in Texas, Lancaster County, Pa., and in Maryland. Owing to its rich green color, and the ease with which it can be cut, it has been used to a

[1] Cf. Building and Ornamental Stones of the United States, by George P. Merrill, Pop. Sci. Monthly, Vol. 27, p. 520.

[2] Contributions to Mineralogy (1876).

SERPENTINE

Color	Locality	Silica	Alumina	Ferric Oxide	Ferrous Oxide	Manganous Oxide	Nickel Oxide	Lime	Magnesia	Water	Hardness	Specific Gravity	Analyst
Theoretical Composition	...	43.48	...	...	...	...	...	...	43.48	13.04	...	...	...
Typical Analysis	Apple green	Smithfield, R. I.	42.56	trace	...	0.95	...	...	trace	43.15	12.84	...	...	J. L. Smith.[1]
Analysis	Dark green	Newburyport, Mass.	41.76	trace	...	4.06	...	...	...	41.40	13.40	...	2.084	T. Petersen.[2]
"	Apple green	Texas, Penn.	45.40	8.50	...	...	...	trace	...	33.60	12.50	4.5	2.59-2.64	Shepard.[3]
"	...	...	41.60	trace	...	3.24	...	0.50	...	41.11	12.70	...	...	J. L. Smith.[4]
"	...	...	42.60	trace	...	1.62	...	0.40	4.25	41.90	12.70	5.0	2.57	J. L. Smith.[5]
"	Apple green	Smithfield, R. I.	44.69	0.56	...	1.75	trace	...	trace	34.63	13.42	...	...	Bowen.[6]
"	"	42.20	trace	...	1.36	...	...	1.90	42.50	13.28	...	...	J. L. Smith.[7]
"	"	42.10	trace	...	1.11	...	...	...	41.23	12.77	...	...	J. L. Smith.[8]
"	Massachusetts	43.20	...	...	5.24	...	...	...	40.00	11.42	...	...	Lychnell.[9]
"	Blanford, Mass.	40.00	...	...	2.70	...	...	...	41.40	15.67	...	...	Shepard.[10]
"	Roxbury, Conn.	42.60	...	...	8.30	...	...	0.93	35.30	13.00	...	...	Jackson.[11]
"	Middletown, Conn.	44.08	0.30	...	1.17	...	...	0.37	40.87	13.00	...	...	Burton.[12]
"	New Haven, Conn.	44.05	...	...	2.53	...	...	...	39.24	13.49	...	...	Brush.[13]
"	Smithfield, R. I.	42.29	...	...	1.21	...	...	0.63	42.29	12.96	...	...	Smith & Brush.[14]
"	Syracuse, N. Y.	40.67	5.13	...	8.12	...	...	...	39.61	12.77	...	...	Hunt.[15]
"	Montville, N. J.	42.53	0.38	...	1.96	...	...	...	42.16	14.22	...	...	Manice.[16]
"	"	42.62	0.66	...	0.27	...	...	...	42.67	14.25	...	...	Reakirt.[17]
"	Hoboken, N. J.	42.32	...	1.64	1.28	...	...	...	42.23	13.80	...	...	Garrett.[18]
"	"	41.67	...	...	...	...	...	...	41.25	13.86	...	...	Lychnell.[19]
"	East Goshen, Penn.	43.89	...	...	1.38	...	0.69	...	40.48	13.45	...	...	Sharples.[20]
"	Texas, Penn.	44.25	4.90	...	3.67	...	0.50	...	34.00	12.32	...	...	Brewer.[21]
"	"	41.60	...	...	3.24	...	0.40	...	41.11	12.70	...	...	Smith & Brush.[22]
"	"	42.60	...	...	1.62	...	0.90	...	41.90	12.70	...	...	Smith & Brush.[23]
"	Bare Hills. Md	44.59	0.75	...	1.39	...	...	...	39.71	12.75	...	...	Hermann.[24]
"	"	42.69	...	1.16	...	...	...	...	40.00	16.11	...	...	Vanuxem.[25]
"	Webster, N. C.	40.95	1.50	...	10.05	...	...	...	34.74	12.60	...	...	Thomson.[26]
"	Texas, Penn.	43.87	0.31	...	7.17	...	0.27	...	38.62	9.55	...	...	Genth.[27]
"	...	43.79	...	...	2.05	...	...	...	41.03	12.07	...	...	Rammelsberg.[28]

1 J. L. Smith, Am. J. Sci. (2), 15, 207; 16, 41; 18, 372. Erdm.
J. pr. ch. 92, 111; 61, 172.
2 T. Petersen Jahresb. über Forts. der Ch., 1866, 932.
3 C. U. Shepard, Am. J. Sci. (2), 6, 349.
4 J. L. Smith, Am. J. Sci. (2), 15, 207; 16, 41; 18, 372. Erdm.
5 Geo. F. Bowen, Am. J. Sci. (2), 5, 113.
6 Same as (4) (5).
7,8 Lychnell, Dana, Mineralogy, 5th Ed., p. 466.

10 Shepard's, Mineralogy I., 291.
11 C. T. Jackson, Proc. Bost. Soc. Nat. Hist., 1856.
12 Burton, Dana, Mineralogy, p. 467, 5th Ed.
13 G. J. Brush, Dana, Mineralogy, p. 467, 5th Ed.
14 Smith & Brush, Am. J. Sci. (2), 15, 211.
15 T. S. Hunt, Rep., Geol. of Can., 1853-1856.
16 Manice, Dana, Mineralogy, 5th Ed., p. 467.
17 Reakirt, Am. J. Sci. (2), 18, 410.
18 Garrett, Dana, Mineralogy, 5th Ed., p. 467, 1850, 690.

19 Lychnell, Dana, Mineralogy, 5th Ed., p. 467.
20 Sharples, Am. J. Sci. (2), 4, 77.
21 Brewer, Dana, Mineralogy, p. 467, 5th Ed., p. 467.
22,23 Smith & Brush, Am. J. Sci. (2), 15, 211.
24 Hermann, J. pr. ch. ...
25 Vanuxem, J. Acad. Sci. Phil., 3, p. 13.
26 Thomson, Phil. Mag. xx. 391.
27 F. A. Genth, Am. J. Sci. (2), 23, 201.
28 Rammelsberg, Dana, Mineralogy, 5th Ed., p. 467.

limited extent in jewelry for charms and other ornaments. It is usually of a more pleasing color than jade, and varies from a dark green to light apple-green and emerald-green shades. Williamsite is one of the handsomest known opaque or transparent stones, rivaling in richness many of the varieties of green jade. The grayish-green serpentinous substance found at Pelham, Mass., named pelhamine by Prof. Charles U. Shepard, admits of a very good polish, which produces a curious effect.

Microlite has been found at the Allen Mica Mines, in Amelia Court House, Amelia County, Va., in beautiful crystals, some of which weigh 4 pounds each, but are opaque. The finest of these,

MICROLITE

	Hyacinth Red.
Color	
Tantalum Oxide...	86·45
Columbium Oxide...	7·74
Tungstic Acid..	0·30
Stannic Acid (Binoxide of Tin).................................	1·05
Lime...	11·80
Magnesia...	1·01
Beryllia...	0·34
Uranium Oxide..	1·59
Yttria...	0·23
Cerium Oxide...	} 0·17
Didymium Oxide...	}
Alumina..	0·13
Ferric Oxide...	0·29
Soda...	2·86
Potassa..	0·29
Fluorine...	2·85
Water..	1·17
Hardness...	6·0
Specific Gravity...	5·656
Locality...	Amelia County, Va.[1]

[1] Analyst, F. P. Dunnington. Am. J. Sci. (3), 22, 82 and 25, 335.

a transparent specimen, is in the cabinet of Clarence S. Bement; it is about ⅝ of an inch long, and in part a rich honey-yellow, having all the color of topazolite, with a higher lustre. Some crystals are of sufficient transparency to afford gems ranging in color from an essonite-red to a rich spinel-yellow and are of remarkable brilliancy. Microlite has the highest specific gravity of any known gem, being about 6.[1]

Meerschaum, or sepiolite, has occasionally been met with in compact masses of smooth, earthy texture in the serpentine quar-

[1] See A Transparent Crystal of Microlite, by William E. Hidden, Am. J. Sci. III., Vol. 30, p. 82, July, 1885.

ries of Mest Nottingham Township, Chester County, Pa. Only a few pieces have been found, but they were of good quality. It also occurs in grayish and yellowish-white masses in the serpentine in Concord, Delaware County, Pa. Masses of pure white material, weighing a pound each, have been found in Middletown, in the same county, and of equally good quality at the Cheever Iron Mine, Richmond, Mass., in pieces over an inch across; also in the serpentine at New Rochelle, Westchester County, N. Y.

Apatite is found in such remarkably perfect and fine-colored crystals in the tourmaline locality of Auburn, Me., that the hill on which it occurs has been named Mount Apatite. The crystals are transparent green, pink, and violet, and so closely resemble tourmaline as to have been mistaken for it. Some of the local collectors attempted to cut them, but without success, for the hardness is too low for a transparent gem.

Beryllonite was first found near Stoneham, Me., in 1886, and this is still the only locality known. Owing to the great transparency and brilliancy of the mineral, as well as its form of crystallization, it at first suggested topaz, and was for a time overlooked, but Prof. E. S. Dana on examination found it to be a new species, to which he gave the name beryllonite.[1] It was analyzed by Horace L. Wells, of the Sheffield Scientific School, who found that it had the following composition :

Phosphoric Acid	55·86
Beryllium Oxide	19·84
Soda	23·64
Moisture	0·08
	99·42

From which the formula $Na_2O.2BeO.P_2O_5$, or $NaBePO_4$ was deduced.

Its hardness was found to be 5·6 to 6 and its specific gravity 2·84. From the great number of its cavities filled with water or carbon dioxide, its lustre and the iridescence of the crystals when viewed from the pyramid face, it strikingly resembles the white topaz of Stoneham, Me. The transparency and brilliancy of this mineral fit it for a mineralogical gem.

[1] See Description of the New Mineral, Beryllonite, by Edward S. Dana and Horace L. Wells. Am. J. Sci. III., Vol. 37, p. 23, Jan. 1889.

Lazulite is found in dark-blue crystals and crystalline masses at Crowder's and Chubb's Mountains in Gaston County, and at Coffee Gap, Sauratown, Stokes County, N. C. At Graves Mount, Lincoln County, Ga., however, are found the finest sky-blue and dark-blue crystals known, often measuring from ¼ an inch to 2 inches in length, and quite compact, and of good color.

LAZULITE

	COLOR.	LOCALITY.	Phosphoric Acid.	Alumina.	Ferrous Oxide.	Magnesia.	Silica.	Water.	Hardness.	Specific Gravity.
Theoretical Composition	....	....	46·80	34·00	....	13·20	....	6·00	......	....
....	Blue......	North Carolina [1]...	43·38	31·22	8·29	10·06	1·07	5·68	5·0-6·0	....
....	Blue......	North Carolina [2]...	44·15	32·17	8·05	10·02	1·07	5·50	5·0-6·0	....
Typical Analysis.......	Blue......	Keewatin, Canada [3]	46·39	29·14	2·09	13·84	2·83	6·47	5½	3·044

[1], [2] Analysts, Smith & Brush. Dana, Mineralogy, 5th Ed., p. 572.
[3] Analyst, C. Hoffman. Hoffman, Report of Geol. of Can., 1878-79, p. (1-6).

Its hardness is 6, and its specific gravity is 3·122. This mineral would make an opaque gem or an ornamental stone, as the color, though lighter, is often as rich as that of lapis lazuli, for which it was mistaken when first found.

Cassiterite has not been observed in fine crystals, what has been found being clear enough to cut only small transparent gems. The wood-tin of Durango, Mex., is used to a very limited extent on the Pacific coast, the stone being simply polished flat, and strikingly resembles a dark wood. The finer crystals of cas-

CASSITERITE

COLOR.	LOCALITY.	Stannic Oxide. (Binoxide of Tin.)	Silica.	Tantalum Oxide.	Ferric Oxide.	Lime.	Magnesia.	Water.	Specific Gravity.
Brownish white to reddish brown.	Rockbridge Co., Va.[1]	94·895	0·760	0·327	3·418	0·244	0·027	0·385	6·536

[1] Am. Chem. Jour., Vol. 6, p. 187.

siterite found at Hebron, Auburn, Norway, and Paris, Me., would afford mineralogical gems. The claims in the Temescal Range, in San Bernardino County, as well as the locality near San Diego, Cal., may yet yield specimens of this mineral equal to that from Durango, Mex. The important occurrence at the Broad Arrow Mines, two miles from Ashland, Clay County, Ala., may produce both the crystals and the stream-tin. No transparent crystals have been found at the Black Hills, Dak. William P. Blake mentions finding on Jordan Creek, Owyhee County, Idaho, a very fine specimen of wood tin, $\frac{1}{8}$ to $\frac{1}{4}$ inch across, and of a very pure and clear material. Cassiterite has also been found in large quantities at King's Mountain, N. C., and in Rockbridge County, Va., though none of these places has yielded a single fine gem, or has as yet been worked with commercial success for tin.

With the exception of small, richly-colored pieces that have been discovered near Gainesville, Ga., hematite is rarely compact enough for cutting, although one of the most abundant ores of iron, and found in many localities in the United States. Most of the gems that are sold in the United States come from abroad, where the mineral and labor of cutting are inexpensive. The foreign material used is the straight, compact, fibrous variety, and is usually cut in the form of small balls, which are supposed to resemble black pearls, but their lustre is higher and more metallic. It is also cut into cubes, into various charms and intaglio cane-heads.

Lodestone, or native magnet, is the iron oxide that possesses magnetic properties. Although not used as a gem at present, it was worn centuries ago for the power it was supposed to possess and for the charm it was presumed to give the wearer. Large quantities of it are found at Magnet Cove, Ark. It is estimated that several tons are sold annually to the southern negroes to be used by the voudoos, who employ it as a conjuring stone. In July, 1887, an interesting case was tried in Macon, Ga., where a negro woman sued a conjurer to recover $5 which she had paid him for a piece to serve as a charm to bring back her wandering husband. As the market price of the magnet was only seventy cents a pound, the judge ordered the money refunded.

Rutile is pure titanium oxide. Specimens from Graves Mountain, Lincoln County, Ga., and from Alexander County, N. C., rival any that have been found; the former for beauty of color, polish and sharpness of crystals, as well as for their great size, and the latter for their perfection, wonderful polish, and fine color. At Graves Mountain, rutile occurs with lazulite.

The rutile crystals are in nearly all cases imbedded in a compact red oxide of iron that can be readily removed by hydrochloric acid, or by means of some sharp instrument, leaving on the surfaces a mirror-like polish. The crystals vary in length from ¼ an inch up to 5 inches, and are in single crystals, twins, and vierlings, often in fine groups. The rutile from this locality has realized at least $20,000 for cabinet specimens, and has supplied the collections of the world through the perseverance of Prof. Charles U. Shepard. The finest small brilliant geniculated crystals are found at Millholland's Mills, White Plains, near Liberty Church, and near Popular Springs, in Alexander County, N. C. These have furnished some of the finest cut black rutile, more closely approaching the black diamond in appearance than any other gem. Some of the lighter colored ones furnish gems strongly resembling common garnet. Beautiful long crystals, at times transparent red, ranging from the thickness of a hair to ¼ and in some instances ⅜ inch across, and from 1 inch to 6 inches in length, often doubly terminated and very brilliant, have been found at Taylorsville, Stony Point and vicinity, North Carolina. Fine crystals are also found in quartz as well as loose in the soil in Sadsbury Township, Pa., for seven miles along the valley, especially near Parkesburgh, where double geniculations forming complete circles have been found, some weighing over a pound each. This is the "money stone" of the inhabitants of the district, who search for it because they can obtain money for it from the collectors; hence the name. Some of the finer stones, as well as the beautiful geniculated nigrine from Magnet Cove, would well serve as natural ornaments. As early as 1836, the rutile of Middletown, Conn., was cut by Prof. Charles U. Shepard into gems that were almost ruby-red in color. On St. Peter's Dome, in the Pike's Peak region, jet-black rutile occurs as black tetragonal crystals about ⅜ inch (10 millimeters) long.

Octahedrite is reported as occurring in small crystals at Dexter's lime rock, Smithfield, R. I., and in flat, tabular, glassy crystals of a pale-green color and very brilliant, in the gold sands of Brindletown Creek and elsewhere in Burke and the adjoining counties of North Carolina. These would probably afford small gems that would compare favorably with the beautiful blue crystals from Brazil, which are so brilliant as to have been mistaken for diamonds.

When the lead mines at Ellenville, Ulster County, N. Y., were worked in 1858, some remarkable flat ruby-red crystals of brookite were found on the quartz crystals; and at Magnet Cove, Ark., brilliant crystals of the variety of this mineral known as arkansite are found in great profusion, at times transparent and of honey-yellow color. The mineral does not readily admit of polish, and hence has little use as a gem.

Compact titanic iron admits of a high polish, especially the porphyritic menaccanite from Cumberland, R. I., in which the included quartz crystals form a very pretty contrast with the deep black color of the polished titanic iron. It has been cut for ornaments to some extent by Edwin Passmore, of Hope, R. I., and resembles a dark black porphyry. At Magnet Cove, Ark., ilmenite or titanic iron is found in fine bright crystals, which take a brilliant polish and form natural ornaments of considerable beauty.

Titanite or sphene is met with abundantly in black and brown crystals in St. Lawrence and Orange Counties, N. Y.

TITANITE (SPHENE)

	COLOR.	LOCALITY.	Silica.	Titanic Oxide.	Manganous Oxide.	Ferric Oxide.	Lime.	Magnesia.
Theoretical Composition	....	....	30·30	41·41	....	....	28·28	....
Analysis................	....	Lehigh Co., Penn.[1]	34·87	43·41	....	....	21·75	....
"	Yellowish white, Vitreous	Statesville, N. C.[2]	20·45	38·33	trace	1·61	39·11	trace

[1] Specific Gravity, 3·45. Analysts, E. B. Knerr and E. F. Smith. Knerr and Smith, Am. Chem. Jour., Vol. 6, 411.
[2] Ignition, 0·60; Specific Gravity, 3·477. Analyst, F. A. Genth. F. A. Genth, Proc. Am. Phil. Soc., 1886, 23, 30.

Some remarkably fine crystals of titanite have been found at Bridgewater, Bucks County, Pa. Certain of these, over an inch long, and very transparent in parts, are of rich greenish-yellow and vitreous golden shades, equaling in color the finest from Tyrol, and would afford gems weighing from 10 to 20 carats each, that would show a play of colors rather adamantine than opalescent. Fine crystals from this locality are now in the cabinet of Clarence S. Bement, the William S. Vaux Cabinet, Academy of Natural Sciences, Philadelphia, and the Peabody Museum, New Haven. Many yellow crystals over an inch long

have been found in the hornblendic gneiss on the Schuylkill
River, near Philadelphia, and in yellow crystals with sunstone,
in Kennett Township, Chester County, Pa. Some small yellow
crystals were found at Fort George, N.Y., by William Niven, one
of which was cut into a transparent gem weighing ⅛ of a carat.
Diana, Lewis County, N. Y., was a famous locality thirty years
ago, but crystals from there are now scarcely mentioned since
the large dark-brown ones have been discovered at various
places in Canada.

Malachite, although occurring in many localities in the
United States, and occasionally in considerable abundance, as
one of the ores, or associated with other ores or copper, is
obtained in gem form only in Arizona, chiefly at the Copper
Queen Mine, at Bisbee. One mass weighing 15 pounds is now
in the State Museum at Albany, N. Y., and others, nearly as
large, and equaling the Russian in quality, have been found,
which, by piecing, will furnish table-tops. One of the finest
specimens of the velvety form of crystals is a piece from the
side of a large cavity, over a foot across, in the American
Museum of Natural History, New York City. It is one side
of a geode filled with stalactites coated with the richest deep-
green, velvet-like crystals of malachite. Many of the stalactites
at Bisbee are over a foot long, an inch across, and are often
curiously entwined. Veins of this mineral from 1 to 4 inches
thick have also been found there. It is to be regretted that
thousands of tons of this beautiful mineral have been put into
the furnace for the copper it contains. One very fine, compact,
fibrous mass of dark-green malachite from the McCullock Mine,
that would cut into a cube an inch square, is in the cabinet of
Clarence S. Bement. Hoffmann mentions malachite in massive
concretions in Copper Cañon, Galena district, and at Mineral
Hill, Nev. At Ducktown, Tenn., some fine, radiated masses
have been found that would polish well. At the Jones Mine,
Berks County, Pa., very dark-green and finely mottled malachite was found that would cut into gems over 2 inches across.
Some of the finest of these specimens are in the cabinet of William W. Jefferis. The material from this locality equals that
from Arizona, but the supply is very limited. Malachite is

found in North Carolina, in Guilford, Cabarrus, and Mecklen-burgh Counties. The fibrous variety has been observed at Silver Hill and at Conrad Hill, in Davidson County, and in a number of other localities in North Carolina, but is rarely of any gem value. In the Torrey Collection at the United States Assay Office, in New York City, are a few fine gem pieces of malachite from the Copper Knob Mine in Ashe County, N. C.

MALACHITE

	COLOR.	LOCALITY.	Copper Oxide.	Ferrous Oxide.	Carbonic Acid.	Water.	Specific Gravity.
Theoretical Composition.	....	....	71·90	....	19·90	8·20	...
Analysis..............	Green....	Wheatly Mine, Penn[1]....	71·46	0·12	19·09	9·02	4·06

[1] Analyst, J. L. Smith. J. Lawrence Smith, Am. J. Sci. (2), 20, 249.

At Morenci, Ariz., there have been found masses of azurite and malachite resulting from the alteration of azurite. These masses are botryoidal in form, so that if the tops of the spheres are cut across, the two minerals are shown in distinct alternate layers (often two to four layers of each) and bandings, forming most beautiful ornamental stones, which are often from 1 to 6 inches across, and admit of a very high polish, that produces

AZURITE

	COLOR.	LOCALITY.	Copper Oxide.	Carbonic Acid.	Water.	Specific Gravity.	Analyst.
Theoretical Composition....	....	....	69·20	25·60	5·20	....	....
Analysis..........	Deep blue..	Wheatly Mine, Penn..	69·41	24·98	5·84	3·88	J. L. Smith.[1]

[1] J. Lawrence Smith, Am. J. Sci. (2), 20, 250.

a novel and pleasing effect. If it were found in sufficient quantity, it would make a valuable ornamental stone. Ruskin has likened this combination of colors to the " green of the fields and the blue of the sky," and notwithstanding the strong contrast, the blending makes it a harmonious one. The association is entirely new and one of the most beautiful ever found. There is little or no demand for this stone outside of Russia, where clocks, jewel-caskets, mantels, table-tops, and doors are covered with a thin veneer of carefully-pieced mala-chite, cemented on slate or marble, not made of solid blocks as is often supposed.

With the malachite at Copper Queen Mine is a variety which has proved on examination to consist of equal parts of carbonate of lime and carbonate of copper. This is slightly harder than malachite, and the name, calcomalachite, indicating its composition, has recently been suggested for it. Like malachite, it admits of a fine polish and is susceptible of similar uses.

A beautiful compact chrysocolla, mixed with quartz, is found at the Allouez Mine, Houghton, Lake Superior region. Some of the specimens would furnish fine, rich, bluish-green gems half an inch square. Specimens of chrysocolla from the Copper Queen Mine, Ariz., coated with quartz and chalcedony, furnish beautiful gems when the polish on the layer of quartz chalcedony is thin enough to allow the chrysocolla to show through. In one case, these markings resembled a human head.

Aragonite (carbonate of lime) or "satin spar," from near Dubuque, Iowa, especially that from Rice's Cave, and in the remarkably fine forms known as the "floss ferri" variety, from near Rapid City, Dak., would admit of the same uses as common satin spar. The satin spar gypsum or sulphate of lime, while made so extensively into ornaments and sold at Niagara Falls and many tourists' resorts, is, almost without exception, imported from Wales, though some few of the common white gypsum ornaments sold at Niagara are cut from the gypsum found in the vicinity. On Goat Island large masses of gypsum are found, and occasionally even under the Falls, where the material for all the ornaments sold there is supposed to be found. Beautiful selenite occurs there, but no satin spar.

The dark amber-colored and brown aragonite (California onyx) from California is extensively used as an ornamental stone, but not as a gem stone. Many thousands of dollars' worth are annually used by marble workers and for decorative purposes.

In the Luray and other American caves are found calcareous concretions called cave pearls, which consist either of pieces of stalagmite worn round by falling water or of similar pieces forming nuclei on which successive layers of carbonate of lime have been deposited.

Fossil corals, consisting of carbonate of lime, often possess great structural beauty and are very compact and susceptible of high polish. Along the shores of Little Traverse Bay, at Petoskey, Mich., are found water-worn pieces of fossil coral of various species, ranging from fragments the size of a small pebble to masses of 2 or 3 pounds weight. The spaces or cells of these corals are entirely filled with carbonate of lime, and being very compact, they take a fine polish. In color they are of various shades of gray, and many of them are exceedingly handsome. Visitors to Petoskey, which is a popular summer resort, gather the corals, and to show the structure keep them in bottles of water or give them a coat of varnish. The lapidaries of the place cut and polish these corals, and at present probably $4,000 or $5,000 worth are annually sold by them, either polished on one side, or in the form of seals, charms, cuff-buttons, paper-weights, and other ornaments. They are first ground on a Berea grindstone, then a polish is put on with four successive grades of emery, and they are finally polished on Spanish felt moistened with oxalic acid and lead ashes. The fossil corals found near Dubuque and Iowa City, Iowa, are magnificent in color and structure, and fine pieces often exceed a foot in width. They have been used to some extent in jewelry, shaped into stones for cuff, shirt, and vest buttons, the light cream-color making a very quiet, rich stone for this purpose.

Pyrite or sulphide of iron is found in many localities in the United States, and one variety occurs in crusts or groups of small, brilliant crystals with slate in the coal regions. These crusts are trimmed and cut into ovals, squares, and other shapes, and sold for mounting as scarf-pins, lace-pins, ear-rings, and ring-stones, as well as other ornaments. Fine single crystals are also sold for ornaments, principally at Mauch Chunk and the summit of the Switchback Railway, and by the local jewelers at Ashland, Shenandoah, and Mahanoy City, in Schuylkill County, who obtain their finest specimens from the Raven Run Mine, six miles from Mahanoy City, Pa. Magnificent groups and fine single crystals with a very high polish have been found at Black Hawk and other mines in Colorado and sold for ornaments just as they are found, principally at Denver, Colorado Springs, and other places in the West. Perhaps $1,000 worth a year is disposed of in this way. The Colorado crystals are compact enough to cut into the faceted gem known in Europe as " marcasite," which was extensively mounted in gold during the last century, but has been almost entirely superseded by the introduction of bright steel jewelry.

Cobaltite, a sulphide of cobalt, is occasionally cut abroad to be used as a gem and then resembles a flesh-colored pyrite. It is not found of fair quality anywhere in the United States.

Little amber of commercial value has been found in the United States. Though its occurrence in several places has been noted, the specimens are believed to be derived from a species of tree quite different from those which yield the Baltic amber. The earliest description of amber found in this country is given by Dr. Gerald Troost,[1] who describes two varieties, one opaque, the other translucent, which had been discovered at Cape Sable, Magothy River, Anne Arundel County, Md. Both these varieties showed a mixture of the various shades of yellow, gray, and brown, the colors being sometimes arranged in nearly concentric zones displaying the most beautiful tints, and sometimes in alternate bands, spots, dots, and clouds, as in agate or jasper. Some of this amber was also wax- or honey-yellow. The transparent variety, which in external appearance resembled colophony gum and had a high lustre, was sparingly found. The opaque variety is described as " very dull." Both varieties broke easily and exhibited a perfectly conchoidal fracture, their hardness being identical with that of the amber found near the Baltic Sea. Their specific gravity varied from 1·07 to 1·08, the difference being due, in the opinion of Dr. Troost, to small particles of pyrites with which the cavities were sometimes lined. Some of the specimens were only slightly electrical, while others exhibited this quality in a greater degree. Dr. Troost also described a variety of amber which occurred in fragments or friable, porous masses, of about the size of a walnut, mixed with iron pyrites, and having a dull, earthy aspect. These fragments, which were all found in an alluvial formation,

[1] Am. J. Sci. I., Vol. 3, p. 8, Jan. 1821.

could be crumbled easily when rubbed between the fingers, and in external appearance resembled clods of loam or of stiff soil. They were of a gray or yellowish-gray color, and when burned gave out the odor, and indeed seemed to possess the other properties, of melted amber. But few specimens of the amber described by Dr. Troost are found in the collections of this country; his conclusions, however, are accepted as correct. A small specimen of the Magothy River amber came into the writer's possession from a collection made about fifty years ago. This resembles the Baltic amber more closely than does that from New Jersey. The specimen is a fractured piece, transparent, rich reddish-brown and yellow in color, like some of the beautiful amber from Catania, Sicily. It is believed that further search in this vicinity would lead to other discoveries. Dr. Philip R. Uhler is authority for the statement that amber has been found in a lignite bed about twenty-five miles from Baltimore, but in very small quantities. In New Jersey it has been found in a great number of localities. As early as 1762, John Bartram, in a letter to Dr. Elliot, states that amber was found in New Jersey near the Delaware " in pieces nearly a pound in weight, and fitted to make a good cane-head." Prof. George H. Cook, State Geologist of New Jersey, says [1] that amber is found irregularly distributed in all parts of the marl region. Marl-pits in every county of the region have furnished specimens, but the finding of one specimen does not insure the finding of others in the same locality. Pieces enough to have filled a barrel are said to have been taken from one marl-pit at Shark River, about the year 1856, but since that time, in looking over many hundreds of tons of marl, not a fragment was found. The mineral is yellow in color, but is not so compact or so lustrous as good specimens of foreign amber.

Prof. Henry D. Rogers, State Geologist of Pennsylvania, mentions the occurrence of amber twice.[2] At Vincentown, Burlington County, N. J., it was found with asphaltum in the cretaceous marl above the green sand. The locality was reported by

[1] The Geology of New Jersey (Newark, 1868), p. 283.
[2] Description of the Geology of the State of New Jersey (Philadelphia, 1840).

Dr. E. Goldsmith to the Philadelphia Academy of Natural Sciences, and the specimens, described as having a specific gravity less than 1, fusing so as to be quite mobile, were regarded by him as related to the variety of succinite called "krantzite." Dr. Charles C. Abbott[1] mentions having several times found, in the bed of Cresswick's Creek, small grains or pebbles of amber which he gave to William S. Vaux of Philadelphia, and which are now in the Academy of Natural Sciences. One of these pieces measures 1 x 4 x 5 inches in thickness. He suggests that they are derived from beds of clay which are exposed in the bluff forming the southern bank of the creek. There are cretaceous clays near Trenton, in which occurs much fossil wood, in and upon which the occurrence of grains of amber is not unusual. These grains are usually very small and difficult to detect. The wood is soft and recent in appearance, burning with an uncertain, flickering flame, and the amber is evidently the fossilized sap of the wood found in these deposits of clay. This same locality is referred to in Comstock's "Mineralogy" (Boston, 1827). Dr. Nathaniel L. Britton has observed traces of amber near Camden, in the cretaceous deposits. In February, 1883, the writer described[2] a mass of amber 20 inches long, 6 inches wide, and 1 inch thick, weighing 64 ounces, that had been found on Old Man's Creek, near Harrisonville, Gloucester County, by Joseph B. Livezey. A quarter-inch section showed a grayish-yellow color, while a similar section, 1¼ inches thick, showed the color to be a light, transparent yellowish-brown. The entire mass was filled with botryoidal-shaped cavities filled with "glauconite" or green sand and traces of vivianite. Its hardness was very nearly the same as that of the Baltic amber, but it was perhaps slightly tougher, cutting more like horn, the cut surface showing a curious pearly lustre, differing in this respect from any other amber yet examined. The lustre is not produced by the impurities, for the clearest parts show it best, and the amber admits of a good polish. The specific gravity of a very pure piece of this amber was found to be 1·061. This figure may be attributable to internal cavities, amber usually ranging from

---

[1] Science, Vol. 1, p. 594.
[2] Am. J. Sci. III., Vol. 25, p. 234, March, 1883.

1·065 to 1·081. The specimen examined was found at a depth of 28 feet, in a six-foot stratum of the middle marl-bed, in and under 20 feet of cretaceous marl. In 1886, a piece of amber was found on the southwest branch of Mantua Creek, near Sewell, Gloucester County, N. J., in the lower marl bed. Prof. Washington C. Kerr mentions the finding of succinite (amber) in lumps of several ounces weight, in Pitt County and elsewhere in the Tertiary marl beds of the eastern counties of North Carolina. It is also found along the Chesapeake and Delaware Canal in Kent County, Del. Dr. Edward Hitchcock[1] refers to one or two masses of amber weighing a pound that had been found in Martha's Vineyard, Gay Head, and Nantucket, and states his belief that they were from the Tertiary formation. In February, 1883, the writer exhibited and described before the New York Academy of Sciences an elongated and twisted mass of opaque, rich yellow-colored amber, weighing 12 ounces, that had been found on the shore at Nantucket, Mass. This specimen, which was evidently from the Tertiary deposit, is now in the Amherst College Cabinet. Other specimens have been found in this locality. The discovery of specimens of amber in one of the Union Pacific coal mines of the Laramie Beds, in Wyoming, was reported by F. F. Chisholm in 1885; but at that time the tests were not completed, so that its genuineness could not be asserted. The material that was brought to Denver was hard, highly electric, and of a good clear yellow color; the fusing point was a little low, and the odor of an ignited fragment slightly resembled that of burning india rubber. In places, the substance was found two inches thick.[4]

Amber, according to Charles G. Yale, is common in the lignite deposits on the peninsula of Alaska. It is also obtained in the alluvium of the delta of the Yukon River, and in the vicinity of most of the Tertiary coal deposits on the Fox Islands, being everywhere an article of ornament among the natives, who carve it into rude beads. The discovery of amber in large quantities in America would be of the greatest interest, for here, as in Europe, it would contain fossil remains that would

[1] Am. J. Sci. I., Vol. 22, p. 50, July, 1832.
[4] Mineral Resources of the United States for 1885, p. 442.

greatly increase our knowledge of the fauna and flora of past ages. Dr. Herman A. Hagen, of Cambridge, Mass., a native of Königsberg, in East Prussia, whence the principal supplies of amber are obtained, writes : " When I first saw the shores of the Lakes Huron and Michigan and the Island of Mackinaw, I was so struck by their resemblance to the shores of my native country, the very locality where amber is found, that I could not help thinking that here also amber would be found." [1] It has been shown by Goppert that amber has been derived from eight species of plants besides the Pinites succinifer. He enumerates 163 species as occurring in amber. No one species has been observed in American amber.

AMBER

CHEMICAL COMPOSITION AND PROPERTIES.	Theoretical Composition.[1]	LOCALITY, South Amboy, N. J.[2]	Analysis by Schrötter.[3]
Carbon......................	78·94	70·68	78·82
Hydrogen....................	10·53	11·62	10·23
Oxygen.....................	10·53	7·77	10·95
Hardness....................	.....	2·0–2·5	.....
Specific gravity.............	.....	1·080–1·085	.....

[1] Text Book of Mineralogy (1871), by E. S. Dana, p. 393.
[2] Mineralogy of New York (1842), by L. C. Beck, p. 185.
[3] Phillips' Mineralogy (1852), p. 630.

Jet occurs in the Wet Mountain Valley, Trinchera Mesa, southeast Colorado, and in the coal seams of most coal-bearing rocks of Colorado. Some specimens a foot long and from 4 to 5 inches wide and an inch thick have been found. It is sold only as mineral specimens, although it admits of as fine a polish as the finest jet from Whitby, Eng., where a large industry in the working of this material is carried on. The beautiful jet, rivalling any jet known, found in El Paso County, Col., is sold extensively as mineralogical specimens, but is little if at all used for ornamental purposes. This is chiefly owing to the fact that it has been almost entirely superseded by black onyx in the United States, owing to the hardness of the onyx and the cheapness with which it is furnished from Oberstein and the Idar. This onyx is colored black by allowing the chalcedony, which is porous, to absorb some carbonaceous sub-

[1] On Amber in North America, Proc. Boston Soc. of Nat. His., Vol. 16, p. 296, Feb., 1874.

stance, such as sugar, molasses, blood, etc., and then putting it into the sulphuric acid, which chars the organic substance into dead black.

Anthracite, one of the hard varieties of coal, is found in many places in eastern Pennsylvania, but the variety used for ornaments is procured from Mountain Top, near Glen Summit; at the Franklin Mine, in Ashley; the Spring Tunnel Mine, the Summit Mine, and Nanticoke in Luzerne County. It is used as jewelry, and for ornamental purposes is carved into various trinkets, such as compass-cases, boots, hearts, anchors, and other small charms. It is also turned into cups, saucers, vases, candle-sticks, and paper-weights. The best work is done by a one-armed man at Glen Summit. Anthracite, like jet, could be made into beads and round ornaments to be used for scarf-pins, lace-pins, bracelets, and similar articles. The objects made often retain a ridge or ridges of the rough coal, while the other portions, being highly polished, form a striking contrast. These articles are sold at Scranton, Wilkesbarre, Pittston, Mauch Chunk, and the Summit Hill Station on the Switchback Railway, from $2,500 to $3,000 being expended for them annually.

The following minerals found in the United States, when fibrous or cut en cabochon across the cleavages, will show the cat's-eye ray:

A dark-brown, almost black, crystal of corundum from Ellijay Creek, Macon County, N. C., when cut en cabochon, furnishes gems two-thirds of an inch across, and showing the cat's-eye ray distinctly. The chrysoberyls of Stow, Peru, and Canton, Me., would cut into inferior cat's-eyes. The milky beryls found at Stoneham, Me., and Branchville, Conn., and some of the North Carolina beryls, especially the fibrous, green, opaque beryl from Alexander County, would furnish cat's-eyes, although not very fine. The so-called "Thetis' hairstone," described by Dr. Charles T. Jackson, found at Cumberland, R. I., is really quartz cat's-eye with acicular crystals of actinolite, and cat's-eyes of good quality have recently been cut from it by Edwin Passmore, of Hope, R. I., one of them nearly two-thirds of an inch long, and equal to many from Hoff, Bavaria. Prof. Frederick A. Genth states that quartz cat's-eye has been observed in several localities in Pennsylvania. A hexagonal crystal with the pyramid of greenish color, resulting from very fine fibers of actinolite disseminated through it, came from York County, Pa. It is found also five miles east of Bethlehem in the allanite locality, but not of gem quality. A curious, dark-gray piece of quartz, obtained from the West Shore Railway tunnel at Weehawken, N. J., was filled with what seemed to be byssolite, but really may be an altered pectolite; it would cut a cat's-eye of fair quality. A fibrous black hornblende from near Chester, Mass., and a white, compact, fibrous pyroxene from Tyringham, Mass., afforded imperfect cat's-eyes. Some of the labrador spar, when filled with included minerals and impurities, will show the cat's-eye ray; this is especially the case with the mineral found in Orange County, N. Y., and in the northern part of the State. Hypersthene, bronzite, and enstatite, when fibrous and cut across the fiber, produce the effect, and are sold abroad as cat's-eyes to a limited extent. Limonite from Salisbury, Conn., Richmond, Mass., and other American localities, can at times be cut into gems showing the cat's-eye ray. Aragonite and gypsum (satin spars) both give the cat's-eye effect.

Catlinite or "pipestone" was stated by Dr. Charles T. Jackson to be a variety of steatite, but it is now regarded by James D. Dana as a rock and not a definite mineral species. It is found in large beds in the upper Missouri region, in Pipestone County, Minn., and at several points in Dakota, Minnesota, and Wisconsin, notably at Flandreau and Sioux Falls, Dakota; Blue Earth River and Sac County, Iowa; Pipe Stone, Cottonwood, Watonwan, and Nicollet Counties, Minn., and in Barron County, Wis. In color it ranges from a deep red to an ashy tint. Reference is made to pipestone by Jacques Marquette, the Jesuit missionary, whose name is linked with the exploration of the upper Mississippi. He smoked the pipe of peace with the Illinois Indians as early as 1673, and gives the following exact description of that important utensil, the bowl of which consisted of red pipestone: "It is made of polished red stone, like marble, so pierced that one end serves to hold the tobacco, while the other is fastened on the stem, which is a

stick two feet long, as thick as a common cane and pierced in
the middle.   It is ornamented with the head and neck of different birds of beautiful plumage ; they also add large feathers of
red, green, and other colors, with which it is all covered."[1]
Carver tells us that near the Marble River " is a mountain from
whence the Indians get a sort of red stone, out of which they
hew the bowls of their pipes,"[2] and adds that individuals belonging even to hostile tribes met in peace at the "Red Mountain,"
where they obtained the stone for their pipes.[3] Loskiel[4] and
Dupraty[5] both refer to it in their works.   George Catlin was
the first white man that the Indians permitted to visit the locality.   He not only described the spot very fully, but also painted
a picture of it in 1836.[6] He says : " The place where the Indians get the stone for their red pipes, the mineral, red steatite,
a variety differing from any other known locality, is a wall of
solid, compact quartz, gray and rose color, highly polished as if
vitrified.   The wall is two miles in length and thirty feet high,
with a beautiful cascade leaping from its top into a basin.   On
the prairie, at the base of the wall, the pipeclay (steatite) is dug
up at two and three feet depth.   There are seen five immense
granite boulders, under which there are two squaws, according
to their tradition, who eternally dwell there—the guardian
spirits of the place—and must be consulted before the pipestone can be dug up.   The position of the pipestone quarry is
in a direction nearly west from the Falls of St. Anthony, at a
distance of 300 miles, on the summit of the dividing ridge between the Saint Peter's and the Missouri Rivers, being about
equidistant from either.   This dividing ridge is denominated by
the French the 'Coteau des Prairies,' and the pipestone quarry
is situated near its southern extremity and consequently not

[1] Discovery and Exploration of the Mississippi Valley, by J. G. Shea (New York, 1852),
p. 35.
[2] Travels through North America (Dublin, 1779), p. 95.
[3] Cf. Ancient Aboriginal Trade in North America, by Charles Rau. Report of the Smithsonian Institution for 1872, p. 23 of reprint.
[4] Missouri der Evangelischen Bruder unter den Indianern in Nordamerika (Barly, 1789),
p. 106.
[5] Histoire de la Louisiane (Paris, 1758), Vol. I, p. 326.
[6] Eight Years Amongst the North American Indians (New York, 1841), plate No. 270, Vol.
2, p. 164. See also Report of the Smithsonian Institution for 1885, part 2, p. 240.

exactly on its highest elevation, as its general course is north and south and its southern extremity terminates in a gradual slope. Our approach to it was from the east, and the ascent, for the distance of fifty miles, over a continued succession of slopes and terraces, almost imperceptibly rising one above the other that seemed to lift us to a great height. The singular character of this majestic mound continues on the west side in its descent towards the Missouri. There is not a tree or bush to be seen from the highest summit of the ridge, though the eye may range east and west almost to a boundless extent, over a surface covered with a short grass that is green at one's feet and about him, but changing to blue in distance, like nothing but the blue and vastness of the ocean." Of his struggles with Indians to visit the place, he relates: "We were persisting in the most peremptory terms in the determination to visit their great medicine (mystery) place, where, it seems, they had often resolved no white man should ever be allowed to go. They took us to be 'officers sent by Government to see what this place was worth.' As 'this red stone was a part of their flesh,' it would be sacrilegious for white men to touch or take it away—'a hole would be made in their flesh and the blood could never be made to stop running.' My companion, Robert S. Wood, and myself were in a fix, one that demanded the use of every energy we had about us. Astounded at so unexpected a rebuff, and more than ever excited to go ahead and see what was to be seen at this strange place, in this emergency we mutually agreed to go forward, even if it should be at the hazard of our lives." He says, concerning the quarry itself: "The thousands of inscriptions and paintings on the rocks at this place, as well as the ancient diggings for the pipestone, will afford amusement for the world who will visit it, without furnishing the least data, I should think, of the time at which these excavations commenced, or of the period at which the Sioux assumed the exclusive right to it." Mr. Catlin tells of the many superstitions about smoking among the Indians, and says: "The red stone of which these pipebowls are made is, in my estimation, a great curiosity, inasmuch as I am sure it is a variety of steatite (if it be steatite) differing from that of any known European locality, and also from any place known in America other than the one from which all these pipes come, and which are all traceable, I have found, to one source, and that source as yet unvisited, except by the red man, who described it everywhere as a place of vast importance to the Indians, as given to them by the Great Spirit for their pipes and strictly forbidden to be used for anything else." Specimens of the mineral were sent to Dr. Charles T. Jackson, of Boston, who was then "one of our best mineralogists and chemists." He gave it the name "catlinite," and pronounced it a new mineral compound, not steatite, harder than gypsum and softer than carbonate of lime.

This locality was visited and referred to by Dr. Charles A. White[1] and subsequently described by Dr. Ferdinand V. Hayden. He says: "On reaching the source of the Pipestone Creek, in the valley of which the pipestone bed is located, I was surprised to see how inconspicuous a place it is. Indeed, had I not known of the existence of a rock in this locality so celebrated in this region, I should have passed it by almost unnoticed. The pipestone layer, as seen at this point, is about 11 inches in thickness, only about 2¼ inches of which are used for manufacturing pipes and other ornaments. The remainder is too impure, slaty, fragile, etc. A ditch from 4 to 6 feet wide and about 400 yards in length, extending partly across the valley of the Pipestone Creek, reveals what has thus far been done in excavating the rock."[2]

Longfellow's lines commemorate the Indian legend:

> "From the redstone of the quarry
> With his hands he broke a fragment,
> Moulded it into a pipe-head,
> Shaped and fashioned it with figures;
> From the margin of the river
> Took a long reed for a pipestem,
> With its dark-green leaves upon it."

Whether catlinite has been used to make pipes for any very great length of time is difficult to decide. According to Dr. Hayden, "the quarry belongs to a comparatively recent

---

[1] American Naturalist, Vol. 2, p. 644, Feb., 1869.
[2] Am. J. Sci. II., Vol. 43, p. 19, Jan., 1867.

period."[1]   On the other hand, Edwin A. Barber, who has reviewed the subject very thoroughly, believes that the stone of Côteau des Prairies and the adjacent territory must have been employed by native sculptors for several centuries at least, and in all probability for a much longer period.[2]   Catlin, who studied the subject with much care, has published numerous drawings of the red pipes.   These are shown in Thomas Donaldson's very elaborate memoir,[3] and bear testimony to the skill and patience of their makers, who in most cases possessed no other implements than the knives and files obtained from the traders.   The cylindrical or conical cavities in the bowl and

CATLINITE

CHEMICAL COMPOSITION AND PROPERTIES.	LOCALITY, Palisades, Minnehaha Co., Dak. Analyst, Dodge.[1]	LOCALITY, Columbia River, Oregon. Analyst, Thomson.[2]	LOCALITY, Columbia River, Oregon. Analyst, Thomson.[3]	LOCALITY, Coteau du Prairie, Upper Missouri Region. Analyst, Jackson.[4]
Color....................	White to Yellowish.	Gray Blue.	Gray Blue.	. . . . .
Silica.....................	50·40	55·62	55·60	48·20
Alumina...................	33·30	17·21	17·42	28·20
Ferric Oxide..............	2·80	7·61	6·31	5·00
Manganous Oxide.........	. . . . .	. . . . .	. . . . .	0·60
Carbonate of Lime.........	. . . . .	. . . . .	. . . . .	2·60
Lime......................	0·60	2·26	2·08	. . . . .
Magnesia..................	0·17	0·11	0·29	6·00
Potassa...................	0·60	. . . . .	. . . . .	. . . . .
Soda......................	3·50	12·16	12·80	. . . . .
Water.....................	9·60	4·60	4·57	8·40
Hardness..................	. . . . .	1·50	1·51	. . . . .
Specific Gravity...........	. . . . .	2·61	2·61	. . . . .

[1] James A. Dodge, Tenth Annual Report of the Geology and Natural History of Minnesota, 1881, p. 203.
[2,3] Th. Thomson's Mineralogy, p. 288.
[4] Chas. T. Jackson, Am. J. Sci. I., 35, 388.

neck of these pipes are drilled with a hard stick and sharp sand and water.   It has been suggested that the manufacture of stone pipes, necessarily a painful and tedious labor, may have formed a branch of aboriginal industry, and that in ancient times the skilful pipe-carver may have occupied among the Indians a rank equal to that of the experienced sculptor in our days. Even among modern Indians, pipemakers have sometimes been met with.   Thus Dr. Kohl speaks of an Ojibway pipemaker whom he met near Lake Superior.   "There are persons among

[1] Am. J. Sci. II., Vol. 43, p. 19, Jan., 1867.

[2] Catlinite, Its Antiquity as a Material for Tobacco Pipes.   Am. Nat., Vol. 17, p. 745, July, 1883.

[3] Report of the Smithsonian Institution for 1885, part 2.

them," he says, "who possess particular skill in the carving of pipes, and make it their profession, or at least the means of gaining part of their livelihood. He inlaid his pipes very tastefully with figures of stars and flowers of black and white stones. His work proceeded very slowly, and he sold his pipes at from $3 to $5 each. The Indians sometimes pay much higher prices."[1] Dr. Daniel Wilson mentions[2] an old Ojibway Indian, "whose name is Pababmesad, or the Flier, but who, from his skill in pipemaking, is more commonly known as Pwahguneka —'he makes pipes.'" The stone is still worked into a large variety of ornamental pipes, that are sold at prices ranging from $1 to $10 each, and at times as high even as $20 for very large pieces of carving. Catlinite is also worked into a number of ornaments and into small charms of different kinds, which are offered to visitors at Minnehaha Falls, Lake Minnetonka, various hotels in St. Paul and Minneapolis, and in Dakota as far west as Fort Sully, and find a ready sale. The amount sold annually is perhaps $10,000 worth. This stone, on account of its compactness, easy working, and the fine polish that it admits of, should find a more extended use. One curious spotted variety, red with white and gray spots, is very beautiful, and would make a good contrast with the common red pipestone in decorative work.

[1] Kitschi-Gami Oder Erzählung Von Obern See (Bremen, 1856), Vol. 2, p. 82.
[2] Prehistoric Man (London, 1862), Vol. 2, p. 15.

CHAPTER XII.

Pearls.

PEARLS are lustrous concretions, consisting essentially of carbonate of lime interstratified with animal membrane, found in the shells of certain mollusks. They are believed to be the result of an abnormal secretory process caused by an irritation of the mantle of the mollusk consequent on the intrusion into the shell of some foreign body, as a grain of sand, an egg of the mollusk itself, or perhaps some cercarian parasite. It has also been suggested that an excess of carbonate of lime in the water may cause the development of the pearl. Accepting the former theory as the more probable one, it is easy to understand how this foreign body, which the mollusk is unable to expel, becomes encysted or covered as by a capsule, and gradually thickens, assuming various forms—round, elongated, mallet-shaped—and is sometimes as regular as though it had been turned in a lathe. Charles L. Tiffany, who has given considerable attention to this subject, suggests that the mollusk continually revolves the enclosed particle in its efforts to rid itself of the irritation, or possibly that its formation is due to a natural motion which is accelerated by the intruding body.

In regard to the formation of pearls, the following general statements may be made: Whatever may be the cause or the process of their production, these interior concretions may occur in almost any molluscan shells, though they are confined to certain groups, and their color and lustre depend upon those of the shell interior, adjacent to which they are formed. Thus the pink conch of the West Indies yields the beautiful rose-colored pearls, while those of the common oyster and clam are dead white or dark purple, according to their proximity to the part of the mantle which secretes the white or the dark material of the shell. The true pearly or nacreous iridescent interior belongs to only a few families of the mollusks, and in these alone can pearls proper be formed at all, while in point of fact they are actually obtained only from a very few genera.

According to William H. Dall,[1] none of the air-breathing mollusks (the land snails) produce a nacreous shell; and among fresh-water mollusks, none are pearl-bearers except certain of the bivalves, notably those belonging to the groups appropriately called the Naiades, of which the common river-mussel (Unio) is a typical example. The soft internal parts of these mollusks are covered by a thin, delicate membrane called the mantle, from the surface, and particularly from the outer edges of which, material is excreted to form the inner layers of the shell. The shell consists of two parts, the epidermis and the shell proper, the latter composed of numerous layers. The epidermis, which resembles horn, is chiefly composed of a substance called "conchioline" and is soluble in caustic alkalies.

The families with iridescent interior layers are the following: Among cephalopods, the Nautilus and Ammonites, the latter wholly fossil. In both these groups the removal of the outer layers of the shell reveals the splendid pearly surface beneath. Modern nautilus shells are often "cleaned" with dilute acid to fit them for use as ornaments, and frequently this is done partially, elaborate patterns being formed by leaving parts of the white middle layers to contrast with the pearly ground. Among the fossil Ammonites, the same effect is produced, very often naturally by decay of the outer layers; and no artificial pearl-work can compare with the richness of color—

[1] Pearls and Pearl Fisheries, American Naturalist, p. 17, pp. 579 and 731, June and July, 1883.

literally "rainbow-hued"—that is presented by many of these fossils from Jurassic and cretaceous deposits. Among the gasteropods, the pearly groups are the Turbos and Haliotes, in both of which, but especially in the latter, there is a frequent occurrence of green iridescence. Shells of both these families are "cleaned" with acid for use as ornaments, and the exquisite green Haliotis material is extensively used in the arts, as described further on.

The pearls of commerce, however, are almost wholly obtained from bivalve (lamellibranch) shells, of which the following families have a nacreous lining: Aviculidæ, Mytilidæ, and Unionidæ, the latter being wholly fresh-water shells, also known as the Naiadæ. A few genera of other families are also brilliantly pearly, but need not be discussed. The true pearloyster (Meleagrina) of the Pacific and Indian Oceans belongs to the first of these groups, and has from time immemorial yielded the bulk of commercial pearls, while its large and thick shell furnishes the mother-of-pearl for countless ornamental purposes. (The Naiades are of particular interest in this country, as it is in North America that this group is most abundant.) Several hundred species of Unio, Anodon, etc., have been found in our great rivers and lakes, and the Mississippi basin teems with them; and for the most part the forms are quite distinct from those of the Atlantic watershed and of the Old World. The Unios, while all iridescent, vary greatly in tint, exhibiting all the delicate shades of pink, brown, purple, etc., as well as white. The rivers of Europe, of Mesopotamia, and of China also yield large numbers of Unios.

The peculiar artificial devices for pearl production employed by the Chinese with Dipsas plicatus are described hereafter in this chapter, as well as similar experiments upon Unios in Germany. Other genera (Hyria and Castalia) represent the family in the Amazon basin of South America.

The same causes and operations that result in the production of pearls or free nacreous concretions in the soft animal substance of the pearl oysters or mother-of-pearl shells also produce in a modified way the tuberculose or knoblike protuberances and irregularities of surface that are frequently seen on the pearly inner surface of the valves and projecting therefrom. The flatter or less pronounced form of these nacreous excrescences are often called "blister pearls," because of their resemblance to vesicular eruptions, or water-blisters caused by burns. These protuberant or vesicular excrescences, as the case may be, are induced in two ways. First, and perhaps more commonly, by the perforation of the shell, from outside to inside, by some species of boring parasite, pholads and lithophagi among the bivalve mollusks (Acephala), also by certain sponges (Clione) and boring worms. For the most part, these are not really parasites, as they do not derive their nutriment from the substance of the pearl oyster, as leeches and ticks do from the blood of their victims; the term "domiciliaire"[1] gives a clearer idea of the relation of these forms to that upon which they fasten or to which they attach themselves. These mollusks, sponges, and worms simply make their residence or domicile, according to their habit, upon or in the shell of the pearl oysters.[2] The boring species are quite small during the early adolescent stage when they first attach themselves, but with increasing growth they have necessarily to increase the size of their burrows, until at last, to the great inconvenience and annoyance of the pearl oyster, the tunnelers have pierced through its shell, and the oyster, in order to maintain the privacy of its own domicile, is forced, as it were, to plaster over the holes with a coating of nacre. This process is repeated and continued as long as the tunneling goes on, until finally a

[1] Name given by Robert E. C. Stearns in paper cited.

[2] In addition to the particular species of fish, Fierasfer dubius, figured in the plate, the occurrence of which had previously been made known, Dr. Stearns has detected another, apparently belonging to the Oligocottus, a form quite different from Fierasfer. The latter is a long, slender, eel-like form, while the other is a shorter, chunky fish, with a squarish head and rather prominent though stumpy spines. The Oligocottæ are small, bull-headed fishes that "usually inhabit rock pools between tide-marks," and are peculiar to the North Pacific waters. The Fierasfers inhabit tropical or semi-tropical regions, and have been reported from Florida Keys to Cuba and Panama. The specimen illustrated was probably from the Gulf of California, as well as the Oligocottus, the occurrence of which as a parasite or domiciliaire had not before been made known. (See Colored Plate No. 8.) It is highly probable that still other species of the ichthyological section of the animal kingdom may yet be discovered occurring under similar conditions, for it would seem that small fishes of many species might occasionally be chased into the gaping valves of the oysters when pursued by some predaceous member of the finny tribe. The Fierasfers, however, exhibit the parasitic habit, as has been pretty well ascertained, not only through its occurrence in the pearl oysters, as before shown, but also through similar relations to the Echinoderms.

nacreous knob or lump of pearl, of greater or less size, results from this defensive and protective action on the part of the oyster. This walling out of intruders can hardly be regarded as an indication of instinct or intelligence in the oyster, analogous to the repairing of a damaged web by a spider or the retunneling of a filled-in gallery by ants : it is a pathological rather than an intelligent action, induced by irritation at the point of intrusion. Secondly, knobs, protuberances, and blister pearls are the result, indirectly, of some intrusive particle, or, it may be, of an organism which has in some way worked in between the delicate tissues of the mantle or sac, or some part thereof, and the interior surface of the shell. This, as may be easily conceived, produces an irritation, as a rough particle of dust on the surface of the human eye, and induces a secretion followed by a flow and deposit of nacreous lymph at the point irritated, and the cause of the irritation, whether an organic form or an inorganic particle, is coated with nacre, and plastered down to or upon the inner surface of the shell. It is rarely the case—but such instances have been known—that a small fish, having entered the shell when the valves were partially open, and having worked its way between the mantle and the smooth surface of the shell up to the region where the adductor muscles are attached (the muscles by which the valves are opened and closed), has here had a stop put to further explorations into the anatomy of the oyster, the latter not only clothing the unfortunate intruder in a pearly shroud, but also burying him in a nacreous tomb.

The disturbance of the muscular economy of the oyster at the point named, it may be assumed, would induce immediate and extreme protective activity in the nacreous deposition.

The report of the United States National Museum for 1886, page 339, contains a paper by Robert E. C. Stearns, "On Certain Parasites, Commensals, and Domiciliares in the Pearl-oysters, Meleagrinæ," and the colored plate (No. 8) (from a painting[1] made for that paper) which illustrates this chapter, as well as the notes and comments herein embodied, have kindly been placed at the author's disposal by Dr. Stearns.

[1] Department of Mollusca, United States National Museum, No. 73,934.

A hundred pearls have been found in a single shell; but as a rule these have little or no value. Very curious nacreous groups made of many small pieces are at times found attached to the hinge, but these are generally without sufficient lustre to be of value, and are rarely collected. These groups are caused by the conglomeration of many small ones cemented by a deposit of nacre, and are often half an inch across. The white and the pink pearls are exceedingly beautiful, and the finest, owing to their delicate sheen or layers, are at times more lustrous than even the best oriental pearls. This lustre is increased by their greater transparency, and a really fine white, pink, yellow, or iridescent pearl is often found quite translucent. In color, the Unio pearls present an extended series of shades from dead opaque white, having but little value, through various tints of pink, yellow, and salmon, passing through a more decided form of these colors, or a faint purple, into a bright red, so closely resembling a drop of molten copper as almost to deceive the eye. Some are very light green and brown, others rose color, and still others are pale steel-blue or russet and purplish brown. In addition to their color and lustre, they are beautifully iridescent. They are found in many odd and remarkable shapes. (See Illustration.) Elongated fish forms found near the hinge of the shell, and called hinge baroque pearls, are abundant. Others, with but a slight addition of gold and enamel, seem to represent human and animal heads, bat and bird wings, and similar objects. Mallet-shaped pearls are found with fine color and lustre at each end, though generally with opaque sides; also grouped or bunched masses of the pearly nacre, made up of from one to over one hundred distinct pearls in fanciful shapes, are of occasional occurrence. Feather-like forms with curiously raised points, and an odd, rounded variety with raised, pitted markings, are quite abundant. A pearl was mounted in this country that strikingly resembled the bust of Michael Angelo, and a number of unique designs have been made of baroques, similar to those mounted by Dinglinger and exhibited in the Green Vaults at Dresden. Although the pearls used here have not been as large as those shown in Dresden, greater taste has been employed in mounting them. The variety of the forms being so great, an artist has a wide field for imagination. The pearls, however, have but slight value unless they are beautiful and lustrous.

Frequently pearls have an opaque appearance and seem to be worthless, but on the removal of their outer layer are found to be clear and iridescent. This outer layer may be removed by dipping them in a weak solution of acid, which dissolves the opaque coating, or it may be peeled with a knife, although sometimes the pearl is not of the same material throughout, and cannot be restored. The story is told of a New York lady who purchased a button-shaped Unio pearl that had a black, diseased appearance on one side. It was so set that the imperfection was all below the mounting. When applauding at the opera one evening, the pearl was broken, and on examination it was found to consist of a very thin nacreous layer, inside of which was nothing but a hard, white, greasy clay. (See Illustration.) Whatever be the method of their formation, it would seem that pearls can be formed only at the expense of the shell, for every substance necessary to their growth is drawn from sources which normally secrete the shell. Hence the presence of the pearl can usually be detected on the outside of the shell. Normal appearing shells rarely contain pearls, while on the other hand those that are deformed often contain pearls of great beauty. There are three indications on which pearl-fishers rely for detecting from the outward aspect of the shell the presence of pearls. These are, first, the thread, that is, the recess or elevation extending from the vertex to the edge; second, the kidney-shape of the shell, that is, an indentation on the ventral side; and third, the contortion of both shells toward the middle plane of the animal.

Much interesting information concerning the structure and quality of the shells of fresh-water pearl-bearing mussels was obtained at the International Exhibition held at Berlin in 1880. The shells were found to consist of three strata: first, the outer yellow or brown conchioline (cuticula or epidermis); second, the prism stratum, consisting of layers formed of minute prisms arranged vertically to the layers and the shell surface; and third, the interior nacre layer, composed of finely folded leaves parallel to the shell surface. The last two strata consist chiefly of carbonate of lime. These formations were illustrated  by transverse cuttings and microscopic sections. (See Fig. 8.) When a wound had been received by the animal in any soft part, the tissues became moistened with a lime-like material and especially with the nacre-substance. This often happens in the muscles which serve to close the shell, and the irregular concretions thus formed are called "sand pearls." When the growth of the pearl is abnormally strong, the pressure which it exerts on the outer wall of this tissue-pocket becomes so powerful that the pocket is absorbed on the side toward the shell, bringing the hard pearl directly against it. It then becomes impossible for the pearl to grow any more at the point of contact, for there is no tissue to secrete the lime substance; but it grows on the rest of the surface, and the thickening layers, as they are formed, pass directly into the nacre layers on the inside of the shell, and thicken the shell itself. Through these over-layers, the pearl is connected with the shell as though by different layers of covering cloths. At first it clings to the shell at one point only, afterward enlarging the area of its adhesion. In this manner twin or united pearls are formed.

FIG. 8. SECTION OF ROUND AND PEAR-SHAPED PEARL SHOWING STRUCTURE.

The most important marine pearl-fishery on the American continent is that of Lower California, the central point being at La Paz. Here the true pearl oysters, Meleagrina or Margaritophora, are found, on the eastern shores of the Gulf of California, from Cape St. Lucas to the mouth of the Colorado River, taking in about 1,500 miles of coast, including the gulf islands. They are also found from La Barra de Ocoz, which is the boundary line between the republics of Guatemala and Mexico, to Mazatlan, a distance of 2,000 miles, making for the pearl fisheries a total extent of 3,500 miles.

These fisheries have recently been confirmed to the Pearl Shell Company of San Francisco, by special franchise from the Mexican Government. The beds were first discovered some three centuries ago by Hernando Cortez when he crossed to the Pacific and discovered Lower California, and the name of California, derived from "calidus," hot, and "fornius," a hearth, it is believed, is due to this journey, having been given by Cortez, who found the heat intense when he first touched California soil. He took possession of the fisheries, and sent a number of fine pearls to the Queen of Spain, subsequently requiring all fishers to send to the Blessed Virgin one-tenth of all they found, and one-tenth to the King of Spain. After some intermittent work, the fisheries, about 150 years ago, were again worked, with system and with great success, by one Juan Ossio, who took from them yearly from 300 to 500 pounds of pearls, actually packing them on mules and selling them by the bushel. The shells were all brought up by head divers, and pearls were taken from them so plentifully that they became of comparatively small value. This heavy drain had the effect of rapidly diminishing the supply, and it is only of late years that fishing has again been carried on systematically. At present numerous beds are known and worked, at Loreto, off Point Lorenzo, the Island of Cerrabro, the harbors of Picheluigo, La Paz, and in fact the whole west coast of the Gulf of California from La Paz to above the island of Loreto, and in the east the island of Tiburon, and the land above and below that island. All these places have been famous for their pearls.

A late authority writes that the beds of the pearl oyster are found on the coast of the Gulf from Cape San Lucas to the twenty-eighth degree north latitude, including the northern islands. The shells are also found on the southern coast at points which are known, but further exploitation has been abandoned on account of the lack of harbors for the protection of vessels used in these fisheries. The pearl oysters seem to prefer well-sheltered bays or harbors where fresh water empties, and in such localities the finest pearls have been found.

According to the report of an expert who visited the district in 1860, the season lasts from June to December, and the time for diving is three hours a day, one hour and a half before low water and one hour and a half after. On an average, one day in every week is a fast-day, on which, as well as on Sundays, no work is done. A good day's work for one diver is to procure ten dozen oysters, though some of the best men frequently get as many as fifteen dozen. Of course a great deal depends upon the locality. The shells average about 7,000 to the ton, and calculating the season at 150 days, each man procuring 15,000 oysters, the total of shells procured by 450 men is about 2,000 tons. Formerly, on the independent system, the divers generally preferred to sell the oysters unopened for about twice the price that they would receive for the shells only, the price of shells averaging $4.50 a thousand. They went out in canoes, three, four, and sometimes five or six to each canoe, but seldom in greater number than four.

The rise and fall of the tides is about twenty feet. The currents run very swiftly among the islands except just before and just after low water, and just before and just after high water ; but before and after high tide the water is too deep for divers, except in the shallowest places, which, of course, are generally exhausted, as they are always accessible. The divers claim that they can easily reach a depth of twelve fathoms when not hindered by the currents, and can remain there from a minute to a minute and a half. On reaching the place where they intend to dive, the canoe is allowed to float, or is paddled slowly by one of the men, while the others, with their heads close to the water, are watching the bottom. Notwithstanding that the bottom is more or less rocky, they can distinguish an oyster at a depth of fifty feet. When one is observed, the diver goes down, and if there are several in the place (it is said that there are always two) he brings up all he can secure during the minute or two he is down. If a spot is found where the oysters are abundant, a basket is sunk by means of stones, having a rope attached, and the diver can sometimes fill it in a few minutes, coming up occasionally to take breath. Those in the canoe take turns in diving, in paddling, and in resting, so that of the three or four in a canoe, not more than one dives at a time. The divers take no food whatever on the day they intend to dive, unless the hours for diving are to be very late in the day, when they take a little broth in the morning. They go down with stomachs as nearly empty as possible, so that the action of the lungs may not be interfered with.

In 1860, in order to conduct pearl gathering in a more scientific manner, the owner of the Mexican grants, Señor Navarro, procured from San Francisco, Cal., a number of expensive schooners, surf-boats, professional divers, and costly apparatus. After several years' experience he found that his experts, with their expensive outfit, were no more successful than the naked Indian divers, while the exorbitant wages demanded by them so diminished his profits that he wisely went back to the primitive methods followed by his ancestors. At present those ship-owners who undertake the fisheries on a large scale use apparatus imported from France and England, by means of which each man is able to bring up daily 300 pearl oysters. The men employed are powerful Mexicans, and every diver has five assistants. Four men work the air pumps for the suited diver, and the fifth attends to the life-line, letting down the diver and hauling him up, as well as hoisting up the nets or baskets full of shells and lowering the empty ones. The pump-men are fed and housed, and receive $15 a month; the life-line man is similarly looked after, and receives $25 a month; the diver receives $45 a month, and one-tenth of all he brings up, netting him as high as $500 a month, if he is fortunate. Connected with each fishing party is a schooner of from 60 to 200 tons burden, and two or three small boats. The men live on the schooner during the entire six months. In addition there are numerous divers who work independently, and who show wonderful skill and aptness in their work. Generally, with no other appliance than a heavy stone attached to the waist, they plunge naked to the bottom, select suitable bivalves, and gather them into a bag, remaining under water as long as two minutes. The shells containing the pearls vary in diameter from 2 to 8 inches, 6 inches being the average size. They are found on hard rocks or on sandstone at the bottom of the sea, usually in bunches, holding to the rocks by a fibrous beard (byssus), the circular opening being on top and the shells usually a little open. The oysters are vertical, not lying on the flat. Each diver has a knife, with which he cuts a bunch loose and places them in a basket or net by his side; this is hoisted up when full, an empty one descending at the same time. On rising to the surface, the fisher empties his bag into one of the waiting surf-boats, which crafts, under careful guard, deliver their loads to a well-armed schooner, the latter vessel running into shore at nights to discharge the accumulated cargo. Occasionally, during all the time he is under water, a man may not send up a single shell containing a pearl; at other times there may be $10,000 worth in twenty shells. A very strict police system is necessary to prevent serious thefts, which, despite the utmost vigilance, are of daily occurrence. On land the cargo is turned over to keepers, and the mass is surrounded by guards armed to the teeth. The shells are pried open with a flat knife, and the mussel is separated from each shell. A gristly substance attaches the body of the oyster to the shell, and covers about one-fourth of its area, the remainder being occupied by the pearl-bearing membrane, a black, jelly-like coat, and of course a part of the living shell-fish. (See Illustration.) The shells are handed over to another man, while the opener takes the separated fish, and examines the inside of the black membrane for the pearls he is in search of, and finally closes his fist over the fish to squeeze out any pearl which may be lodged in the interior, after which the pearls found are examined by experts, their value estimated, and a settlement made at once with the divers. Usually their wages amount to twenty-five per cent. of the total find, and they are paid by an allotment of the pearls taken during the day. On the outside the shells are covered with seaweed or other submarine growths, and look not unlike a Tam O'Shanter cap. All this growth is removed, and the shells are cleansed and picked, finding a ready market in Liverpool, London, and Hamburg at prices of from ten to twenty cents a pound. The profit from these fisheries is not as large as might be imagined, because the expenses are very heavy, and there is always involved a very considerable element of chance.

About 1863 a company organized in New York City for the purpose of gathering pearls and pearl shells on the Pacific Coast, and secured the use of a submarine boat, the peculiarities of which were that it carried a large supply of fresh air condensed within its walls and was provided with a means of purifying the air in the working chamber, thus dispensing with the necessity of communicating with the surface as it furnished an atmosphere in which men could work for a whole day with perfect ease.   The company procured a lease of property at the island of Tiburon, hoping, with their facilities, to secure unusual returns; for, with their submarine boat, they would have the advantage of exploring, locating, and working beds where divers could not go.   Presumably their efforts were not successful, for the company soon went out of existence.

During the subsequent summer a new company obtained the concession for the Lower California pearl fisheries, and they decided that all the fisheries on the Gulf of California should in the future be worked by Chinamen.

For more than 300 years these fisheries have been in the possession of private grants dating back to the days of the conquest.   The Mexican Government has in recent years annulled the old grants and leased the fisheries to the highest bidders. The house of Gonzales & Ruffo, having offices both in La Paz and the City of Mexico, secured a concession for sixteen years permitting them to work the fisheries around the Espiritu Santo and La Paz Islands, which are considered the best of the beds. The Government has recently granted to a single firm the exclusive right to raise the mother-of-pearl shells, and for the reproduction of such oysters the system used in the State of Maryland will be followed.   The fisheries, which constitute one of the leading industries of Lower California, are now diminishing yearly, and, owing to the continued exploitation, many of the ship-owners find themselves losers at the end of the season.

In the year 1831, according to T. J. Farnham,[1] more than $40,000 worth of pearls were taken from the coast of Sonora. The pearls from this fishery at one time brought from $150,000 to $200,000 a year.   As the search has been so actively carried on, the Government has deemed it necessary to prohibit fishing the second time for a period of two years.

Robert A. Wilson,[2] in speaking of pearls, says: " Their abundance is one of the first things to strike a stranger on entering

[1] Scenes on the Pacific, p. 307.
[2] Mexico, its Geography, its People, and its Institutions (New York, 1846), p. 307.

Mexico.   Every woman above the rank of a peasant must have at least one pearl to ornament the pin that fastens her shawl or mantilla upon the top of her head." It is common to see girls with strings of pearls around their necks that would bring a large price in London, and there are women in La Paz who have pearls of extraordinary value, but are so poor that they have not means to buy food.

The pearls of the Countess de Regia, of the Marquesa de Guadeloupe, and of Madame Velascó are from these fisheries and are remarkable for their size and value.   The great pearl presented to Gen. Guadeloupe Victoria, while President, was from the same locality.   The pride of the Spanish regalia is an enormous Mexican pearl which was secured near Loreto by a Mexican diver.   This most perfect pearl weighs 400 grains. In the Bay of Muleje a pearl was taken weighing 400 grains and as large as a small egg.   During 1883 several notable specimens were found, among them a light-brown pearl, flecked with darker shades, weighing 260 grains, and valued at $8,000. It was sent to Paris.   Another one was pear-shaped, white, with dark specks, weighed 176 grains, and sold for $7,500.   About the same time the pearl merchants of La Paz secured a pearl from some unknown Indian diver for which they paid $10, and received for it $5,500 in France.   It was oval-shaped, of a light sandy color, perfect in contour and brilliant in lustre, and weighed 32 carats.   In 1881 a black pearl, weighing 162 grains, was sold in Paris for $10,000.   During 1884 two other pearls, weighing respectively 140 and 124 grains, and of surprising lustre, brought $11,000.   Recently a pearl from these beds, weighing 48 grains, was sold in London for $7,500.   A black pearl weighing 108 grains, taken from the San Lorenzo Channel, was sold for $3,000.   A year later one of the principal shipowners found a pearl weighing 300 grains, and in the same year another weighing 180 grains was sold in Paris for $1,000. More recently a fine pearl was found in the Bay of Guaymas that weighed 372 grains.   At the World's Fair held in Paris during 1889 there was exhibited a set of seven black pearls from this district valued at $22,000.   The poorer pearls are sold in Germany, the finest in the United States and France.

The largest and finest black pearls (for it is the black pearls which are the specialty of these fisheries) that have been found weigh from 120 to 140 and even 240 grains each. A pearl of 12 grains, which is perfect in beauty, color, and shape, may be worth $200, but very slight defects will reduce the price to one-tenth of that sum. The best black pearls found come from these fisheries, though peacock-green, blue, green, gray, and white ones are also found. In shape they vary greatly, being spherical, pear-shaped, egg-shaped, conical, in the shape of a little round loaf, or a wax match. Frequently pearls are found attached to and forming a part of the inside of the shell, instead of being in the membrane, when they are of little value, because they are difficult to remove, and are usually imperfect.

Most of the pearls from this place are sent to market by way of San Francisco. A letter to the author from a leading firm there contains the following: "The pearl fisheries average about 5,000 carats a year, which represent a value of $200,000, to which you must add about 800,000 pounds of pearl shells representing a value of about $180,000. The cost amounts to about $100,000." During 1887 it is believed that more than $50,000 worth of pearls were found. The total product of the fisheries has amounted to as much as $250,000 in a single year, and the sale of the shells to as much more. From November, 1868, till September, 1869, $26,000 worth of pearls were purchased from this locality by one New York house. These were of various sizes, including four that weighed over 20 grains and one of 49 grains. In color, the pearls from this locality vary from pure white through gray and brown to black. The latter have become so fashionable in late years that their value has increased tenfold. One black pearl weighing 50 grains was valued at $8,000.

Fresh-water pearls are found, as before stated, in various species of the Unios, more frequently, according to Dr. Isaac Lea, in the Unio complanatus, but also in the following: U. Blandingianus, U. Buddianus, U. costatus, U. Elliotti, U. fragilis, U. globulus, U. gracilis, U. Mortoni, U. nodosus, U. orbiculatus, U. ovatus, U. torsus, U. undulatus, and U. Virginianus. Not one pearl in a hundred from Unios is of good shape, and probably not more than one in a hundred is really fine, therefore, as the worth of a pearl depends on lustre and form, the greater number obtained from this source are of slight value. Rev. Horace C. Hovey, however, is credited with having found a pearl half an inch across in the shell of a Unio ovatus, near Cincinnati, Ohio. Unio pearls have been sought since the settlement of this country, and the narratives of early voyagers abound with references to them. In an ancient catalogue[1] of objects of natural history, made in 1749 by John Winthrop, F. R. S., the following items are mentioned:

" 30. Unripe pearls which in time would have become [31].

31. Bright pearls which are produced in the same shells [30].

32. Some of the larger sea pearl shells which are often found in deeper waters three times as large and bear larger pearls.

N. B.—Almost all the lakes, ponds, and brooks contain a large fresh-water clam which also bears pearls. The Indians say they have no pearls in them at certain seasons, but at the season when they grow milky, the pearls are digested in them, which causes their milkiness."

Dr. Samuel P. Hildreth writes: "Some of the fresh-water shells produce very fine pearls. I have one taken in the waters of the Muskingum, from the shell known as the Unio nodosus of Barnes. It is a thick, tuberculated shell, with the most rich and pearly nacre of any in the western rivers. The specimen is perfect in form, being plano-convex on one side and a full hemisphere on the opposite. It is nearly $\frac{1}{4}$ inch in diameter across the plane face, and $\frac{3}{8}$ inch through the transverse diameter, and of a very rich pearly lustre. Set in a gold watch-key and surrounded by facets of jet, it makes a beautiful appearance and is by far the largest and finest pearl I have ever seen. Several others have been found, but none to be compared to this."[2]

The greatest find of these Unio pearls was in a mound in the Little Miami Valley explored by Prof. Frederick W. Put-

[1] Am. J. Sci. I. Vol. 47, p. 284, Jan., 1845.

[2] Am. J. Sci. I., Vol. 25, p. 257, April, 1834. Ten Days in Ohio, from the Diary of a Naturalist.

nam and Dr. Charles L. Metz, who procured over 60,000 pearls, nearly two bushels, drilled and undrilled, undoubtedly of Unio origin, all of them, however, decayed or much altered, and of no commercial value. (See Illustration.) In 1884 these scientists examined the Marriott Mound, where they found nearly 100 Unio shells, and among other objects of special interest six canine teeth of bears, that were perforated by a lateral hole near the edge at the point of greatest curvature of the root, and by passing a cord through this, the tooth could be fastened to any object or worn as an ornament. Two of these teeth had a hole bored through near the end of the root on the side opposite the lateral perforation, and the hole countersunk in order to receive a large spherical pearl, about ⅜ inch in diameter. When the teeth were found, the pearls were in place, although chalky from decay. Upward of 250 pearl beads were found, concerning which they say: "The pearl beads found in the several positions mentioned are natural pearls, probably obtained from the several species of Unios in the Ohio rivers. In size they vary from ¹⁄₁₆ inch to ¼ inch in diameter, and many are spherical. They are neatly drilled, and the larger from opposite sides. These pearls are now chalky and crumble on handling, but when fresh they would have formed brilliant necklaces and pendants."[1]

One of the most singular circumstances connected with the New Jersey "pearl fever" was the discovery of several shells which proved that the local savants had experimented on the pearl-bearing Unios by dropping mother-of-pearl buttons inside the shell, hoping that the mussel would cover them with its secretions. The specimens found had apparently been experimented on over thirty years previous, a time when European scientists were greatly interested in shells received from China, containing small images of Buddha. These images had been moulded in tin and placed between the mantle and the shell. The mussels were then returned to their natural environment, and after several months the layer of mother-of-pearl became of sufficient thickness, and the images were removed.

[1] Explorations in Ohio from the Eighteenth Report of the Peabody Museum (Cambridge, 1886), p. 449.

In a shell of the Unio in the Lea Collection of the Unionidæ, which has recently been presented to the United States
National Museum, an oval piece of white wax, flat on the
lower side and rounded on the upper, which had been inserted
in the valve near the hinge, is entirely coated with a beautiful
pink nacre. It has been broken out of the shell, the pearly
nacre of the lower or flat side remaining in the shell, whereas
the dome-shaped piece is covered with this material.

FIG. 9.
IMBEDDED INSECT IN UNIO FROM LONG ISLAND.

At the International Fisheries Exhibition held in Berlin
during 1880, there were shown the results of experiments undertaken in Germany toward the production of artificial pearls
from Unios, in a manner similar to that practiced by the
Chinese. Flat tin figures, usually of fish, were introduced between the mantle and the shell. Similar experiments were conducted in the Royal Saxon pearl fisheries. Either small foreign
bodies were introduced into the mantle, in order to furnish the
nucleus for the free pearl formation, or the Chinese method of
inserting such bodies between the mantle and the shell was
followed. From the second method successful results were
shown. The foreign bodies that had been introduced, poor
pearls from other mussels, pieces of grain, or china buttons,
were entirely covered with nacreous substance. The shape of
these objects makes it impossible for the mantle to fit closely
around them, and hence the nacre covers them so irregularly that
it is quite out of the question to make any use of them. From
specimens exhibited, it was shown that German oysters could
be made to cover a plain relief with nacre as well as those of China.

Efforts to make the river pearl-mussel available in another way met with better success. In 1850, Moritz Schmerler conceived the idea of making small fancy articles of the shells themselves, and succeeded so well that the Government allowed him to take from the royal beds the shells he needed for his manufacturing business. Large numbers of pearl-shell pocket-books and hand-satchels have been made since then. The almost faultless white and reddish tinted "rose-pearl mussels" are specially prized for this purpose, as the material may be cut so thin that a photograph pasted on the inside can be seen through the shell, conveying the appearance of being produced on the shell itself. Other manufacturers engaged in the business as soon as its success became apparent, and many hundred thousands of pearl-mussels are now annually used at Adorf, where the business is chiefly carried on. The principal sources of supply are brooks in Bavaria and Bohemia that are owned by private persons.

Some of the earliest American pearls, that were found, came from near Waynesville, Ohio, $3,000 worth being collected in that vicinity during the pearl excitement of 1878. At that time, Israel H. Harris, of Waynesville, began what has since become one of the finest and best known collections of Unio pearls in this country, purchasing, during many years, every specimen of value that he could find in that part of the State. Among his pearls was one button-shaped on the back and weighing 38 grains, also several almost transparent pink ones, and an interesting specimen showing where a pearl had grown almost entirely through the Unio. His collection contained more than 2,000 pearls, weighing over 2,000 grains, and is in all probability the last collection that will be made from that district. It was exhibited in the jewelry department at the World's Fair held in Paris during 1889.

Large and valuable Unio pearls have been obtained in New Jersey. In 1857 a pearl of fine lustre, weighing 93 grains, was found at Notch Brook, near Paterson. It became known as the "Queen Pearl," and was sold by Tiffany & Co. to the Empress Eugenie of France for $2,500; it is to-day worth four times that amount. (See Colored Plate No. 8.) The news of this sale created such an excitement that search for pearls was started throughout the country. The Unios at Notch Brook and elsewhere were gathered by the millions and destroyed, often with little or no result. A large round pearl, weighing 400 grains, which would doubtless have been the finest pearl of modern times, was ruined by boiling open the shell.

During the early part of the summer of 1889 a quantity of magnificently colored pearls was found in the creeks and rivers of Wisconsin, in Beloit, Rock County ; Brodhead and Albany, Green County ; Gratiot and Darlington, La Fayette County ; Boscobel and Potosi, Grant County; Prairie du Chien and Lynxville, Crawford County. Of these pearls, more than $10,000 worth were sent to New York within three months; including a single pearl worth more than $500, and among them were pearls equal to any ever found for beauty and coloring. The colors were principally purplish red, copper red, and dark pink. A fine, very round pink pearl of 30 grains was found in a Unio near St. Johns, N. B., and now belongs to George Reynolts of Toronto, Canada.

The lumbermen, while sailing down the Canadian rivers on their rafts, collect Unios for food, by fastening bushes to the rear of the raft, so that when they pass through the mussel shoals, where the rivers are shallow, the bushes touch, the Unios close on the leaves and thin branches, holding to them securely ; and at intervals the bushes are taken out and the Unios removed. Many brooks and rivers, among them the Olentangg, at Delaware, Ohio, and a number of streams near Columbus, have been completely raked and scraped, often in a reckless manner, and consequently with little result. The general method of collecting shells was for a number of boys and men to wade into the mill-race or into the river to their necks, feeling for the sharp ends of the Unio, which always project. When one was discovered in this manner, the finder would either dive after it or lift it with his feet. It was the custom at that time to open the shells in the water, and once during the process a pearl the size of a pigeon's egg is said to have been dropped into the water and never recovered.

At the United States National Museum in Washington,

D. C., there is a very interesting collection of pearls and the mollusks which bear them, including the Unio and the common conch, the common clam, mussel and the Trigona crassatelloides. The common mussel (Mytilus edulis Linn.) secretes small pearlaceous bodies, somewhat resembling those from the common oyster, but they have no value. Trigona crassatelloides of Conrad also secretes small pearls possessing neither lustre nor value.

Within one year pearls were sent to the New York market from nearly every State. In 1857 fully $15,000 worth, in 1858 about $2,000, in 1859 about $2,000, in 1860 about $1,500, in 1860–1863, only $1,500. The excitement abated until about 1868, when there was a slight revival of interest, and since then many Little Miami River pearls have been found. Since 1880, pearls have come from a comparatively new district, the supply from which is apparently on the increase. At first, few were found, or rather few were looked for, west of Ohio, but gradually the line has extended, and now Kentucky, Tennessee, and Texas are the principal pearl-producing States, and some pearls are sent North from Florida.

Of single pearls, one from Montpelier, Vt., valued at $300; one from Waynesville, Ohio, valued at $200; one from Boston, Tex., valued at $250; one pink pearl, 19½ grains, from Murfreesborough, Tenn., valued at $80, another at $150; one from Llano, Tex., valued at $95, have been sold in New York.

The production during recent years has been as follows:

September, 1881, to 1882	210 lots worth		$7,500
September, 1882, to 1883	72 "	"	5,000
September, 1883, to August, 1884	71 "	"	5,000

That so few American conchologists have paid attention to American pearls is probably accounted for by the fact that the pearls are contained in old, distorted, and diseased shells, which are not so desirable for collections as the finer specimens. Collectors who have opened many thousands of Unios have never observed a pearl of value. Pearls are usually found either by farmers, who devote their spare time to this industry, and, if no result is obtained, suffer no loss, or by persons in country villages who are without regular occupation, but are ever seeking means for rapid increase of fortune. Many shells that do not contain pearls are destroyed. In order to obviate this wholesale destruction, it would be well to use

FIG. 10.

METHOD OF OPENING MUSSELS IN SAXONY IN ORDER TO SEE IF THEY CONTAIN PEARLS.

instruments like those that have been employed in Saxony and Bavaria. In the former country a thin, flat iron tool with a

FIG. 11.

INSTRUMENT USED IN SAXONY TO OPEN MUSSELS WITHOUT KILLING THEM.

bent end is inserted in the shell. The handle is then turned to 90°, and the shell is opened without injury to the animal. (See Fig. 10.) Another implement is a pair of pliers with sharp-pointed jaws and a screw between the arms, which is turned by the hand until the valves of the shell are sufficiently distended to see whether it contains a pearl. (See Fig. 11.) If it does not, the animal is returned to its former haunts, perhaps to propagate more valuable progeny. This wholesale destruction, together with the depredations of hogs, which have exterminated whole shoals of Unios when the brooks were low, and the elements introduced into the water by manufacturing industries, have no doubt exhausted many varieties of these shells. The more eastern States are so densely populated and the streams so contaminated with sewerage and refuse from factories, that animal life is rapidly disappearing from the water-courses in many localities.

It is probable that the existence of carbonate of lime in excess where mussels abound influences the secretion that causes the growth of the pearl. In limestone regions, where the waters are polluted by products of decompositions that are acid, these unite with the lime and form other compounds, that are precipitated or are carried away with the impurities of the water. There can be no doubt that this cause would tend to decrease the amount of lime which the shell would receive, thus not only retarding the growth of the pearl, but often eventually leading to the extermination of the Unio itself. At nearly all the marine pearl fisheries, coral-banks abound, and it may be that these have more or less influence on the development of the pearl in the shell. In Vermont, New Jersey, and Ohio, where pearls were formerly found, a fine one is now rarely obtained. In gathering the shells, only those that are full-grown, old, and distorted by disease should be taken, so that the fisheries may be preserved, and the shells should be opened as soon as taken from the water, and not allowed to open by decay for this discolors the pearls ; and particularly, they must never be opened by boiling, as this dims the lustre and lessens the value of the pearl.

The common clam (Venus mercenaria) secretes pure white pearls, scarcely distinguishable from ivory buttons, as well as others faintly tinted with a purplish blue, passing at times to a reddish purple and a purplish black. · The white pearls are worthless, the tinted ones of very little value, but those of darker color are often from $\frac{1}{4}$ to $\frac{3}{8}$ of an inch in diameter, and the finest ones bring from $20 to $100. The supply is limited and there is very little demand, for unless the color is exceptionally good they possess little beauty, lacking the lustre peculiar to other pearls ; still, when mounted with diamonds, the appearance of the darker ones is much improved.

It would seem from an article on wampum, written by Dr. Samuel L. Mitchell in 1825, that clam pearls were of much more frequent occurrence in the early part of the century than they are now. "To form an opinion," says Dr. Mitchell, " of the frequency of their occurrence, I mention a circumstance that happened on Long Island. A man desirous of making a collection of clam pearls gave notice through the neighborhood that he would pay a quarter of a dollar each for those of proper size; and in the course of two months he received two dozen. The clam-mongers in the city save the pearls they find on opening the shells, and sell them to persons who come to the stalls in the market to purchase them." He himself possessed a purplish one weighing 69 grains, which surpassed all that he had ever heard of.

The manufacture of wampum to be sold or traded to the Indians is an old American industry, and the manufacture is still in the hands of the Campbell family, who originated it. The first to engage in this industry was John Campbell, who was succeeded by Abraham Campbell, and by the survivors of the four sons of Abraham, the youngest of whom is now over seventy-five years old. Mrs. Erminnie A. Smith described the manufacture,[1] and took a series of the beads, to represent the industry, to the New Orleans Exposition. She says :

" Originally the grandfather of the Campbells, who resided at Tea Neck, N. J., would make trips to Rockaway in a boat which, when they returned, was loaded with clams (Venus mercenaria), the meat of which was given to the country people in return for opening the shells. The blue 'heart' of the clam, as it was called, was cut out and made into beads used for the groundwork of the wampum belts. At one time this industry flourished so that thousands of dollars were paid out weekly to buy the beads made by the white country-people who manufactured them at the time. The hole of the bead was made with an arm-drill and they were polished or ground on grindstones. The white beads were not made from the clam, but from conch-shells (Strombus gigas), which they have always imported from the West Indies. The young clams cannot be used, and the old have so decreased in number that this branch of the industry has been greatly reduced."

When Mrs. Smith visited the Campbells she had with

[1] Science, Vol. 5, p. 3.

her an Iroquois wampum belt, bearing the marks of age, which they immediately pronounced to have been made after their manner. Although they had been familiar with Indians, they had never known of their making the beads. They had always depended upon the trappers for their market, and related incidents connected with their dealings with "fur companies," etc. The conch-shell is used also in the manufacture of "pipes," beads, rosettes, etc. The "pipes" vary in length from 2 to 6 inches, and resemble a tobacco-pipe stem with bulging sides; those of 6 inches in length are quite rare, and are highly prized. The rosettes consist of a concentric series of round, flat disks placed on them, secured one to the other by means of a string passed through the holes drilled in the center.

The common oysters (Ostrea borealis and Ostrea Virginica) occasionally secrete one or more pearly bodies, always dead-white in color. The reflections produced by their fibrous, radiated structure is similar to that observed in the common conch. The "skin" of these pearls is never smooth or lustrous, and consequently they have no value. Rev. Horace C. Hovey, in a letter to the author states that he had found twenty-nine pearls in a single common oyster

FIG. 12.
CURIOUSLY-SHAPED PEARL
FROM COMMON OYSTER.

(O. borealis) at New Haven, Conn. In the Smithsonian report for 1881 it is stated that Charles E. Ash took forty-five pearls from a single oyster in Providence Bay. A curiously formed oyster pearl is shown in Fig. 12.

CONCH PEARLS.—That is, the concretions found in the common conch (Strombus gigas), are not nacreous, and therefore cannot be considered true pearls. They are usually a little elongated or oblong in form, rarely round, and most of them are very beautiful, owing to the reflections produced by their fibrous stellated structure causing the light to play over the surface, but giving a different effect from the cat's-eye or that of satin-spar. They are almost always pink in color and the fine ones are wonderfully lustrous.

James R. Curry, of Key West, Fla., states that there and at Tortugas, fully 15,000 of these shells are used annually for food and for bait, being sold at the rate of three for ten cents, uncleaned.   He has paid some attention to pearl-collecting, and has never observed more than one in a shell.   He instances one as large as a small hazel-nut.   A few found by him were really finer in color than those from the West Indies, although not so regular in form.   The principal shades are canary, salmon-pink, pink, and pure white.   The value of none was over $50.   Conch pearls from the West Indies have occasionally been observed half an inch in length and of very fine quality, and are sometimes worth $1,000 apiece; yet the taste for pink pearls is on the increase, although for years the demand has been somewhat limited in the United States.   A necklace of these pearls valued at $4,000 has been collected, which is worthy of mention.

The pearls of the queen conch vary in color from a rich yellow to a yellowish-brown shade, and if anything are more highly polished than those of the Strombus gigas, or pink conch.   Cassis cornuta, C. tuberosa, C. Madagascarensis, C. rufa, also contain pearly concretions, varying from yellow to brown, somewhat similar to those from the common conch, but no large ones have as yet been observed.

The Abalone (Haliotis or Ear-Shell), the principal species of which are Haliotis splendens and Haliotis rufescens (called ormer in the Channel Islands, fuh-yu in China, awabi in Japan, and abalone in California), also secretes pearls.   (See Illustration.)   The nacreous portion of the shell itself is used for ornamental purposes, such as buttons, etc., and surface ornamentation in lacquer work, papier-maché, etc.   The mollusk itself, called "mutton-fish" by the New Zealanders, has long been known to the Indians of the Pacific coast as a valuable article of food, and it is much sought after by the Japanese and Chinese for the same purpose.   The former take only the very smallest fish, and eat them when freshly caught with cayenne pepper and vinegar, while the Chinese seek out the largest, and eat them only after they have been dried.[1]

The fishing is conducted at low tide, the principal grounds on the coast being along the Catalina and Santa Rosa Islands, in the Santa Barbara Channel, and from Monterey to San

[1] From an article on this subject by Charles R. Orcutt.

Diego, although a large number are gathered in Halfmoon Bay and from the rocks that line the shore of Mendocino County. The ear shells attach themselves to the rocks by means of their large muscular disk-shaped foot (so called) which acts like a sucker or exhaust cup. Just before the tide leaves them on the ebb and just after it has reached them on the flow, the abalones keep their shells slightly raised above the surface of the rock with the feelers drawn in. Then the fisherman, with either a long, broad knife or a spade-like instrument—both are used—gives a quick lift to the sucker or foot, letting in the air. The suction is destroyed and the fish falls off, when it is seized and thrown into a boat or basket, before it can fasten itself afresh. If the fish lies below water, a sort of grappling iron is let down, and after the point is inserted under the shell a vigorous wrench pulls it away. All this has to be done quickly and quietly, for if the abalone closes down on the rock, it cannot be drawn off, so great is its power of adhesion, and it will be broken into fragments before it releases its hold. When caught, the abalones are thrown on the beach, and the fish is pulled from the shell with a flat, sharp stick, and stripped of its curtain, boiled, salted, and strung on long rods to dry in the air. This process is very disagreeable, and that of stripping and cleaning so offensive that none but Chinese will undertake it. The abalones must be as hard as sole-leather when properly dried, and they are then packed in sacks, and sent to China. The price of the meat is from five to eight cents a pound in San Francisco, or from seven to ten cents a pound in China. When cooked, it is cut into strips and boiled, the taste being similar to that of the clam, but with a more meatlike consistency.

The trade in this dried meat is considerable. In 1866 there were exported from San Francisco by steamer 1,697 sacks, valued at $14,440, and in 1867 the exports had risen to 3,713 sacks, valued at $33,090. At present there are exported upwards of 200 tons a year, which at $175 a ton would amount to $35,000. At San Diego, Cal., the dried meat is quoted at $110 a ton. The shells vary from almost microscopic size to eight or ten inches in diameter. Before they were found to be of any marketable value, they were thrown away. One heap a little south of San Diego, containing over a hundred tons of shells, from exposure to the rain and the sun was converted into lime on the outside, but this was broken into and many fine shells were found.

The shell in its natural state is no more attractive than that of the oyster; it is rough on the outside, looking much like a piece of dried brown clay, and is frequently covered with a growth of barnacles, seaweeds, etc. Commercially there are five varieties, the green, the black, the red, the pink, and the mottled; but considering them from an ornamental standpoint, the shells may be grouped under three heads, red, black, and green, so-called, of course, from their prevalent color. The black, which is the smallest and least valuable, is found from Monterey down to the Gulf of California; the red, which is next in value, but the largest in size, is found from Mendocino to Monterey; while the green comes from below San Diego. The black seldom exceeds 6 inches in diameter, the green rarely goes beyond 9, while the red runs as high as 12 or 14 inches. The black is not beautiful on the outside, even when cleansed of lime and marine parasites, but inside there lies a small patch of the most beautiful opalescent tints, and this is sawn out, and made into brooches and lockets. The red is of a general mother-of-pearl appearance, with stripes and mottles of a rich burnt umber. The green, both within and without, is full of fire and color, some interiors being fully as vivid and of much the same prevailing color as a peacock's neck. This variety is principally used for jewelry, and is worked into every kind of ornament, from a table-top, inlaid with representations of flowers and butterflies, to the smaller varieties of jewelry. The Pueblo, Zuni, and Navajo Indians, and all the Indians of the California coast as far north as Alaska, have made it into charms and have used it for ornamentation for ages. It has been used as an applied decoration on silver objects, and was exhibited at the World's Fair held in Paris during 1889.

The play of colors is attributed by Sir David Brewster to minute striæ or grooves on the surface of the nacre alternating with the grooves of animal membrane. These laminæ decompose the light in consequence of the interference caused by the reflection from the two sides of each film, as may be seen in soap-bubbles. The nacreous laminæ when magnified are seen to be of minute cellular structure.

The first adaptation of the abalone shell to ornamental purposes was made by an English worker in mother-of-pearl who went to San Francisco more than twenty years ago. He saw the possibilities of the wonderful, brilliant shell, and began a business which now requires the services of more than ten men. The little trifles made of this shell are considered by the eastern visitor and the European tourist as distinctively Californian as a piece of big-tree bark. The incrustations were formerly removed by soaking the shells in a bath of muriatic acid, but it was found that this process injured the texture, and they are now cleaned and polished by friction lathes. Twenty years ago abalone shells were considered so worthless that freight steamers would not transport a bag of them without advance payment for the freight. Now they are worth $150 to $175 a ton in New York and Liverpool. The shells are shipped first to San Francisco, where they are assorted and the damaged ones thrown aside, about three tons of merchantable shells being procured from five tons of material as it comes from the abalone hunters. These shells are quoted (1889) in San Diego at $20, $25, and $35 a ton according to quality, and in consequence of such low prices the trade is comparatively dull. The output of the shells during 1888 was estimated at 300 tons. The amount of shells made into jewelry in San Francisco is very small compared with that consumed by the button-makers of France, England, Germany, and New York. Orders for abalone shells are constantly received from these places, and there are times when the export reaches as high a figure as 100 tons a week. The collector of customs at San Francisco furnishes the information that for the fiscal year 1887–1888 the export of abalone shell amounted to $185,414, which, together with $35,000, the value of the dried meat annually exported, makes this quite an important industry. These shells secrete very curious pearly masses, sometimes of fine lustre and choice enough to deserve a place among pearls. A pearl measuring 2 inches in length and from $\frac{1}{4}$ to $\frac{1}{3}$ inch in width has been found. A necklace made in California from the finest specimens was valued at over $2,000. A pearl over half an inch long and of good color cost $30 and was used as the body of a jeweled fly. The abalone pearls from the coasts of Korea and Japan are often very beautiful. In a lot of about one hundred shells only five were found bearing pearls, two with three pearls each, two with two pearls each, and one with a single pearl.

The history of American pearls dates back to the discovery of the New World. Arthur Helps[1] says:

"It is strange that this little glistening bead, the pearl, should have been the cause of so much movement in the world as it has been. There must be something essentially beautiful in it, however, for it has been dear to the eyes both of civilized and uncivilized people. The dark-haired Roman lady, in the palmiest days of Rome, cognizant of all the beautiful productions in the world, valued the pearl as highly as ever did the simple Indian woman, and a love for these glistening beads came upon the Spaniards from two quarters, from the Romans who had colonized them, and from the Moors they had conquered. The perilous nature, however, of his submarine possessions was not yet visible to the poor innocent Indian on the coast of Paria or Cumana, and it was with childish delight that he threw the strings of pearls, strung in a way that would have driven the jewelers of Europe wild with vexation, on the smooth brown arm or rich brown neck of his beloved."

Of Columbus[2] it is said that the natives of Paria possessed such quantities of fine pearls that the most sanguine anticipations were roused in him. Remembering the assertion of Pliny, that pearls were generated from drops of dew which fell into the mouths of oysters, he deemed no place so propitious as this coast for their growth and multiplication. When nearing the island of Cubagua, this admiral, Charlevoix tells us, beheld a number of Indians fishing for pearls, who at the approach of the strangers at once made for the land. A boat being sent to communicate with them, one of the sailors noticed many strings

---

[1] The Spanish Conquest of America (London, 1855), Vol. 2, p. 89.
[2] Life and Voyages of Columbus and his Companions, by Washington Irving (New York, 1849), Vol. 2, p. 123.

of pearls around the neck of a woman. Having a plate of Valencia-ware, a kind of porcelain painted and varnished with gaudy colors, he broke it, and presented the pieces to the Indian woman, who gave him in exchange a considerable number of her pearls. These he carried to the admiral, who immediately sent persons on shore well provided with Valencian plates and hawks'-bells, for which, in a little time, he procured about three pounds' weight of pearls, some of which were of a very large size, and were sent by him afterward to the sovereigns as specimens.

At the time of the Spanish invasion, the pearl was held in high esteem by the Mexican people as an ornament, and, upon occasions of state, its beauties were invoked to increase the magnificence of the apparel and lend additional lustre to the pomp of royalty. When Montezuma alighted from his regal palanquin, "blazing with burnished gold" and overshadowed by a "canopy of gaudy feather-work powdered with jewels and fringed with silver," to grant personal audience to Cortez, his cloak and golden-soled sandals were sprinkled with pearls and precious stones.

To Vasco Nuñez de Balboa, Tumaco gave jewels of gold and 200 pearls of great size and beauty, although they were somewhat discolored. Observing the value which the Spaniards set upon them, the cacique sent a number of his men to fish for them. The largest pearls were generally found in the deepest water, sometimes in three or four fathoms, and were sought only in calm weather. The smaller pearls were taken at a depth of two or three feet, and the oysters containing them were often driven in quantities on the beach during the violent storms. The party of pearl divers sent by the cacique consisted of thirty Indians, with whom Balboa sent six Spaniards as eye-witnesses. A number of the shell-fish were driven on shore, from which they collected enough to yield pearls to the value of twelve marks of gold. They were small but exceedingly beautiful, not having been injured by heat like those collected by the Indians, who opened the shells by putting them in a fire, and many of these pearls were sent to Spain as specimens.[1]

[1] Life and Voyages of Columbus and his Companions, by Washington Irving (New York, 1849), Vol. 3, p. 181.

Oviedo, the Spanish historian, commemorates the circumstance that this cacique, Tumaco, subsequently furnished Balboa with a canoe formed from the trunk of an enormous tree and managed by a great number of Indians. The handles of the paddles were inlaid with small pearls, a fact which Balboa caused his companions to testify before the notary, that it might be reported to the sovereigns as a proof of the wealth of this newly discovered sea. In another bay of the Pacific coast, this bold navigator saw groups of islands abounding with pearls, many of them as large as a man's eye.

Barnard Shipp states, "The first Spaniards who landed on terra firma found savages decked with necklaces and bracelets of pearls, and among the civilized people of Mexico and Peru, pearls of a beautiful form were generally sought after. The Indians of Virginia wore pendants in their ears, and round their arms chains and bracelets of pearls."[1]

When the King of Spain made Hernando De Soto Governor of Cuba and conqueror of Florida, with the title of Adelantado, his concession provided that one-fifth of all the gold and silver, precious stones and pearls, won in battle, or entering towns, or obtained by barter with the Indians, be reserved to the Crown. It was further stipulated that the gold and silver, stones, pearls, and other things which might be found and taken, as well in the graves, sepulchers, ocues or temples of the Indians, as in other places where they were accustomed to offer sacrifices to idols, or in other concealed religious precincts or buried houses, or in any other public place, "should be equally divided between the king and the party making the discovery."[2]

It is evident that among the valuable trophies of this expedition, precious pearls were confidently anticipated, and that the Spaniards were not disappointed in this expectation the early narratives abundantly testify. These establish beyond all controversy that pearls were used as ornaments among the Indians of Florida and the South.

It is related how, near the Bay of Espiritu Santo (now Tampa Bay), in Florida, the followers of De Soto came upon

[1] The History of Hernando De Soto and Florida (Philadelphia, 1881).
[2] Antiquities of the Southern Indians, by Charles C. Jones (New York, 1873), p. 467.

the town of an Indian chief called Ucita. His house stood near the beach, and at the other end of the town was a temple, on the top of which perched a wooden fowl with gilded eyes. Within these eyes were pearls such as the Indians greatly value, piercing them for beads and stringing them to wear about their necks and wrists.

When the Indian queen welcomed the Spanish adventurer to the hospitalities of the Cutifachiqui, she drew from over her head a long string of pearls, and throwing it around his neck, exchanged with him gracious words of friendship and courtesy. Observing that the Christians valued these pearls, the cacica told the governor that if he would order some sepulchers to be searched that were in the village, he would find many ; and, if he chose to send to those which were in the uninhabited towns, he might load all his horses with them. The Spaniards did examine and rifle of their contents the sepulchers in Cutifachiqui ; and, upon the authority of the Knight of Elvas, obtained from them 350 pounds' weight of pearls, some of which were formed after the similitude of babies and birds. If the truth were known, or if an Indian had written this account, we should feel assured that De Soto and his companions, in their eager quest for treasures, violated the graves without permission and plundered the receptacles wherein were gathered the most costly possessions of the natives. As a proof that the Indians did not willingly part with these ornaments, but suffered the pillage through fear of these strange and wanton men, we are informed that when the cacica, whom De Soto compelled to accompany him with the intention of taking her to Guaxule, which was the farthest limit of her territory, succeeded in making her escape, she carried back with her a cane box filled with unbored pearls, the most precious of all her jewels.

Luys Hernandez de Biedma says that the governor, while at this town, opened a "mosque" in which were interred the chief personages of that country. "From it we took a quantity of pearls of the weight of as many as 6½ or 7 arrobas, though they were injured from lying in the earth and in the adipose substance of the dead." In the estimate of the relator, one of the saddest losses encountered by the expedition in the bloody affair at Mauilla was the destruction of the pearls which the Spaniards had been sedulously collecting during their wanderings in this strange land.

The most minute and interesting description of the manner in which the Indians obtained pearls and converted them into beads is that furnished by Garcilasso Inca de la Vega. During the time when De Soto remained in the town of Ichiaha, which was probably located at or near the confluence of the Etowah and Oostanaula Rivers, and possibly the very spot now occupied by the village of Rome, Ga., the following circumstances occurred: "The cacique came one day to the governor, bringing him a present of a string of pearls, five feet in length. These pearls were as large as filberts, and had they not been bored by means of fire, which had discolored them, would have been of immense value. De Soto thankfully received them, and in return presented the Indian chief with pieces of velvet and cloth of various colors, and other Spanish trifles held in much esteem by the natives. In reply to the demand of De Soto, the cacique stated that the pearls had been obtained in the neighborhood. He further told him that in the sepulcher of his ancestors was amassed a prodigious quantity, of which the Spaniards were welcome to carry away as many as they pleased. The Adelantado thanked him for his good will, but replied that, much as he wished for pearls, he never would insult the sanctuaries of the dead to obtain them, adding that he only accepted the string of pearls from the chieftain's hands.

"De Soto having expressed a curiosity to see the manner of extracting pearls from the shells, the cacique instantly despatched forty canoes to fish for oysters during the night. At an early hour next morning, a quantity of wood was gathered and piled up on the river bank, and being set on fire was speedily reduced to glowing embers. As soon as the canoes arrived, the oysters were laid upon the hot coals. They quickly opened with the heat, and from some of the first thus opened, the Indians obtained ten or twelve pearls as large as peas, which they brought to the governor and the cacique, who were standing together looking on. They were of a fine quality, but somewhat discolored by the fire and smoke. The Indians were apt also to further injure pearls thus obtained by boring them with a heated copper instrument.

"De Soto, having gratified his curiosity, returned to his quarters to partake of his morning meal. While thus engaged a soldier entered with a large pearl in his hand. He had stewed some oysters, and in eating them, felt the pearl between his teeth. Not having been injured by fire or smoke, it retained its beautiful whiteness, and was so large and perfect in its form that several Spaniards, who pretended to be skilled in those matters, declared it would be worth 400 ducats. The soldier would have given it to the governor to present to his wife, Doña Isabel de Bobadilla, but De Soto declined the generous offer, advising him to preserve it until he should arrive at Havana, when he could purchase horses and other necessaries with it; moreover, as a reward for his liberality, De Soto insisted upon paying the fifth of the value due the Crown."[1]

During the course of the weary march of the expedition through the mountains of Upper Georgia, the following circumstance is related by the same historian.

"A foot-soldier, calling to a horseman who was his friend, drew forth from his wallet a linen bag in which were six pounds of pearls, probably filched from one of the Indian sepulchers. These he offered as a gift to his comrade, being heartily tired of carrying them on his back, though he had a pair of broad shoulders capable of bearing the burden of a mule. The horseman refused to accept so thoughtless an offer. 'Keep them yourself,' said he. 'You have most need of them. The governor intends shortly to send messengers to Havana, when you can forward these presents and have them sold, and obtain three or four horses with the proceeds, so that you need no longer go on foot.' Juan Terron was piqued at having his offer refused. 'Well,' said he, 'if you will not have them, I swear I will not carry them, and they shall remain here.' So saying, he untied the bag, and whirling it around, as if he were sowing seed,

---

[1] The foregoing is taken from Theodore Irving's Conquest of Florida under Hernando De Soto (London, 1835), Vol. 2, p. 14, and is from Pierre Richelet's translation made in 1831. De la Vega's entire work, translated from the same source, appears in the History of Hernando De Soto and Florida, by Barnard Shipp (Philadelphia, 1881).

scattered the pearls in all directions among the thickets and herbage. Then putting up the bag in his wallet, as if it was more valuable than the pearls, he marched on, leaving his comrades and other bystanders astonished at his folly. The soldiers made a hasty search for the scattered pearls and recovered thirty of them. When they beheld their great size and beauty, none of them being bored or discolored, they lamented that so many of them had been lost; for the whole would have sold in Spain for more than 6,000 ducats. This egregious folly gave rise to a common proverb in the army, 'There are no pearls for Juan Terron.' The poor fellow himself became an object of constant jest and ridicule, until at last, made sensible of his absurd conduct, he implored them never to banter him further on the subject." [1]

Fontaneda states that at the place where Lucas Vásquez went, seed-pearls were found in certain conchs, and that between Havalachi and Olagale is a river called by the Indians Guasaca-esqui, which means in the Spanish language Rio de Canas (river of canes), which is an arm of the sea, and along the adjacent coast, pearls are procured from certain oysters and conchs. These are carried to all the provinces and villages of Florida, but principally to Tocobaja, the nearest town. The Indians of the town of Abalachi asserted that the Spaniards hanged their cacique because he would not give them a string of large pearls which he wore around his neck, the middle pearl of which was as big as the egg of a turtle-dove. Ribault frequently alludes to the possession of pearls by the natives of Florida, and on one occasion saw the goodliest man of a company of Indians with a collar of gold and silver about his neck from which depended a pearl " as large as an acorn at the least." [2]

David Ingram, during the " Land Travels " of himself and others in the year 1568–1569, from the Rio di Minas in the Gulf of Mexico to Cape Breton in Acadia, made the following observation: " There is in some of those Countreys great abundance of Pearle, for in every cottage he founde Pearle, in some howse a quarte, in some a pottell, in some a pecke, more or lesse,

---

[1] Conquest of Florida under Hernando De Soto, by Theodore Irving (London, 1835), Vol. 2, p. 7.

[2] The Whole and True Discovery of Terra Florida, by Thomas Hackett (London, 1563).

where he did see some as great as an acorn, and Richard
Browne, one of his companions, found one of these great pearls
in one of their canoes, or Boates, wch Pearle gaue to Mouns
Champaine, whoe toke them aboarde his Shippe, and brought
them to Newhaven in ffruñce." [1]

The English were quick to note the presence of pearls in
this country, and it is interesting to find that, centuries before,
Suetonius states that Cæsar undertook his British expedition
for the sake of finding pearls, and Pliny and Tacitus report his
bringing home a buckler made of British pearls, which he dedi-
cated to Venus Genetrix [2] and hung up in her temple. An account of the pearl fisheries in Ireland was published, stating
that oysters were found set up in the sands of the river-beds,
with the open side from the torrent. About one in one hundred would contain a pearl, and one pearl in one hundred would
be tolerably clear. Between the years 1761 and 1764 the river
Conway in Scotland supplied the London market with pearls
to the value of £10,000 sterling, and fine Scotch pearls are still
sold in London. The rivers of Cumberland, the Conway in
Wales and the Tay in Scotland, have yielded pearls that were
noted for their beauty in time past.

Father Louis Hennepin assures us that the Indians along
the Mississippi wore bracelets and ear-rings of fine pearls, which
they spoilt, having nothing to bore them with but fire. He
adds: " They gave us to understand that they received them in
exchange for their calumets from nations inhabiting the coast of
the great lake to the southward, which I take to be the Gulph
of Florida." [3]

A member of the expedition of Sir Walter Raleigh collected from the natives of Virginia 5,000 pearls, " of which number he chose so many as made a fayre chaine, which for their
likenesse and uniformity in roundnesse, orientnesse and
pidenesse of many excellent colors, with equalitie in greatness,
were very fayre and rare." [4]

[1] Documents connected with the History of South Carolina, edited by Plowden Charles Jennett
Weston (London, 1856), p. 8.

[2] Transactions of the Philosophic Society for 1693.

[3] New Discovery, etc. (London, 1698), p. 177.

[4] A Briefe and True Report of the New Found Land of Virginia (Frankfort on the Main,
1590), p. 11.

In the plates illustrative of the " Admiranda Narratio" and the " Brevis Narratio," the natives both of Virginia and Florida are represented in the possession of numerous strings of pearls of large size; and in his description of the "treasure of riches" of the Virginia Indians, Robert Bevery says : " They likewise have some pearls amongst them, and formerly had many more, but where they got them is uncertain, except they found them in the oyster banks which are frequent in this country."[1]

Wilson asserts that he saw pearls "bigger than Rouncival pease," and perfectly round, taken from oysters found on the Carolina coast.[2]

The existence of shell-heaps may also be traced to the making of wampum and of shell beads in general, which formed a trade among the tribes inhabiting the sea coast. This labor required much time, and promised success only to those who, by long practise, had attained skill in the operation. The supposition gains some ground by an observation of Roger Williams, who states that " most on the Sea side make Money and Store up shells in Summer against Winter whereof to make their money." He further observes : " They have some who follow onely making Bowes, some Arrowes, some Dishes (and the women make all their Earthen vessels), some follow fishing, some hunting."[3]

Kjoekkenmoeddings on the St. John's River, Florida, consisting of river shells, were examined and described by Prof. Jefferies Wyman. He saw similar accumulations on the banks of the Concord River, in Massachusetts, and was informed by eye-witnesses that they are numerous in California.[4]

Charles Rau[5] says : " The term 'wampum' is often applied to shell-beads in general, but should be confined, I think, to a certain class of cylindrical beads, usually $\frac{1}{4}$ of an inch long and drilled lengthwise, which were chiefly manufactured

[1] History of the Present State of Virginia (London, 1705), Book 3, p. 59.

[2] An Account of the Province of Carolina (London, 1682), p. 12.

[3] A Key into the Language of America, reprinted from the London edition of 1643 (Providence, 1827), p. 1331.

[4] Cf. Fresh-Water Shell-Heaps of the St. John's River, East Florida (Salem, Mass., 1868), p. 6.

[5] Ancient Aboriginal Trade in North America, in the Report of Smithsonian Institution for 1872, p. 32 of Mr. Rau's reprint.

from the shells of the common hard-shell clam (Venus mercenaria). This bivalve occurring, as every one knows, in great abundance on the North American coasts, formed an important article of food of the Indians living near the sea, a fact demonstrated by the enormous quantity of cast-away clam-shells, which form a considerable part of North American Kjoekkenmoeddings (as these heaps are called). The natives used to string the mollusks and to dry them for consumption during the winter. The blue or violet portions of the clam-shells furnished the material for the dark wampum, which was held in much higher estimation than that made of the white part of the shell or of the spires of certain univalves. Even at the present time, places are pointed out on the Atlantic sea-board, for example on that of Long Island, where the Indians manufactured wampum, and such localities may be recognized by the accumulations of clam-shells from which the blue portions are broken off."

Wampum beads formed a favorite material for the manufacture of necklaces, bracelets, and other articles of ornament, and they constituted the strings and belts of wampum which played such a conspicuous part in Indian history. Loskiel says on this point : "Soon after their arrival in America, the Europeans began to manufacture wampum from shells, very neatly and in abundance, exchanging it to the Indians for other commodities, thus carrying on a very profitable trade. The Indians now abandoned their wooden belts and strings and substituted those of shells. The latter, of course, gradually declined in value, but, nevertheless, were and still are much prized."[1]

According to Albert J. Pickett, the oyster alluded to by Garcilasso was identical with the mussel so common in all the rivers of Alabama. "Heaps of mussel shells," he says, "are now to be seen on our river banks wherever Indians used to live. They were much used by the ancient Indians for some purpose, and old warriors have informed me that their ancestors once used the shells to temper the clay with which they made their vessels. But as thousands of the shells lie banked up,

[1] Mission der evangelischen Brüder unter den Indianern in Nordamerika (Barby, 1789), p. 34.

some deep in the ground, we may also suppose that the Indians in De Soto's time, everywhere in Alabama, obtained pearls from them. There can be no doubt about the quantity of pearls found in this State and Georgia in 1540, but they were of a coarser and more valueless kind than the Spaniards supposed. The Indians used to perforate them with a heated copper spindle and string them around their necks and arms like beads."[1]

Sufficient historical evidence has been given to show that pearls were in general use among the southern Indians; that the choicest of them were the prized ornaments of the prominent personages of the tribes; that the fluviatile mussels were collected and opened for the purpose of procuring them; that the marine shells of the Atlantic, the Gulf of Mexico, and the Pacific, yielded generous and beautiful tribute to the labor, skill, and taste of numerous and well-trained pearl divers; and that these pearls were found, not only in the possession of the living, but also in large quantities in the graves of chieftains and the sepulchers of the undistinguished dead. A present of pearls from the caciques to the conquerors was an earnest token of consideration and the most acceptable pledge of friendship that he could offer.

Doubtless, however, the accounts that have reached us from the pens of the historians of these expeditions and voyages are somewhat extravagant with regard to the quality and quantity and size of the pearls seen in the possession of the natives. From these interviews between the Europeans and the natives, it appears that the Indians obtained their supplies of pearls both from marine shells and from fresh-water mussels. Some of the oysters in Georgia and Florida are margaritiferous and many of them contain seed-pearls. Specimens symmetrical in shape, as large as pepper-corns, and not wanting in beauty, have been observed by Charles C. Jones, who says: "Some were quite big enough to have been perforated in the rude fashion practised by the Indians. They were, however, of a milky color and opaque. Neither in size nor quality did they answer the description spoken of in the Spanish narratives."[1]

[1] History of Alabama (Charlestown, 1851), Vol. 1, p. 12.
[2] Antiquities of the Southern Indians (New York, 1873), p. 481.

Perforated pearls were found by Dr. Edwin H. Davis[1] on the hearths of five distinct groups of mounds in Ohio, and sometimes in such abundance that they could be gathered by the hundred. They were generally of irregular form, mostly pear-shaped, though perfectly round ones were also found among them. The smaller specimens measured about ¼ of an inch in diameter, but the largest had a diameter of ¾ of an inch.

According to this same authority, pearl-bearing shells occur in the rivers of the region whose antiquities are described, but not in such abundance that they could have furnished the amount discovered in the tumuli, and the pearls of these fluviatile shells, moreover, are said to be far inferior in size to those recovered from the altars. The latter, it was erroneously thought, were derived from the Atlantic coast and from that of the Mexican Gulf.

The Indians of Carolina, Georgia, Florida, Alabama, and the other Southern States, subsisted largely on oysters, clams, and conchs, as shown by the numerous refuse piles and shell heaps that abound upon the salt-water creeks. It is not matter of surprise that the Indians, as they opened these shells, should have carefully watched for pearls, and from the vast numbers examined, should have accumulated a store. If the shores of Carolina, Georgia, and Florida did not afford the larger and more highly-prized pearls, it is not improbable that pearls from the islands and lower portions of the Gulf of Mexico, and even from the Pacific Ocean, may have found their way into the heart of Georgia and Florida and into more northern localities, to be there bartered away for skins and other articles. The replies of Indians to Father Hennepin and others, and the presence in remote localities of beads, ornaments, and drinking-cups made of marine shells and conchs, still peculiar to the Gulf of Mexico, confirm the truthfulness of this suggestion.[2]

But marine shells are not the only source whence the southern Indians derived their pearls. The fluviatile mussels contributed perhaps more freely than other shells to the treas-

---

[1] See Ancient Monuments of the Mississippi Valley, by Ephraim G. Squier and Edwin H. Davis (Washington, 1848), p. 232.

[2] Ancient Aboriginal Trade in North America, by Charles Rau. Report of the Smithsonian Institution for 1872 (Washington, 1873).

ures of these early people. At various points along the southern rivers, relic beds are found, composed of the fresh-water shells native to the streams. The inland lakes of Florida show similar evidences of occupancy of their shores by aborigines, and even some ponds in middle Georgia and Alabama exhibit along their banks signs of ancient refuse piles where lacustrine shells abound. These heaps are common in the South, and several of them on the banks of the Savannah River, above Augusta, are fully described by Charles C. Jones.[1] He says: "In these relic beds no two parts of the same shell are, as a general rule, found in juxtaposition. The hinge is broken, and the valves of the shell, after having been artificially torn asunder, seem to have been carelessly cast aside and allowed to accumulate."

Thus, in addition to the historical evidence given, physical proof is adduced of the pearl fisheries of the aboriginal tribes of the South. In order to ascertain the precise varieties of shells from which the southern Indians obtained their pearls, Mr. Jones invited an expression of opinion from the following scientists, whose pursuits rendered them familiar with the conchology of the United States. They throw considerable light upon this inquiry.

Dr. William Stimpson, of the Chicago Academy of Sciences, considered the statements of the early Spanish historians with regard to the size of the pearls (as large as filberts) exaggerated. He says: "The pearls of the Aviculæ, our only margaritiferous marine genus, are very small, and those of the oyster valueless. The Indians must have obtained their pearls from the fresh-water bivalves (Unio and Anodon) which abound in the rivers of Georgia, etc. These are usually small, but in very rare instances examples have occurred reaching in diameter $\frac{1}{4}$ of an inch."

"Most of the fresh-water mussels," writes Prof. Joseph Le Conte, "contain small pearls now and then. By far the best and largest number I have seen were taken from the Anodon gibbosa (Lea), a large and beautiful shell abundant in the

---

[1] Antiquities of the Southern Indians (New York, 1873), p. 483; also Monumental Remains of Georgia (Savannah, 1861), p. 14.

swamps of Liberty County, Ga., at least in Bulltown and Ala-tamaha Swamps.   Some of the pearls taken from this species are as large as swan-shot.   Of the salt-water shells, I know not if any produce pearls except the oyster (Ostrea Virginica). Pearls of small size are sometimes found in them."

Prof. William S. Jones, of the University of Georgia, says that he has seen small pearls in many of the Unios found in Southern Georgia.

Prof. Jefferies Wyman, after a careful and extensive series of excavations in the shell heaps of Florida, failed to find a single pearl.   "It is hardly probable," he remarks, "that the Spaniards could have been mistaken as to the fact of the orna-ments of the Indians being pearls, but in view of their frequent exaggerations, I am almost compelled to the belief that there was some mistake; and possibly they may not have distin-guished between the pearls and the shell beads, some of which would correspond with the size and shape of the pearls men-tioned by the Spaniards."

Prof. Joseph Jones, whose investigations throw much valu-able light upon the contents of the ancient tumuli of Tennessee, says: " I do not remember finding a genuine pearl in the many mounds which I have opened in the valleys of the Tennessee, the Cumberland, the Harpeth, and elsewhere.   Many of the pearls described by the Spaniards were probably little else than polished beads cut out of large sea-shells and from the thicker portions of fresh-water mussels, and prepared so as to resemble pearls.   I have examined thousands of these, and they all pre-sent a laminated structure, as if carved out of thick shells and sea conchs."

Charles M. Wheatley was confident that there were "splendid pearls in southern Unios."   He instances the Unio Blandingianus and the large old Unio Buddianus (Buckleyi) from Lakes George and Monroe in Florida, as pearl bearing. "In Georgia," he continues, "the large, thick shells of the Chattahoochee, such as the Unio Elliottii, would be the most likely to contain fine ones; but there is no positive rule, as an injured shell of any species will doubtless afford some, irregular in most cases and of no value, but in some instances worth from $50 to $100." He also mentions that he has received from the Tennessee River, in Alabama, fine round pearls both white and rose colored.

John G. Anthony writes: "I never have collected in Florida and but little in Georgia, but what I can say about Ohio I presume will hold good in other States, that the Unios of various species furnish them tolerably abundantly there. They are not confined to any one particular species, but are generally found in the thicker and more ponderous shells, though even the thinner shells often have small ones, especially such as are found in canals, ponds, and places which seem to be not so healthy for the animal on account of stagnant water. I recollect taking over twenty small ones out of the mantle of one specimen of Unio fragilis, U. gracilis (Barnes), which I found in the Miami Canal; and almost every old shell there had more or fewer pearls in it. U. torsus (Raf.), U. orbiculatus (Hildreth), U. costalus (Raf.), and U. undulatus (Barnes), also produce them in Ohio. I have seen about half a pint of beautiful pearls, regularly formed and pea size, which were taken in one season and in one neighborhood; so you may judge of their frequency, though, as I hinted before, it is probable that a kind of disease caused by impure water may govern their production somewhat. No doubt the southern waters are given to making pearls, as well as Ohio streams. I have seen protuberances of the pearl character in southern shells, and have no doubt that one collecting them with the animal in them would find pearls. I particularly recollect Unio globulus (Say) and U. Mortoni (Conrad), both Louisiana species, as having these protuberances in their nacreous matter. Georgia Unios are generally too thin to produce any excess of pearly matter and form pearls, but the Louisiana shells from Bayou Techa, which I have seen, have a remarkably pearly nacre, quite thick, reminding one very much of the marine shell Trigonia, as to nacre. No doubt the bayous, which have in general no current at all, would make first-rate places for pearl breeding."

Dr. Charles Rau[1] writes: "I learned from Dr. Samuel G.

[1] Ancient Aboriginal Trade in North America, Report of the Smithsonian Institution for 1872, p. 38 of the author's reprint.

Bristow, who was surgeon of the Army of the Cumberland during the Civil War, that mussels of the Tennessee River were occasionally eaten 'as a change' by the soldiers of that corps, and pronounced no bad article of diet. Shells of the Unio are sometimes found in Indian graves, where they had been deposited with the dead, to serve as food during the journey to the land of spirits."

Dr. Brinton saw on the Tennessee River and its tributaries numerous shell heaps consisting almost exclusively of the Unio Virginianus (Lamarck). In every instance he found shell heaps close to the water-courses, on the rich alluvial bottom lands. He says: "The mollusks had evidently been opened by placing them on a fire. The Tennessee mussel is margaritiferous, and there is no doubt but that it was from this species that the early tribes obtained the hoards of pearls which the historian of De Soto's exploration estimated by bushels, and which were so much prized as ornaments."[1]

A source has recently been pointed out whence small pearls, and perhaps some fine specimens, could have been obtained by the Indians of Florida, and in considerable quantities. In the Unios of some of the fresh-water lakes of that State, there were found not less than 3,000 pearls, most of them small, but many large enough to be perforated and worn as beads. From one Unio there were taken eighty-four seed-pearls; from another fifty, from a third twenty, and from several ten or twelve each. The examinations were chiefly confined to Lake Griffin and its vicinity. It is said that upon one of the isles in Lake Okeechobee are the remains of an old pearl fishery, and it is proposed to open the shells of this lake, which are large, in hopes of finding pearls of superior size and quality.

The use of the pearl as an ornament by the southern Indians, and the quantities of shells opened by them in various localities, make it seem strange that it is not more frequently met with in the relic-beds and sepulchral tumuli of that region; but after exploring many shell and earth mounds, Col. Charles C. Jones failed, except in a few instances, to find pearls.[2]

[1] See Artificial Shell Deposits in the United States, in the Report of the Smithsonian Institution for 1866, p. 357.

[2] Antiquities of Southern Indians, p. 486.

A few were obtained in an extensive relic bed on the Savannah River, above Augusta, the largest being $\frac{4}{10}$ of an inch in diameter, and all of them blackened by fire. Many of the smaller mounds on the coast of Georgia do not contain pearls, because at the period of their construction the custom of burning the dead appears to have prevailed very generally; hence, it may be that the pearls were either immediately consumed or so seriously injured as to crumble out of sight.

This absence of pearls tends somewhat to confirm the opinion that beads and ornaments made from the thicker portions of shells that were carved, perforated, and brilliant with their primal covering, were regarded by the imaginative Spaniards as pearls. More minute investigation, however, will doubtless reveal the existence of pearls in localities where the pearl-bearing shells were collected. Perforated pearls have been found in an ancient burying-ground located near the bank of the Ogeeche River, in Bryan County, Ga.; and many years ago, after a heavy freshet on the Oconee River, which laid bare many Indian graves in the neighborhood of the large mounds on Poullain's plantation, fully a hundred pearls of considerable size were gathered.

There can be no doubt that what were regarded as pearls by the early Spanish voyagers were really such, although it is well known that shell-beads have been found in mounds in connection with pearls; but the numbers found in Ohio, and which have been mentioned by Prof. Frederick W. Putnam and by others, leave no room for doubt in this matter. That the Indians of the South also had these pearls, both drilled and undrilled, is beyond question. Notwithstanding the intimacy existing between remote Indian tribes, as shown by many authorities, and the fact that Pacific coast shells have been carried to Arizona, and that clam shells have been found in Zuni cities by Lieut. Frank H. Cushing, it is likely that these pearls came, not from the pearl oyster of the Pacific coast, but from the marine shells of the Atlantic coast and the fresh-water shells of the eastern part of the continent. It is more than probable that the Indians opened the shell to secure the animal, which they valued as an article of food; that the shells of some varieties, such as the common clam and the conch, were made into wampum; and that the pearls found in the shells were used as ornaments, whether lusterless pearls from the common oyster or lustrous ones from the Unio. The great shell-heaps along our coasts bear evidence that many pearls must have been found, and that, though the Spaniards who invaded the country may indeed have obtained great hoards of pearls, all of them, perhaps, were not of great value.

CHAPTER XIII.

In the Dominion of Canada.

ALTHOUGH Canada can scarcely be called a gem-producing country, and no mining for precious stones is carried on there, still it furnishes some stones that are of more than passing interest to the mineralogist and of some little value in jewelry and the arts. A number of gem minerals, not of gem quality, are here obtained of such wonderful size and perfection that they have been given places in the cabinets of the world, and are even more prized as specimens than cut stones from other localities. Their mineralogical value gives them commercial importance. For instance, the zircon crystals, occurring as individuals up to 15 pounds in weight, many fine ones weighing nearly a pound, and the beautiful twin crystals of the same mineral; the dark brown titanite in simple and twinned crystals up to 70 pounds each; the quantities of amethyst from Lake Superior; the ouvarovite or green chrome-garnet from Orford, Ont.; and the white garnet crystals from near Wakefield, Que., are among the most notable Canadian minerals. Not the least wonderful are the apatite crystals (one weighing over 500 pounds), which are found of such size and beauty that the rich green variety might be worked into ornaments similar to those made from fluorite.

Corundum, in red and blue crystals, has been found in limestone near Burgess, Ont., also disseminated through a rock made up of feldspar, quartz, calcite, and titanite, in contact with the crystalline limestone. These grains vary from light rose-red to sapphire-blue color, and are of no gem value, nor in quantity sufficient for commercial use.

In the seigniory of Daillebout, Que., translucent octahedrons of blue spinel are found in micaceous limestone; and from Wakefield, Que., come pink and dark-bluish spinels in rounded cubic crystals and opaque light-blue cubes nearly an inch in diameter. Very interesting black spinels in brilliant crystals, 1 to 2 inches in diameter, occur in Burgess and Bathurst Townships, Ont., where a vein of them has been traced for a mile in one direction. They are also associated with fluorite in the township of Ross, Ont. None of these possess gem value.

At the World's Fair held in London, 1862, there was exhibited two so called topazes, from Cape Breton, N. S., one in the rough, and the other, which had been cut at Pictou, N. S., ¼ an inch in length and of a yellow color,—the variety of this mineral peculiar to Brazil. This fact leads to the inference that these stones may have been citrine or artificially decolorized smoky quartz, and not the true mineralogical topaz.

Little if any beryl of value for gems has been discovered in Canada. Pale-green, well-defined crystals have been reported by Dr. Bigsby at Rainy Lake, 230 miles west of Lake Superior; and in Berthier and Saguenay Counties, Que., crystals over an inch in diameter have been found.

The zircons of Ontario, especially those from Lake Clear, and Sebastopol and Brudenell Townships, in Renfrew County, are the most remarkable known for beauty, size, perfection, and richness of color. An occasional crystal top or a small fragment will afford a gem of the hyacinth variety, but they rarely exceed a carat in weight. Some of these individual crystals weigh about 15 pounds, and are more than 4 inches in diameter. One was observed 3 inches in diameter and nearly a foot in length. In Brudenell Township, twenty-five miles west of Eganville, Ont., fine crystals are obtained. The twin zircons from Lake Clear are beautiful and interesting, one of them measuring nearly 4 inches in length; they are of no gem value, but many thousand dollars' worth have been sold as specimens. Short's Claim, on the north shore of Lake Clear, yields the choicest twins. Perhaps the finest twin crystals ever found, and one of the best single crystals, are in the British Museum Collection; while the best series of this mineral is probably that in the collection of Clarence S. Bement, of Philadelphia. An enormous single crystal is in the cabinet of the Academy of Natural Sciences at Philadelphia. In Burgess and adjoining townships fine crystals occur, not so large as those from Renfrew County, but of exquisite polish and highly modified forms; in Templeton and near Grenville, Que., especially four miles north, are found smaller crystals, often cherry-red and transparent, that would yield gems; and many of the crystals are modified and associated with wollastonite and graphite.

Tourmaline in green crystals is found in Chatham Township, Que., and the green and red varieties in Villeneuve Township, Que. Brown tourmalines are frequently met with in the Laurentian limestone. Fine crystals, rich yellowish or translucent brown in color, often occur imbedded in a flesh-red limestone in Ross, Ont., Calumet Falls, Clarendon, and Hunterstown, Que. These furnish an occasional gem. Slender crystals in white quartz occur at Fitzroy, Island Portage, and Lac des Chats, and of inferior color at McGregor's Quarry in Lachute, Ont. Black tourmaline of no gem value is found in a number of localities, principally at Yeo's Island, near the Upper end of Tar Island, one of the Thousand Islands. It occurs in large crystals at Murray Bay, Cape Tourmente, Que., and in white quartz near Bathurst, Ont.; in the granitic veins in Ross, Ont.; on Roche Fendue Channel, on Camping Place Bay, on Charleston Lake in Lansdowne, in Blythfield, on the Madawaska, and at North Elmsley and Lachute, Ont.; and on the west side of the North River at St. Jerome, St. Felix, and Calumet Falls, Que. The velvet-black, fibrous tourmaline found at Madoc and Elzevir, Ont., gives a blue powder and is evidently an indicolite, like the variety from Paris, Me.

Almandite garnets occur plentifully in crystals in mica schist along the Stickeen River in British Columbia. Owing to their perfect form and polish, the faces of these crystals are the most beautiful in the world. Although they are not transparent enough to be of value to cut into gems, yet if obtained in sufficient quantity, they would be useful for watch jewels. Beds of nearly pure red garnet rock, from 5 to 25 feet thick, are sometimes met with in the gneiss at St. Jerome, Que., and in quartzite in Rawdon and Marmora Townships, Ont., and at Baie St. Paul, Que. Some small pieces would afford gems of little value, but the stone is of considerable use in the arts as a grinding material and for sand-paper. The large red garnet, disseminated through a white oligoclase gneiss, at Lake Simon, would not afford gems, but if polished with the rock would afford an ornamental stone. An-dradite garnet is found on Texada Island, B. C., in fair crystals, but not suitable for jewelry. Essonite, cinnamon-colored gar-net, is found in small crystals in Grenville, Que., but not of gem value, and in fine crystals, associated with idocrase, in Wakefield, Que. But few of these would furnish even small gems. Gros-sularite, white lime-alumina garnet, is found in Wakefield and in Hull, Que., in large quantity, in veins lining the crystalline lime-stone, and associated with essonite, idocrase, and pyroxene. This is the most remarkable locality for this mineral, superb crystals 2 inches across having been found there, as well as groups of crystals a foot across. In color the crystals vary from colorless to light yellow and light brown, and some of them are transpar-ent enough to afford colorless gems of from 1 to 2 carats in weight. Melanite, black garnet, is found in Marmora, Ont., but it is not used for jewelry. Ouvarovite, or green chrome-garnet, found in Orford, Ont., yields the most beautiful known specimens of this rare mineral. The crystals, which are transparent dode-cahedrons, rarely over ¼ inch in diameter, and of the deepest emerald-green color, are found lining druses in cavities of crys-talline limestone, often on the chrome pyroxene and associated with millerite. If it were not for the small size of the crystals, it would be a gem of the highest rank. A few crystals have been found in Wakefield, some of which rival in size any that have been discovered, the largest measuring nearly ¼ an inch in diameter. They are of a fine green color, but opaque, and some-times have a yellow center.

Rock-crystal is found in many localities of Canada,

especially in veins with amethyst in the Lake Superior region, but not of sufficient size to afford crystal balls or other art objects. The small, doubly terminated crystals found in the limestone of the Levis and Hudson River formations, and locally called " Quebec diamonds," are sold as souvenirs to tourists. Fine crystals are found in the soil in Lacolle, Que., and beautiful limpid crystals in the cavities of the calciferous formations in many places. Larger crystals have been found with smoky quartz near Paradise Bay, N. S., also in the geodes on agate throughout the entire Bay of Fundy district, and on the Musquash River, N. B., at Cape Blomidon, N. S. Milky quartz is found all through Canada, but it is never of any value in the arts except for porcelain. Rose quartz is also found in many localities, especially at Shelburne, N. S. It has little value in the arts, but has been made into various ornaments and charms. Smoky quartz in fine groups occurs in the same veins with amethyst on both Lake Superior and the Bay of Fundy, so uneven in color, however, as to afford gems of little value. It has been found in immense crystals in the vicinity of Paradise River, also near Bridgetown and Lawrencetown, Annapolis County, N. S., from a light yellow color to the dark, smoky "cairngorm." Dr. How mentions a crystal 13 inches high and 6 inches in diameter. Single crystals, weighing 100 pounds each, have often been obtained from the decomposing granite and have been piled up with the stones from the fields, near Paradise River, and loose in the soil. It occurs in crystals about 2 inches in length at Mill Village, Lunenburgh County, N. S., and at Margaret's Bay, Halifax County, N. S. In King's College cabinet there is a specimen of the almost black variety known as " Morion," with crystals ½ an inch across.

Amethyst is found in some form in nearly every vein cutting the cherty and argillaceous slates around Thunder Bay, on the north shore of Lake Superior. At Amethyst Harbor this mineral constitutes almost the entire vein, and numerous openings have been made to obtain it for tourists who visit the spot. Thousands of dollars' worth are annually sold here, and as much more is sent to Niagara Falls, Pike's Peak, Hot Springs, and other tourists' resorts, as well as to the mineral dealers. Surfaces several feet across are often covered with crystals from ¼ inch to 5 inches long, rich in color, and having a high polish. Sometimes, when large, the crystals have a coating of a rusty brown color, owing to the oxidation of the included göthite. This is one of the famous occurrences of this mineral, regarded as natural specimens, but the purple color is very unevenly distributed, and as the crystals are not transparent like those from Siberia, they afford very few gem-stones of value. In Nova Scotia, fine amethysts occur in bands, veins, and geodes at Partridge Island, Cumberland County, surfaces a foot square being covered with splendid purple crystals 1 inch across. Dr. Gesner mentions a geode that would hold about two gallons, found at Cape Sharp, nearly opposite Blomidon, N. S. Another, lining walls of chalcedony with concentric bandings, and weighing 40 pounds, was found at Sandy Cove, Digby County, N. S. De Monts is said to have taken crystals from Partridge Island to Henry IV. of France, whom they greatly pleased, and a crystal from Blomidon was among the French crown jewels twenty years ago. A bushel of crystals was obtained by Dr. Webster, of Kentville, N. S., in digging a well. Dr. Gesner also states that he had seen a band of amethyst some feet in length and perhaps 2 inches thick, about a mile east of Hall's Harbor, N. S. Other localities are the south side of Nichols Mountain, Cape d'Or, Mink Cove, Scott's Bay, in Nova Scotia, and Little Dipper Harbor and Nerepis in New Brunswick, and elsewhere along the Bay of Fundy. The beautiful masses of straight, concentric, and irregular banded amethyst (banded with quartz and agates) found in Nova Scotia on the Bay of Fundy, are similar to a variety found abroad, and used for ornamental purposes, principally for clock-cases and jewel-caskets. The material is slit into plates so thin that they are often strengthened by cementing them on plates of glass, and the colors are enhanced by setting the plates so that the light can pass through. Dr. How mentions prase, green quartz, as occurring at Kail's Point, N. S. A beautiful hyaline quartz is found at Scott's Bay, N. S. Sagenite (flèche d'amour, or Venus' hair-stone) is reported by Dr. How as having been found at Scott's Bay, N. S.

Agates are found along the coast of Lake Superior in abundance and of considerable size and beauty. The finest are

derived from the trap of Michipicoten Island, Ont. They occur
on St. Ignace and Simpson's Islands, Ont., on the former only as
nodules in the trap. Both chalcedony and agate occur as veins
filling dislocations and cracks which penetrate the trap. In the
Thunder Bay district they are associated with amethysts, occurring
also as pebbles. Although these agates are often of rich color,
and are beautifully veined, they are rarely over 2 inches across.
Many are sold to tourists for ornaments, and a greater number
could probably be disposed of if more attention were given to
cutting and polishing them. As natural agates, their color is exceptionally fine. Nearly all the large agates sold in this region
are from abroad as well as of foreign coloring and cutting. Agate
pebbles known as Gaspé Pebbles are found in the conglomerate
of the Bonaventure formation, on the Baie des Chaleurs, Que.,
and along the shore of Lake Superior, in the vicinity of Goular's
Bay, and especially on the St. Mary's River. Handsome agate
and chalcedony in nodules and veins are of frequent occurrence
on the south shore of the Bay of Fundy, between Digby and
Scott's Bay, N. S. Large masses of agate have frequently been
found on this coast. Gesner mentions a mass of 40 pounds
weight, made up of curved layers of white, semi-transparent
chalcedony and red carnelian, forming a fine sardonyx. A mass
showing distinct parallel zones of cacholong, white chalcedony,
and red carnelian, was found a few miles east of Cape Split, N. S.
When polished it resembles an aggregation of circular eyes, and
hence the name eye-stone, or eye-agate, is applied to it. At Scott's
Bay, N. S., large surfaces of rocks are studded with these minerals.
Specimens are also found at Blomidon and at Partridge Island,
N. S. Fine agates and carnelians occur at Digby Neck, Woodworth's Cove, and at Cape Blomidon, N. S. Agate, chalcedony,
and carnelian are also found in New Brunswick, at Darling Lake,
in Hampton, near the mouth of the Washdemoak River, in Dal-
housie, and on the Tobique River, in Victoria County. A blue
chalcedony, rich brownish-green by transmitted light, is mentioned by How, from Cape Blomidon, N. S. Agate often occurs
in layers, forming an onyx, in the Bay of Fundy and Lake
Superior regions. Beautiful specimens are found at Two
Islands, Cumberland County, near Cape Split, at Scott's Bay

and at Parrsborough, N. S. In the Queen Charlotte Islands, B. C., they occur abundantly at some localities, being derived from the miocene-tertiary rock.

Beautiful moss agates are found at Two Islands, Cumberland County, and near Cape Split, Partridge Island, also at Scott's Bay, Kings County, N. S., where they are exceptionally fine. Chrysoprase of fair color has been found in the Hudson's Bay district, on Belanger's Island.

Silicified woods are found in the northwest Territories and in British Columbia.

Jasper conglomerate exists in mountain masses, along with the quartzite masses of the Huronian series, for miles in the country north of the Bruce Mines, on Lake Superior north of Goular's Bay, on the St. Mary's River about four miles west of Campment d'Ours, on the east shore of Lake George, and on Lake Huron, Ont. It is a rock consisting of a matrix of white quartzite, in which are pebbles, often several inches across, of a rich red, yellow, green, or black jasper, and smoky or other colored chalcedony, which form a remarkably striking contrast with the pure white matrix. It is susceptible of a high polish, and has been made into a great variety of ornamental objects. Some very beautiful mosaics have been produced by using the rock and included pebbles. It occurs in thick bands extending for miles, and in large boulders, scattered along the shores of the lakes and rivers. Within half a mile of the northern extremity of Goular's Bay, Ont., is a ridge containing several varieties. Large quantities of rich, red jasper are found in Hull, Que. Yellow and red occur at Handley Mountains, Annapolis, Pictou, Gulliver's Hope, Blomidon, N. S.; at Belleisle Bay, Kings County, Grand Manan, Darling's Lake, and Hampton, near the mouth of the Washdemoak River; at Red Head and at the Tobique River, Victoria County, N. B.; at Woodworth's Cove, west of Scott's Bay, and all along the shore of the Bay of Fundy from Sandy Cove, N. S. Near the head of St. Mary's Bay lie large blocks of red, yellow, and yellowish-red jasper, often banded, but generally impure.

Heliotrope (bloodstone) in good specimens is of rare occurrence in the North Mountain, Bay of Fundy, N. S.

Dr. Gesner mentions finding two small nodules of opal, of a waxy color, at Partridge Island, N. S. Semi-opal has been found at Partridge Island in fine specimens, also at Grand Manan, N. B., and other localities in that vicinity.

Cacholong has been found associated with chalcedony in Nova Scotia on the Bay of Fundy. The hornstone found at Partridge Island admits of a fine polish and is of some use as an ornamental stone.

Jade (nephrite), in the form of archæological implements, has been found from the Straits of Fuca northward along the entire coast of British Columbia and the northern end of Alaska.[1] At the latter place it is closely allied with other minerals, such as the new form of pectolite, and is found, with other relics of various kinds, about shell heaps and old village sites, in graves, or still preserved, although seldom used, by the natives. It is also found as far inland as the second mountain system of the Cordillera belts, represented by the Gold, Cariboo, and other ranges, principally among remains from Indian graves, and along the lower portions of the Fraser and Thompson Rivers, within the territory of the Selish people. It is less common in the interior of the province, which Dr. Dawson accounts for in part by the facts that adzes or adze-like tools had not been so much employed by the Indians of the interior and by those of the coast, who are pre-eminent as dextrous workers in wood and noted for the size and superior construction of their wooden houses and canoes; and that, previous to the introduction of iron tools among the Eskimos and Indians, the use of jade must have been much more frequent, so much so as to preclude the theory of its having been obtained in trade from remote sources. The Indians of the west coast, although they value the jade, have for it no superstitious or sentimental feeling. The finding of two partly worked small boulders of jade on the lower part of the Frazer River, at Lytton and Yale, B.C., respectively, and the discovery of unfinished objects in old Indian graves near Lytton, make it certain that the manufacture of adzes had been actually carried on there. A series of specimens, numbering sixty-one in all, have been deposited in

[1] On the Occurrence of Jade in British Columbia, by Dr. George M. Dawson. Canadian Record of Science, Vol. 2, No. 6. April, 1887.

the Museum of the Geological Survey of Canada at Ottawa and in the Peter Redpath Museum, McGill College, Montreal. These consist of both jade and pectolite articles, in the form of adzes, drill-points of borers, cut boulders, sockets for fire-drills, mallets, axes, pendants, and burnishers. Of the sixty-one objects found, seventeen show evidence more or less distinct of having been sawn from other pieces. Nordenskiold[1] describes figures and a broken harpoon-point of bone and nephrite, from Point Clarence, 65° north latitude, north of Norton Sound. Dr. Dawson says: "It is among the highly altered and decomposed rocks of the Carboniferous and Triassic that silicates of the jade class might be expected to occur, and I feel little doubt that when these rocks are carefully investigated they will be found to be the sources of the jade." The Indians of the region, however, have usually, if not invariably, obtained their supply from loose fragments and boulders. Jade is also reported from the Rae River and from the Hudson Bay district.

Axinite in fine crystals was reported by Dr. Bigsby from a boulder of primitive rock in Hawksbury, near Ottawa.

Epidote is found in many localities, though not in gem form, except when with flesh-colored feldspar in the amygdaloid trap on Lake Superior. This has been polished to form an odd ornamental stone. At the falls of the Mingam River, Que., and in Ramsay Township, Ont., is found a peculiar, fine-grained, reddish gneiss, traversed by veins of a pea-green epidote. It is very beautiful when polished. Pale-green epidote with quartz is found on the Matane River. The epidote which forms mountain masses in the Shickshock Mountains, Que., is hard, susceptible of a high polish, and would be of value as an ornamental stone.

Amazonstone (microcline) has been found in Sebastopol, Ont., and in Hull, Que., in cleavages of good color.

Labradorite, the most beautiful of all the chatoyant feldspars, exists in great quantities on the coast of Labrador, especially at Nain, and on St. Paul's Island adjacent to it, where the finest known occurs in veins of some size, where for over a century it has been mined for use in the arts. It occurs on Lake Huron, Ont., at Cape Mahul, and at Abercombie, Que., also in cleavages

[1] Voyage of the Vega, Vol. 2, 1882.

several inches in diameter and of rich color, showing blue opalescence, at Morin, Que.

The beautiful variety of albite called peristerite, exhibiting a peculiar bluish chatoyancy or opalescence, is sometimes mingled with pale green and yellow, and called "moonstone." It is found in crystals and in large cleavable masses, containing disseminated grains of quartz, in veins cutting the Laurentian strata at Bathurst, Ont., on the north side of Stony Lake, near the mouth of Eel Creek; in Burleigh, Ont., in crystals, in large, opalescent, cleavable masses of reddish albite, and north of Perth, Ont. It is also reported by Mr. G. Christian Hoffmann, of the Canadian Geological Survey, in specimens showing beautiful blue color, from Villeneuve, Ottawa County, Que.

Perthite occurs in large cleavable masses in thick pegmatite veins, cutting the Laurentian strata, and is often made up of flesh-red and reddish-brown bands of orthoclase and albite, interlaminated. When cut in certain directions, it shows beautiful golden reflections like aventurine, and being susceptible of a high polish, is adapted for an ornamental stone or for use in jewelry. It is also found in considerable quantity at Burgess, Ont., about seven miles southwest of the town of Perth, and likewise near Little Adams Lake.

Sunstone, aventurine feldspar, has been described by Dr. Bigsby in the form of a largely crystallized flesh-red feldspar, constituting part of a granitic vein traversing gneiss, twenty miles east of the French River, on the northeast shore of Lake Huron, and occurs in fine specimens at Sebastopol, Ont.

Obsidian has been found in British Columbia, but it has little value except for the cheaper jewelry, and even then is rarely used for such purposes.

The porphyries which cut the Laurentian limestones in the townships of Grenville and Chatham, Que., form a dike running east and west twenty feet in breadth. They have a dark-green or brownish-black base, homogeneous and compact, containing crystals of red orthoclase, and admitting of a high polish, which strongly recommends the material for ornamental use.

The pegmatite at Montgomery's Clearing on Allumette Lake, five miles above Pembroke, Ont., consisting of a brownish-red orthoclase with white quartz, is a beautiful ornamental stone, and admits of a good polish.

Idocrase occurs in wax-yellow crystals imbedded in limestone, in Grenville, Que., in crystals of remarkable perfection and rich brown color; in a white calcite, near Wakefield, Que.; on Frye's Island, N. B.; and in large brown crystals at Calumet Falls, Que. Some of these would cut small gems, for which there is a slight demand to represent the initial I in sentimental jewelry.

Pyrite is found in many localities, but nowhere in great perfection. It was extensively cut and polished for ornaments a century ago, but has been largely superseded by the more recent introduction of steel jewelry.

Hematite occurs, finely crystallized, at Cape Spencer, and exceptionally perfect and brilliant at Digby Neck, N. S., Sussex, Kings County, and Black River, St. John County, N. B. This fibrous form of red oxide of iron is extensively worked into jewelry in England and Germany; but it has not been found of sufficient value in Canada to warrant working, as it can be cut so much more cheaply abroad. All the hematite jewelry of the Lake Superior region is believed to be foreign, not only in workmanship, but in material.

Although olivine, chrysolite, or peridot, is found in a number of localities as a rock constituent, and often in the form of imperfect olive and amber-colored crystals, $\frac{1}{4}$ an inch in diameter, at Mount Royal, Montarville, Mount Albert, and Rougemont, Que., it has not yet been observed of sufficient clearness and perfection to afford gems of any value.

The andalusite, found on Lake St. Francis, Que., in small, flesh-red prisms not exceeding $\frac{1}{16}$ inch in diameter, and also in black crystals and the variety known as chiastolite, macle, or cross stone, is sold abroad for use in jewelry. It also occurs at Guysborough, N. S., in fair macles.

Of the deep chrome-green pyroxene found at Orford, Que., many fine crystals have been found. Occasionally they are transparent and would afford gems. The lilac-colored variety from Grenville, Que., does not admit of a fine polish.

Staurolite has been found at several localities in Nova Scotia, more especially at Guysborough.   This mineral, when in perfect crosses, finds some sale in Switzerland for charms.

Diopside is found as a rock-constituent in many localities in the Laurentian area.   At Calumet Falls, Que., it occurs in crystals 6 inches long, though not of gem value.

Scapolite or wernerite occurs in large cleavable masses in a limestone at Grenville, Que., and Bathurst, Ont.   When free from the lilac-colored crystals of pyroxene with which it is associated, it admits of a good polish, but is of little value, if any, in the arts.

Ilvaite has been found in a boulder about a foot in diameter in the vicinity of Ottawa, Ont., and is believed to form a bed in the Laurentian series.   It has little value as a gem, but is occasionally used for the letter I in sentimental jewelry.

Sodalite in fine blue grains has been found in the granite of Brome, Que., in seams at Montreal, Que., and in veins several inches wide on the line of the Canadian Pacific Railway, at Kicking Horse Pass, B. C., by Dr. B. J. Harrington.   It is occasionally used in the arts.

Lazulite is reported from the Hudson Bay district, but of little gem value even when of fine color.

Prehnite is associated with native copper and calcite in the Lake Superior region, where it is often of a rich green color, in spherical masses of crystals an inch across, or in aggregations even larger, affording a curious green stone resembling a chrysoprase. Fine specimens occur at Clifton, Clark's Head, and Black Rock, Kings County, N. S.

The titanites of Canada have a world-wide reputation, not only for their color, their polish, and the perfection of the crystals, but also for their great size.   A twin crystal of this mineral has been found on Turner's Island, in Lake Clear, weighing 80 pounds.   They occur abundantly in this region, associated with apatite.   The crystals are generally of such deep brown color as to appear black.   It is rare that even a small transparent gem could be cut from them; as crystals, however, they are unexcelled, and many thousand dollars' worth have been sold as specimens.   The finest are found in Renfrew County, especially in Sebastopol and Brudenell Townships, Ont.

Zonochlorite, said by Hawes to be a chemically impure variety of prehnite, is yet distinctive enough as a gem-stone to entitle it to its name. It occurs in small rolled masses, and in the rock, at Nipigon Bay, Ont., and was described by Prof. A. E. Foote. It is a dark, opaque, green stone, beautifully marked and veined, and admitting of a high polish.

Thomsonite of a red color, compact and fibrous, often banded with green in a number of concentric rings, is found on the northern shore of Lake Superior, Ont., and at Cape Split, N. S. The pebbles vary in size from ⅛ inch up to an inch across, and are quite extensively sold on all sides of the lake as an ornamental stone. The green which Peckham and Hall (see p. 181) described as lintonite, an uncrystalline green variety of thomsonite, often forms the center or band, making an effective gem-stone, and is sold for that purpose.

The ilmenite in the parish of St. Urbain, at Baie St. Paul, sometimes contains grains of a greenish triclinic feldspar, and would furnish an ornamental stone similar to the porphyritic menaccanite found at Cumberland, R. I. It also contains rutile crystals, too small to have value as gems, though adding to the beauty of the material when polished.

Natrolite is found in stout crystals with other zeolites at Peter's Point and other localities on the Bay of Fundy, and at Swan's Creek, Cape Blomidon, and Partridge Island, N. S. When transparent and of sufficient size, it is occasionally used as a gem to represent the initial N in sentimental jewelry.

Apophyllite is often found along the coast of Nova Scotia, on the Bay of Fundy, principally at Cape d'Or, Haute Island, Partridge Island, and Swan's Creek just above Cape Blomidon, in magnificent crystals sometimes an inch or more across. It occasionally occurs on agate and amethyst in the trap rock, and would afford a mineralogical gem, as its pearly lustre produces a curious effect, like that of a fish's eye; hence the name ichthyophthalmite, or fish-eye stone. The color is generally white, but occasionally the crystals have a rich green tinge.

Hoffmann has described a part of a crystal of monazite, weighing 14 pounds, from Villeneuve, Ottawa County, making this one of the most remarkable occurrences known. If transparent, it would afford a hyacinth-yellow gem, rather low in hardness.

Apatite, which has added so much to the mining industry of the Dominion, is found there in greater quantity and in finer crystals than in any other country.   The crystals are often of great size and perfection, one famous crystal from the Emerald Mine, at Buckingham, Que., weighing 550 pounds.   Magnificent crystals are found throughout eastern Ontario, on the shores of Lake Clear, several feet in length and of fine color; at Sebastopol and elsewhere throughout Renfrew County; and at Wakefield, Templeton, Portland, and Buckingham Townships, Ottawa County, Que.   The crystals are often partly transparent, and are of all shades of red-brown, brick-red, and often rich, deep green, especially in Ottawa County, in which case they should be adapted to some of the uses of fluorite as ornamental stones.

Wilsonite is found in Bathurst and Burgess, Ont., and Ottawa County, Que., in masses of some size, associated with scapolite.   The specimens are beautiful, the minerals often passing into each other.   The rich, purplish-red color of this mineral, and the fact that it admits of a good polish, make it one of the most interesting of gem minerals.

Fluorite is occasionally found in purple crystals measuring several inches on a face, associated with and on the Lake Superior amethyst. Green and purple fluor often fills mineral veins in the Lake Superior region, and veins in syenite opposite Pic Island, on the mainland.   On an island near Gravelly Point, in a porphyry, it occurs in green octahedral crystals, with barite; in green cubes associated with calcite and quartz at Prince's Mine, Ont.; and in small, beautiful crystals near Hull, Que.   Fluor spar of a beautiful blue color is found at Plaster Cove, Richmond County, N. S., and also on the west side of the harbor of Great St. Lawrence, Newfoundland.   Small purple crystals of great beauty are occasionally found on pearl-spar in the geodes at Niagara Falls, Ont., and elsewhere in the Niagara formation.   A green, compact variety occurs in white calcite associated with galena, in veins cutting the Potsdam sandstone, at Baie St. Paul and Murray Bay, Que., which would work into an ornamental stone.   It is found frequently all through the Laurentian rocks.   It is rarely cut into mineralogical gems, but when compact, of good color, or beautifully veined, it is worked into vases, cups, and other ornamental objects, known as Blue John, Derbyshire Spar, and similar names.

Malachite of gem value has not been found to any extent in Canada, although the species occurs in nearly every locality where copper and its ores are obtained.   It has also been observed at Sutton, Que.

The agalmatolite found in Canada is not of such quality as to fit it for the uses of the Chinese figure-stone.

Jet is found at Pictou, Pictou County, N. S., in fine pieces.   It has been very generally superseded in jewelry by black onyx, and the little now used is mined at Whitby, Eng., owing to the superior hardness of that found there, and the perfect facilities for working it.

What Canada has produced in precious and ornamental stones was well shown at the Centennial Exhibition, Philadelphia, 1876, and at the Colonial and Indian Exhibition at London, in 1886.   The fine minerals have found their way into the well-arranged collection of the Geological Survey of Canada at Ottawa, the British Museum, the mineralogical collection of McGill College, which contains the cabinet of John G. Miller, and the Provincial Museum of Nova Scotia.   Many of the finest specimens, in full series, grace the cabinets of Clarence S. Bement, at Philadelphia, King's College, at Windsor, N. S., the School of Mines, New York (which contains the collection of Dr. Henry How), Walter G. Ferrier, Montreal, W. J. Wilcox (deposited at the Wagner Institute, Philadelphia), Amherst College, at Amherst, Mass., Prof. Othniel C. Marsh, New Haven, Conn., and the New York State Museum, at Albany, N. Y.

Further reference to this subject can be found in the following works: "Remarks on the Mineralogy and Geology of the Peninsula of Nova Scotia," by Charles T. Jackson and Francis Alger (Cambridge, 1832); "Geology and Mineralogy of Nova Scotia," by Abraham Gesner (Halifax, 1836); "Catalogue of the Mineral Localities of New Brunswick, Nova Scotia, and Newfoundland," Am. J. Sci. II., Vol. 35, 1863; "Mineralogy of Nova Scotia," by Henry How (Halifax, N. S., 1868); "Geology of Canada : Report of the Geological Survey from its Commencement to 1863 " (Montreal, 1863), XXVII., 983 ; " The Mineral Resources of the Dominion of Canada" (Ottawa) ; " Descriptive Catalogue of a Collection of Economic Minerals of Canada at the Philadelphia International Exhibition" (Montreal, 1876) ; " The Woods and Minerals of New Brunswick at the Centennial Exhibition at Philadelphia," by L. W. Bailey and Edward Jack (Fredericton, N. B., 1876) ; "Dana's Mineralogy" (New York, 1875) ; " Descriptive Catalogue of a Collection of the Economic Minerals of Canada at the Colonial and Indian Exhibition, London, 1886," by the Geological Corps, Alfred R. C. Selwyn, Director (London, 1886) ; Mineral Wealth of British Columbia, Dr. George W. Dawson (Ottawa, 1889) ; Mineral Statistics of Canada, 1887, Eugene Coste (Ottawa, Canada).

CHAPTER XIV.

In Mexico and Central America.

OUR knowledge regarding the precious and ornamental stones of Mexico and Central America is very meagre, especially when one considers the extent of the territory and the richness of the mineral wealth that undoubtedly exists there. The fullest information on these subjects is furnished by Santiago Ramirez,[1] from whose book we shall have occasion to take many extracts in the sequel. We have his statement that the diamond has been found in Mexico, but the description is altogether too unsatisfactory to establish it positively. The story he gives, on the authority of Señor Del Moral,[2] is this: General Guerrero, while searching with a few soldiers for a suitable camping-ground, found what appeared to be large pebbles, some of which, on being broken open, proved to be hollow geodes, and to contain loose, brilliant crystals. Two of these, which had been given by the General to a lady friend, and had been mounted as ear-rings, were, about the latter part of 1822, shown by her to Professor Del Rio, who is said to have pronounced them diamonds, octahedral in form, and of a quality not inferior to any from India or Brazil. Señor Guillow, a lapidary of the City of Mexico, is said by the same author to have bought a number of these diamonds that to-

---

[1] Noticia História de la Minerca de Mexico (Mexico, 1884–1885), pp. 237–250.
[2] La Naturaleza, Vol. 2, pp. 257–302, 1873.

gether in their rough state weighed 18 carats.   The largest of
these, which weighed 3 carats, he presented to the Mining College Museum.   Another crystal, that weighed 2 carats, General
Guerrero kept for himself.   As this is the only reference to the
finding of diamonds in that country, it is to be regretted that
nothing further is known regarding it.   It is, however, highly
probable that if, as is represented, the crystals were found in
goedes, they were not diamonds, but quartz crystals.

｜Sapphire and ruby of value as precious stones have never
been reported from Mexico, but among a number of rolled pebbles of jasper, agate, and chalcedony, that were found near San
Geronimo, Oaxaca, Mexico, near the Isthmus of. Tehuantepec,
and brought to the writer by Dr. Knight Neftel, of New York,
for examination as to their gem value, a rolled pebble of sapphire was found. In color it was mottled blue and yellowish-white.
It 'was slightly fissured, and translucent, and did not show crystalline form, although the rolling seemed to have brought out the
cleavage more distinctly, and this reflected a fine pearly lustre.
It weighed 19·223 grams and had a specific gravity of 3·9.   This
is low, but may be due to the impurities in the veining.   From
this single pebble, it would be impossible to decide whether gems
occur in that region, but further investigation will determine this.
It is not improbable that corundum may exist there in quantities
large enough to be of commercial value.   And since it is in this
very State that so many jadeite objects with aboriginal carvings
have been found, it may be that sapphire, perhaps from the
same locality, was used in the slitting, drilling, and cutting of
them.   Ruby is said to occur in Durango, and also in Secom,
near the Falls of California ; but whether the stone so reported is
ruby or only garnet, it has been impossible definitely to determine.
Emerald is found according to D. Ignacio Alcocer, member of
the Scientific Commission of Mexico, in the vicinity of Tulancingo in the State of Hidalgo.   No statement is made as to the
quality, and the name may have been applied to common beryl,
which has been reported from the State of Hidalgo, Tajupilco,
and on the hill of Cerro Gordo in Guanajuato.

Fine pyrope garnets, similar to those found in Arizona and
New Mexico, are found in Chihuahua, especially near Lake Jaco.

These are often gathered by the Comanche Indians. Pyrope garnet is also reported from the State of Sonora. At Trumfo, Lower California, beautiful garnets, in crystals of ¼ to ½ inch in diameter, are said to occur in a white rock. These are probably not of value. A. B. Damour describes essonite, with a specific gravity of 3·57, in light-red dodecahedral crystals in a granular limestone, from Rancho de San Juan.

Iolite or dichroite is reported as associated with beryl in Tajupilco, in the State of Hidalgo, and it is said to occur on the hill of Cerro Gordo in Guanajuato.

The name "jade" is popularly applied to several distinct ornamental stones, but since A. B. Damour's investigations into the character of jade, jadeite, and chlormelanite, the term has been restricted mineralogically to what is specifically known as nephrite. Many bric-à-brac dealers never distinguish between jade and jadeite, calling both simply jade.

The word "jade" is evidently a corruption of the Spanish "ijada," since the mineral is first mentioned under this name in the writings of Monardas in 1565, and was brought from Mexico and Peru under the name "Piedra de ijada," or "Stone of the Loins," in allusion to its supposed curative properties in diseases of the loins and kidneys. Amulets of jade-like minerals have always been highly venerated by the natives throughout Central America, Mexico, and Peru.

Jade, or nephrite, is a silicate of calcium and magnesium. It has a specific gravity varying from 2·7 to 2·9, a hardness of 6·5, and is extremely tough, more so than jadeite. It may be described as a cryptocrystalline variety of hornblende, exhibiting no crystalline form or cleavage, with a splintery fracture, like horn. The color is generally uniform, commonly either white or green, occasionally yellow or brown, very rarely with a bluish or pink tint. Jadeite is a silicate of aluminum and sodium. Its specific gravity ranges from 3·25 to 3·35, and it has a hardness of 7. It has a crystalline structure, is not as tough as jade, and its composition places it nearer to epidote than to hornblende. Its lustre is more brilliant than that of jade. It is generally white, occasionally greenish, with veins or spots of almost emerald-green color, also lettuce-green and sage-green. In 1865, Damour, who originally described jadeite, also described chlormelanite, another substance resembling jade. This material contains a larger percentage of iron than jadeite, and its specific gravity is higher, ranging from 3·4 to 3·65. The color is generally blackish green, spinach or sage green, marked with patches of lighter shades. It is found enclosing garnet with iron pyrite.

Prof. Heinrich Fischer of Freiburg, Baden, devoted his entire life to the study of the literature, archæology, and examination of jade, jadeite, chlormelanite, and allied minerals. The results of this study were published in his "Nephrit und Jadeit" (Stuttgart, 1875), in which he shows that these and other green stones have been called by 150 different names, which he gives, as well as a chronologically arranged table of the literature on the subject from the earliest time to his death. The United States National Museum has a large and fine series of jadeite objects from Mexico, nearly all of which are from the State of Oaxaca. The Museum is also well supplied with jadeite objects from Central America, only a few of which are from Nicaragua and Guatemala. The finest are from Costa Rica. With these jadeites are many articles of softer green stones, and occasionally an object of quartz or chalcedony.[1]

Among the more remarkable jadeite objects of Mexican origin is an adze described by the author, believed to be the largest yet found. On its face is figured a grotesque human figure, and for so hard a material, the workmanship is excellent. It is said to have been found about twenty years ago in Oaxaca, Mexico. It measures 272 millimeters (10 11/16 inches) in length, 153 millimeters (6 inches) in width, and 118 millimeters (4⅝ inches) in thickness, and weighs 229·3 Troy ounces. Across the ears 153 millimeters (6 inches), across the lower axe end 82 millimeters (3¼ inches), height of head to neck 158 millimeters (6¼ inches), height from chin to foot 115 millimeters (4½ inches), and the legs 50 millimeters (2 inches). From the back a piece about 160 millimeters (6¼ inches) long and 50 millimeters (2 inches) wide has been removed. The color is light grayish-green with a tinge of blue, and streaks of an almost emerald-green on the back. In style of ornamenta-

[1] On Nephrite and Jadeite by Prof. Frank W. Clarke and George P. Merrill. Proceedings of the United States National Museum (1888), Vol. 11, p. 115, et seq.

tion it very closely resembles a gigantic adze of granite, 57 centimeters long and 34 centimeters wide, mentioned by A. Chavero,[1] and it has almost a counterpart in the green aventurine quartz adze now forming part of the Christy Collection at the British Museum, and formerly in the possession of Percy Doyle of the British Diplomatic Service, differing from these two objects, however, in having no ornamentation on the forehead, and in having four dull markings on each ear, one under each eye, and one near each hand, which seemingly could have served no other purpose than to hold thin plates or films of gold, which the polished surfaces would not do. If this was so, no trace of the gold can now be seen. From all appearances, this adze was shaped from a boulder, since weathered surfaces, such as appear on all sides of it, would be found only on an exposed fragment. The lapidarian work on this piece is probably equal to anything that has been found, and the polish is as fine as that of modern times. One point of interest, which should not be lost sight of, is the removal of a portion which has weighed fully two pounds. Why was this? Similar removals and divisions have been mentioned. On April 27, 1881, in a paper read before the American Antiquarian Society, Philip J. J. Valentini described two carved jadeites which showed similar treatment. One was the Humboldt celt, a votive adze presented to Humboldt by Del Rio in 1803, and the other, the so-called Leyden plate, which was found by S. A. von Braam near St. Félipe, in Honduras, near the borders of Guatemala, and given by him to the Leyden Museum. Both of these objects are 9 inches in length and 3¼ inches in breadth, the former having a thickness of 1⅜ inches and the latter ¼ of an inch. From the fact that the two, if placed together, face to face, have exactly the same outline, it is highly probable that they were originally part of one and the same celt, and it is quite possible the remaining parts may yet be found. In 1886, Professor Frederick W. Putnam exhibited before the same Society a remarkable series of Nicaraguan and Costa Rican jadeites, which were all ornaments, and showed that they had been made by cutting celts which had been perforated by one or two drilled holes, into halves, thirds, and

[1] Mexico à travès de Los Siglos, Mexico, 1886, p. 64.

quarters, in one instance, two of them fitting together. The explanation offered is, that the supply of the material having become exhausted, recourse was had to division, or a removal of a part from existing objects, evidently for the purpose of making others, perhaps to be buried with some dead chief, or to be bestowed on new branches of the tribe, an object held sacred. Fully one-eighth was removed from the back of this adze, and the manner in which the instrument used in the removal was held has produced a rounded cut on each side, lending probability to the supposition that some abrasive was employed, drawn with a string held in the hands, or stretched across a bow. If the Aztecs knew of the existence of this sapphire, we can more readily understand how they worked so large a mass of tough and hard material. So far as the writer has been able to ascertain, no similar object of equal magnitude and archæological interest exists. Neither the Humboldt celt, the Leyden plate, the Vienna adze, nor the one in the Ethnological Museum at Dresden, which weighs only seven pounds, and is entirely devoid of ornamentation, can compare with it.

A jadeite breastplate was obtained in 1884, by a German engineer, from a tomb near Santa Lucia, Cotzulmaguapa, where Dr. Behrendt had made some extensive excavations and obtained a quantity of large engraved stones and other antiquities from the old temples and tombs of the ancient kings of Quiché, which exist in that neighborhood. It is 16 centimeters (6¼ inches) wide, about 12 centimeters (5 inches) high, 1 centimeter (⅜ inch) thick. The color round the edges is a grayish-green, while on the outside, at the center, it is a light rusty brown, perhaps from burning. By transmitted light, the color is a light apple-green. It has been drilled at two places on the back edges with holes 4 millimeters (⅙ inch) in width, and has been sliced or cut from some boulder, as the back edges show. At one place, there are evidences of an attempt to slit it. According to Dr. Valentini, the cutting represents a human face or mask, or rather the headgear of a man, representing the symbol Achau, meaning " Ross " or " Lord ;" the head of the tribe (one of the most common motives of the Maya, which is found at least a thousand times drawn and colored in the Maya codices) forming the walls and friezes of their structures. Achau is also the name of the nineteenth day of the Maya month. Each of the eyes is represented by a circle with two flattened sides. Below these is a beard or tattooing. A circle with a central dot represents a mouth, and the nose is an oblong between the eyes, extending below the tattooing. From the ears, which are quite natural, are suspended feather pendants. Feathers also cover the top of the head, and probably ornament the chin as well.

In 1879 Mrs. Erminnie A. Smith exhibited an interesting jadeite mask, having a specific gravity of 3·3 at the Saratoga

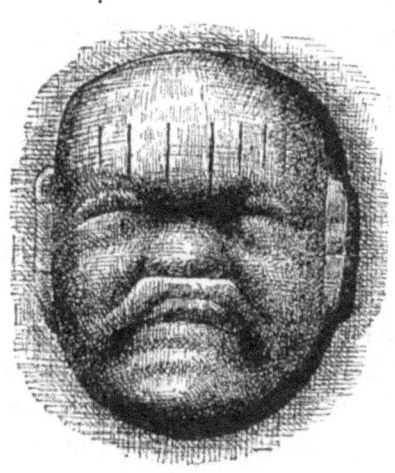

FIG. 13.
JADEITE MASK.

meeting of the American Association.[1] It represented a crying baby-face (see Fig. 13.) and is almost identical with one made of quartzite in the Peabody Museum of Archæology at Cambridge, Mass.

The Codex Mendoza, a copy of the tribute-roll of the ancient Mexican Empire published in Lord Kingsborough's "Antiquities of Mexico" (London, 1830), defines the tax from each district, naming the cities. Strings of chalchihuitl are mentioned as part of the tribute from a number of localities, and refer evidently to small rounded pieces used as beads, and obtained from the sands of streams. Only from one district were large pieces of chalchihuitl demanded. These, three in number each year, were required from Totoltepec, Chinantlan, and other towns situated in the present State of Oaxaca, and principally in the department of Valalta (Zoochila). Mühlenpfordt describes this region as mountainous and wild, inhabited by the Mixe Indians and the Chinantecas.[2]

Dr. Daniel G. Brinton suggests that in Valalta (Zoochila), in the State of Oaxaca, if jadeite exists in Mexico, it may be found in large pieces, and that this is the locality which the explorer

[1] Proc. Am. Asso. Adv. Sci., 1879, Vol. 28, p. 523.
[2] Schilderung der Republik Mejico, Vol. 2, p. 213.

should penetrate if he would discover the locality of the large pieces of Mexican jadeite [1] or perhaps the mineral in situ.

Bernardino de Sahagun [2] gives the following description of chalchihuitl:

The nahuatl (Mexican) name for jadeite is chalchihuitl. This appears to have been applied to any greenish, partially transparent stone capable of receiving a handsome polish. All such were highly esteemed. Specific distinctions were established between such precious minerals by descriptive adjectives, as follows:

Iztac chalchihuitl, white chalchihuitl; of a fine green, quite transparent, without stripes or stains.

Quetzal chalchihuitl, precious chalchihuitl; white, much transparency, with a slight greenish tinge, somewhat like jasper.

Tlilayotic, literally "of a blackish watery color"; with mingled shades of green and black, partially transparent.

Tolteca-iztli, literally "Toltec knife" or "Toltec obsidian"; of a clear, translucent green, and very beautiful.

It is very evident that this is the so-called Mexican onyx, or Tecali marble or onyx, which exists in Tecali in veins, and is in reality an aragonite stalagmite. Great quantities of it were made into Mexican figures, ornaments, and beads, which are found all the way from northern Mexico down to Oaxaca. This so-called onyx is extensively quarried to this day, forming one of our richest ornamental stones. (See Mexican Onyx.)

Quetzal chalchihuitl is precious chalchihuitl, white, with much transparency, and with a slight greenish tinge, something like jasper. Various green stones exist at present, and were used in considerable abundance in ancient Mexico. Among eight green stone objects, sent to the writer at one time as jadeite, four were jadeite, one was laminated serpentine, another a greenish quartz, and two a mixture of white feldspar and green hornblende. In a string of beads were four pieces of jadeite; but all the others were, as are the jadeite beads, in the form of rounded pebbles, drilled from both sides, and there were nearly a dozen different substances in this string. The question is, are these pebbles a part of the tribute mentioned in the Codex Mendoza?

[1] Science, Vol. 12, p. 168, Oct. 5, 1888.
[2] Historia de la Nueva Espana, Book 11, chap. 8.

If so, they must have existed in some abundance; and they have not been reworked from other objects, as are the larger pieces, like the Costa Rican celts. Can it be that the large pieces came from lower Mexico, and, after use as implements, were bartered, but being green stones, which have been given preference the world over by savages and barbarians, and here were considered precious, were made into votive objects? Among other green stones used by the ancient Mexicans were green jasper, green plasma, serpentine, a fine-graded green shale and the Tecali marble, which was often of such a rich green color that at a glance it might be mistaken for jadeite.

Dr. Heinrich Fischer, who gave much time to the study of this subject, endeavored to prove that the jadeite objects found in Mexico and Central America were of Asiatic origin and were brought to this continent by migration. The facts above mentioned, in connection with the slicing and division of the adze and other objects, implying a scarcity of the material, and the further fact that Burmese jadeite, when green, if exposed to a high temperature, assumes the brownish-green color presented by some of the Mexican objects when subjected to the same process, all tend to support Dr. Fischer's theory. On the other hand Dr. A. B. Meyer, Director of the Ethnological Museum at Dresden, and others, firmly believe in the indigenous character of this material. In support of this the following reasons are advanced: First, large objects such as celts, entirely devoid of ornamentation, are occasionally found. Second, that objects sliced from celts and axes have been found which were subsequently carved and ornamented, as if the beauty and durability of the material had been recognized. Third, that in strings of beads one or more made of jadeite, not forming a central ornament, or put in such a part of the string as to show that they were considered of more importance than the others, but apparently selected for their size, have been found. Fourth, it is also probable that the Mexicans never knew of the existence of the true veins of this mineral, these veins, perhaps, occurring on the summits of some of the higher mountains and the material used by the natives being found in the form of boulders and fragments in the valley below, where it had been transported by the mountain streams.   Fifth, the fact that jade has been found elsewhere on the continent although jadeite has not. In 1884, among a quantity of things sent to the National Museum from Point Barrow, were some hammer-heads, supposed to be jade, but, on analysis by Prof. Frank W. Clarke, found to be a new compact variety of pectolite,[1] with a specific gravity of 2·873. (See Pectolite.) Some early writers have attributed Alaskan nephrite to Siberian sources, but four or five years ago it was determined to be of native origin. The native reports assigned as its source a place known as Jade Mountain, about 150 miles above the mouth of the Kowak River, and after several attempts the spot was visited by Lieut. G. M. Stoney, U. S. N., who collected a series of specimens. The material was of a grayish-green color and splintery lamellar structure, one variety being more granular, brownish in color, and highly foliated in form. Sixth, according to Bernardino de Sahagun, all the green stones of the Aztecs were simply varieties of the chalchihuitl, and it is not improbable, as has been supposed by some, that jadeite, like turquoise, was one of the varieties of chalchihuitl, and perhaps the most prized. This theory has been greatly strengthened during the last ten years, and especially since Professor Frederick W. Putnam exhibited his remarkable series of Nicaraguan and Costa Rican jadeites before the American Antiquarian Society, in April, 1886.

Professor Clarke and George P. Merrill concluded their examination of the Jade Collection in the United States National Museum with the suggestions: Confirmation of the theory that the widely scattered jadeite and nephrite objects were derived from many independent sources, and are of no value whatever in the work of tracing the migration and intercommunication of races, lies in the fact that these substances are comparatively common constituents of matamorphic rocks, and hence liable to be found wherever these rocks occur, so that their presence is as meaningless as would be the presence of a piece of graphite. The natives required a hard, tough substance, capable of receiving and retaining a sharp edge and a polish, and took it wherever it was to be found.

Rock crystal has not, in our time at least, been discovered,

[1] Am. J. Sci. III., Vol. 28, p. 20, Jan., 1884.

in Mexico or Central America, of a quality or in sufficient quantity to be of much use in the arts, yet there have been found a number of interesting prehistoric objects made of rock crystal, —skulls from 1 inch to 7 inches in width, crescents, beads, and other articles,—of which the material is excellent, and the workmanship equal to anything done by the early lapidaries. Small skulls are in the Blake Collection at the United States National Museum, the Douglas Collection, New York, the British Museum, and the Trocadéro Museum. (See Fig. 14.) A large skull, now in

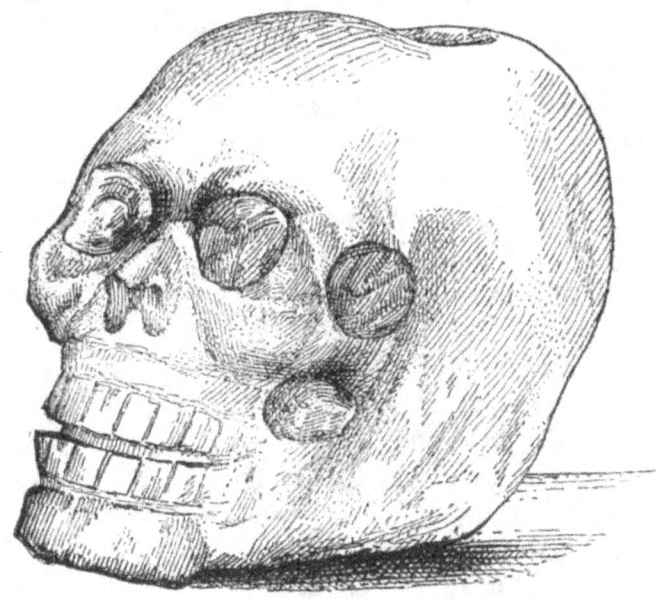

FIG. 14.
ROCK CRYSTAL SKULL.
TROCADÉRO MUSEUM.

the possession of George H. Sisson of New York, is very remarkable. It weighs 475¼ Troy ounces, and measures 210 millimeters or 18⅜ inches in length, 136 millimeters or 15⅜ inches in width, and 148 centimeters or 15¹¹⁄₁₆ inches in height. The eyes are deep hollows ; the line separating the upper from the lower row of teeth has evidently been produced by a wheel made to revolve by a string held in the hand, or possibly by a string stretched across a bow, and is very characteristic of Mexican work. Little is known of its history and nothing of its origin. It was brought from Mexico by a Spanish officer sometime before the French occupation of Mexico, and was sold to an English collector, at whose death it passed into the hands of E. Boban, of Paris, and then became the property of Mr. Sisson. That such large worked objects of rock crystal are not found in Mexico might lead one to infer its possible Chinese or Japanese origin. But it is evident that the workmanship of the skull is not Chinese or Japanese, or nature would have been more closely copied; and if the work were of European origin, it would undoubtedly have been more carefully finished in some minor details. Prof. Edward S. Morse of Salem, Mass., who resided in Japan for several years, and Tatui Baba of Japan, now of New York City, state positively that this skull is not of Japanese origin. Mr. Baba gives as one reason

for his belief that the Japanese would never cut such an object as a skull from so precious a material. In ancient Mexico there was undoubtedly a veneration for skulls, for we find not only small skulls of rock crystal, but real skulls, notably the one in the Christy Collection in the British Museum, incrusted with turquoise, and it may have been one of these that suggested the making of this skull, the one at the Trocadéro Museum, and the smaller one. Two very interesting crescents are

FIG. 15.

CRESCENT OF ROCK CRYSTAL FROM VALLEY OF MEXICO, TROCADÉRO MUSEUM.

known, the one in the Trocadéro Museum (see Fig. 15), the other in the collection of Dr. Maxwell Sommerville, in the Metropolitan Museum of Art, New York City. Beads of this material are sometimes found in the tombs with jadeite and other stone beads. They rarely have a diameter of an inch.

Rock crystal in large masses has been reported from near Pachuca, Hidalgo, in the State of Michoacan, and in veins near La Paz in Lower California; the center of the vein is said to be beautifully pellucid, but the sides are opaque white. It is not known whether the rock crystal used by the aborigines was obtained at a Mexican locality, or whether it came from Calaveras County, Cal., where masses of rock crystal are found containing vermicular prochlorite inclusions identical with those observed in the large skull described above.

The amethysts of Guanajuato, which have a world-wide reputation, are found in large quantities, associated with pink and white apophyllite, and ranging in color from the most delicate pink to the deepest red. The crystals are frequently light in color at the base, but very much darker at the terminations. Groups a foot across, not good enough to cut gems, are frequently found; it is certain that fine amethysts were formerly found at some locality in Mexico, since the collections contain fine objects made by the Aztecs, but not at all resembling the Guanajuato mineral either in color or structure.

Chalcedony, agate, jasper, and the other varieties of quartz undoubtedly exist in abundance at many places in Mexico and Central America, judging from the numbers of objects, such as beads, figures, and ornaments, in the collections. Some finely carved agate figures six inches in length are in the Blake Collection in the United States National Museum, and similar objects exist in the collections of other museums.

The opal,[1] in all its varieties, is found in Mexico and Central America, the noble opal occurring more frequently in Central America than in Mexico. The opal consists principally of silica, differing from quartz, however, in not being crystalline, and in containing from 9 to 12 parts of water in 100. The specific gravity of quartz is 2·65, of opal about 2·2. Quartz has a hardness of 7, and opal of only 6 and even as low as 5·5.

Noble opal is the harder variety, in which the color is uniformly distributed, and ranges from opaque white to almost the pellucidness of glass. Fire opal or girasol is the variety showing flashes of red and yellow, green, and other colors, the opal itself ranging from colorless to white, transparent yellow, reddish-brown to almost opaque, and is usually less hard than the noble opal. The name lechosos is applied by the Mexicans to the variety showing deep-green flashes of color. The name Harlequin is

[1] Whatever may be the origin of the widespread notion of the unluckiness of opal, it is certain that opal was the favorite gem of the Romans, even in their palmiest days. Since it has become known that Queen Victoria is partial to it, the old superstition, which it is said may be traced to Sir Walter Scott's Anne of Geierstein, is slowly yielding, and the gem has gained much public favor during the last ten years.

applied to the variety in which the patches of color are small, angular, variously tinted, but evenly distributed.   Common opal, so-called, exhibits no play of color.   This variety is found of many hues, chiefly, however, milky, rose-colored (when it is called quinzite), and green.

Hyalite, Muller's glass, is the name applied to a colorless, transparent, jelly-like variety, usually occurring in botryoidal masses.   Moss opal is the name applied when the variety contains dendritic marking ; semi-opal when impure ; opal agate and opal jasper when a mixture of either agate and opal or jasper and opal ; opalized wood, when replacing wood ; and hydrophane when it is transparent, or exhibits play of colors on being wet.

The noble opals of Honduras are often exceedingly beautiful, although not as fine or as durable as those from Hungary, which are the finest in the world, and the most valuable.   The Honduras mines are little worked, and the opals only seldom reach the market and generally in an uncut state.   A remarkable specimen of these weighed over a pound, and when cut furnished a quantity of fine stones, some of the finest of which are in the collection of Dr. A. C. Hamlin.

The best description of the Honduras deposits is that given by Dr. John L. LeConte, of Philadelphia, Pa., who, in a report which he made on his return, says : " Extensive beds of common opal and semi-opal are seen along the belts extending through the central part of the Department of Gracias.   The localities worthy of exploration are those in which the opal forms veins, not beds, in compact but brittle trachyte of a dark color.   The veins, as will be seen, are not confined to such rocks, but seem to have their origin in it, and probably are not found except in connection with it.   The best known mines of precious opal are in the Department of Gracias ;   several localities have yielded valuable gems, but they are all remote from the line of road. Some are in the vicinity of the town of Gracias, others near Intibucat ; but the most important ones are at Erandique.   The working is now carried on in a very small way, but the locality is extensive, and in the opinion of Dr. LeConte, mining on a large scale would be attended with profit.

The country near by abounds with common opal, but the

gems occur in somewhat irregular veins running in a northeast and southwest direction, and with a nearly perpendicular dip. The veins are not continuous, but branch off and disappear at short intervals; neither are the contents of uniform quantity, but the valuable parts are usually in belts in the vein, and limited on each side by portions of ordinary opal without play of colors. These lines of light are sometimes numerous and narrow, alternating with the common opal, forming a very beautiful gem. Many again, even of large size, are uniform in structure, and exhibit as brilliant a play of colors as do the finest opals from Hungary. The hill where they are found is about 250 feet high, and two or three miles in length, and for the width of half a mile for its whole length opals have been found wherever excavations have been made. The rock in which they occur is a hard, brittle trachyte with a vitreous lustre, splintering into acute fragments when struck. A bed several feet in thickness overlying this rock, of a gray color and soft consistency, probably a trachyte changed by atmospheric action, also contains opal veins. Other localities within two leagues of Erandique have furnished fine opals, but as they are not now worked, Dr. LeConte did not visit them. Many places between Intibucat and Las Pedras appear favorable to the existence of opal mines, but these can be discovered only by careful scrutiny of a number of explorers. As most worthy of future attention, the vicinity of Le Pasale and of Yucusapa and the ascent of the great mountain of Santa Rosa may be mentioned. Almost certain success will attend the search for opal mines in the valley leading from Tambla towards the pass of Guayoca, nearly on the line of the proposed road. Within half a mile of Tambla are immense beds of common opal of various shades of color. Near Guayoca are banded opals of alternate layers of opaque and semi-transparent white, having the appearance of onyx. They occur in a red, vitreous trachyte, and sometimes in contact with the masses of petrified wood which strew the ground for a considerable distance. Veins of a pearl-colored opal, with red reflections, are also found here. They have no commercial value, but serve as indications of better things in the neighborhood. Between the two localities mentioned, near Tambla and about Guayoca, W. W. Wright, chief assistant of the Survey, by following some obscure indications, discovered a vein of very pretty glassy opals and yellow fire-opals, not of great value, but serving to strengthen the general belief in the ultimate discovery of precious opals in the vicinity. Near Choluteca are found fire-opals, some of which possess merit. One, not of the best, is precisely similar to those obtained by Mr. Wright near Tambla. Within one league of Goascoran is a mine producing opals with a good play of color. Another remarkable deposit of opal was found by Mr. Wright about five miles east of Villa San Antonio in the plains of Comayagua, which though not of high value, may be used for ornamental purposes, being of a fine red color with transparent amethystine bands. It occurs in veins of gray porphyry, sometimes several inches in thickness, and may be procured in large quantities.

Dr. LeConte had a favorable opportunity to purchase a series of fine opals, which he did, and these still remain in the possession of his family. From the fact that parcels of opals are occasionally brought to the large cities by Indians, it may be considered certain that there are many mines in Honduras and other parts of Central America, and future investigation may show that an opal belt exists that extends from Mexico southward perhaps to Central America.

The opals of Mexico are well known throughout the world, although they do not rank in value, and often not in durability, with those from Hungary. Del Rio mentions that in 1802 in Zimapan, near the sanctuary of Guadaloupe Hidalgo, in a red trachytic porphyry, fire-opals were found in abundance, the color of the opal being a hyacinth red; and the same variety of opal is mentioned by Sonnenschmidt as occurring in the mine of Toliman, in a trachytic conglomerate. John Mawe[1] mentions these opals in his work on precious stones, published in 1812, as having been sent to England in quantities at that time. The fire-opal still occurs in its greatest perfection in the porphyritic rocks at Zimapan in Mexico. It is generally of a translucent hyacinth-red or topaz color, and flashes forth dazzling beams of fiery carmine-red, with yellow and green reflections of more or less intensity. When these opals are still in the compact red porphyry, they form ob-

[1] A Treatise on Diamonds and Precious Stones (London, 1812).

jects of remarkable beauty, the flashes of red, green, yellow, and blue color intermingling as the light falls on them. A beautiful opal was exhibited by the Mexican Commission, at the World's Fair held at Philadelphia in 1876, and was very greatly admired by visitors.

An opal from Zimapan, Mexico, was analyzed by Klaproth, with the following result:

Silica	92·00
Peroxide of Iron	0·25
Water	7·75

The noble opals at Esperanza are remarkable for the extent and intensity of their reflections. The harlequin opals are noted for the diversity and the small size of their colored spots, which form beautiful miniature mosaics. One of the most pleasing varieties has a play of red fire like the red variety from Zimapan, and mingled with it flashes of brilliant metallic emerald-green, and occasionally a violet-blue of remarkable intensity. One of the red varieties from the Rosario Mine, on the hill of Jurado, has a violet-blue reflection of peculiar beauty; and the same mine produced a variety with a metallic emerald-green and a dark ultramarine color combined, or rather showing one after the other. The lechosos opals, as those with the red and green reflections are called in Mexico, are very common on the hill of Peineta, and less plentiful in the other mines of Queretaro. The opal mines of Esperanza are situated ten leagues northwest of San Juan del Rio, in the State of Queretaro, and are very extensive, having been traced over a district thirty leagues long and twenty leagues wide. They were discovered in 1835, on the landed estates on which they are situated, by a farm laborer. It was 1870 before a settlement was made on the edge of the mountain Ceja de Leon, by José Maria Siurob, near the present mine of "Santa Maria Iris." In 1873 Dr. Mariano de la Barcena[1] made a special report on this opal district, in which he states that he has discovered ten veins, or mines, as they are called. He says: "The opals of Esperanza are found forming chains more or less regular, on the banks of porphyry in quartz which forms its base, or disseminated through the mass of the

[1] Am. J. Sci. III., Vol. 6, p. 466, Dec., 1873.

same rock.   Veins of porphyry are met with in regular banks,
which in many cases preserve the same constant direction as on
the hill of Ceja de Leon, southeast to northwest.   The porphyry
is of a grayish-red color, although in some parts it is lighter
colored, changing into a reddish-white, even on the surface where
it is altered.   The aspect of the porphyry indicates generally the
class of opals it contains.   Where the rock is brick-red in color,
compact and hard, the varieties with a fiery-red color abound,
also the color combined with red, formed from different change-
able colors, or rather a mixture of colors.   However, where the
porphyry is lighter colored and mottled, noble opals are found more
abundantly, notably in the mines situated on the hill of Peineta."
These mines are remarkable for their richness and the variety of
their product.   In a single piece of rock, from the mine of Sim-
patica, Dr. Barcena found noble opal, fire opal, harlequin opal, and
the lechosos opal.   One of the mines at which the greatest amount
of work has been done is the Jurado.   Here an excavation fully
150 feet deep, 100 feet wide, and several hundred feet long has
been made, and at the depth of 150 feet the porphyritic rock
contains an abundance of hydrated silica and common opal.   A
deposit of opal was discovered in 1851 on the brow of the hill of
the Navajas, at a place called Tepezala, by Juan Orozca and Juan
Hill, two pupils of the Mexican School of Mines.   These were
fire-opals, in a conglomerate consisting for the greater part of
trachytic porphyry.   In the borough of Tepoentitlan, in the
district of San Nicholas del Ora, and near Huitzuco, both in the
State of Guerrero, fire-opals are also found, either light topaz-
yellow in color, with green, red, and yellow reflections, or white
changing into reflections.   Common and fire-opals have also been
observed north of San Luis Potosi, on the ridge of Mount
Mezquitic, and an opal district of considerable extent has been
observed on the Ciervo estate, three leagues south of Cadereita,
and fourteen leagues from Esperanza.   A quantity of semi-opals,
cacholong, and hyalite has been obtained here, but no true opal.
In this locality, the hills are of porphyry, yet the opal-bearing
rock is readily detected by the appearance of the soil which
covers it.   In the district of Amealco, in the State of Queretaro, opals were observed on the Batan, Galindo, and Lallare

properties. Dr. Manuel Gutierrez says of these that on the hills of the El Astillero estate, which is in the jurisdiction of Contepec, and in the State of Michoacan, were discovered some very rich beds of opals, but D. José Maria Siurob of Queretaro, the owner of the Esperanza mines, his informant, did not give him more definite particulars.

At present (1889) only one mine is being worked by the owner of six of the largest mines. The rock containing opal is brought to the city of Queretaro, a distance of twenty-five leagues, and about twenty lapidaries are continually employed in cutting and polishing the stone. The miners receive an average of twenty-five cents a day and the polishers an average of seventy-three cents for their work.

The noble opals found in Mexico generally exist (unfortunately) only in thin layers, between or upon layers of common opal, without any play of fire. Often only one-half or two-thirds of the cavity which contains this variety is filled with opal, and it generally shows stratified layers, like an onyx. A layer of hyalite is often present on the upper layer, or else the opal is very smooth, the opal coating being thin, with a very strong play of color, usually too thin to be polished. Both of these varieties of opal exist in great abundance in Mexico, and many thousand stones are sent to Germany to be remounted in the cheaper class of jewelry. Thousands are annually sold to visitors to the cities of Mexico, Queretaro, and at railroad stations in Mexico, and in Texas, New Mexico, and Arizona in the United States.

The fire-opal is perhaps the most gorgeous of all varieties of the opal, and it is also the most sensitive. It is frequently injured by water or exposure or by sudden atmospheric changes; indeed, so easily affected are fire-opals by the vicissitudes of the weather that they are believed to be brighter in summer than in winter, though this difference may be due to the fact that the light is better and the weather is warmer in summer. Some varieties are not so easily influenced, however, and are not injured by contact with water. Stones have been known to lose their brilliancy even when removed from the influence of atmospheric changes; when wrapped in paper, and placed in a jeweler's iron safe, or in the drawers of a collector's cabinet, they have lost their color, or become entirely filled with fissures, more especially the very limpid varieties with the flames of color. Often the stones, with only a small loss of color, have become entirely flawed, the cracks being such as to render the stones unfit for setting, since they are liable to break.

Dr. Augustus C. Hamlin[1] gives the two following illustrations of the loss of fire in opals: "A traveller from Central America brought home a splendid rough fire-opal which dazzled the eye with its fiery reflections. It was taken to an honest lapidary, who received it with a doubtful look. The next day the opal was returned, having been shaped into the usual oval form, but only faint gleams of any of the colored rays flashed from its surface or the interior. 'Is this the gem which was given you yesterday?' was demanded of the artisan. With a smile the lapidary took the transparent stone, and roughened its finely polished surface upon the wooden wheel. In an instant the lost fire returned, as if directed by magic's wand. The perfect transparency of the gem, with its high polish, had allowed the rays to pass directly through it, and there was but little refraction, but on roughening the surface the light was interrupted, and the peculiar property of the mineral displayed. Unfortunately the lesson was not concluded here. At the last touch of the wheel the beautiful gem flew into two parts, and its glories departed in an instant. Saddened with the day's experience, the two fragments were taken, cemented together, and tossed into a drawer which contained other mineral specimens of no great value. Some months after, when searching for a misplaced mineral, a gleam of light suddenly flashed out as the drawer was again opened. It was the neglected and abused opal, which now gleamed with the energy of a living coal of fire. It had recovered its beautiful reflections, and still adorns, notwithstanding its fracture, a most cherished jewel. Whence this mysterious change? the reader may ask. It can only be said that the complete transparency of the stone had been lessened, and perhaps the change was due to the action of some of the ingredients of the cement with which the fragments of the broken gem were united."

[1] Leisure Hours Among the Gems, 8vo., Boston, 1884.

Many of the Mexican and some of the Central American opals have the reputation of fading and becoming translucent, or opaque, or cracking in course of time, according to the circumstances of the exposure.

Dr. Hamlin's second instance illustrates this : " A few years ago, some Spaniards arrived in New York with a bag of rough opals brought from Central America, but from what particular locality we never learned. The specimens varied from the size of a bean to that of an English walnut, and were extremely beautiful. They had a fresh appearance, as though they had been recently extracted from the mines, and many of them had portions of the soft, sandy matrix still attached to them. They excited suspicions of not having been properly tempered and hardened by exposure. But their beauty, which reminded one of the perfect glowworm, or lumps of phosphorus moistened with oil, did not allow the spectator to hesitate about the purchase of them, especially as they were offered at a moderate price. We invested in the purchase of several charming ones, and never wearied in examining their exquisite effects. Still we felt a vague suspicion of the enduring qualities of our newly acquired treasures. The most beautiful stone, the size of a small almond, we carried in our pocket for a long time, not only for our gratification, but for the purpose of studying the effect of the atmosphere upon its reflections. Soon after the acquisition, we fancied a slight shadow, or nebulosity, appearing in one end of the stone. We carefully watched it, and before long an indistinct cloudiness began to appear, like the dim and distant haze of a summer sky on the commencement of a storm. Even then, we thought it might be mere fancy on our part. But when the shadow changed to opacity, and the transparency of the gem, with its beautiful reflections, vanished never to return, we were compelled to admit that even substances of the mineral kingdom had their diseases, as well as forms of the organic world. This is indeed but one example to illustrate a theory, but most of those we purchased at that time of the Spaniards have altered in appearance, and some of them quite as seriously. Wherefore, we have arrived at the conclusion that recently mined opals should be bought with caution, and that the perfection of a rough opal as a gem cannot be safely estimated until after it has been cut by the lapidary." Possibly these opals had been soaked in oil, a device which is sometimes practised to improve or restore the color.

About one hundred natives work the Esperanza mines in a desultory manner. The opals are nearly all cut at three cutting establishments, in the city of Queretaro. The cutting is done in the rudest manner, by native lapidaries, who neither give the stones a good shape nor polish them properly; hence they rarely show their true beauty, and very few are sent out of Mexico to be cut. Fully 50,000 are cut annually, and this amount could be doubled should the demand exist. Occasionally, when the color is thick enough, they show an intensity of color—often only one color, such as red, green, or yellow—not rivalled even by the Hungarian stones, and the Mexican opal in all its varieties is often purchased with the hope of realizing for them an equal value. The prices asked vary from a few cents to upward of $100. Lots of thousands are often sold for less than ten cents each, occasionally exceptional stones selling for $100, rarely for more. A beautiful series of opals exhibited by the Mexican Commission at the World's Fair held in Paris during 1889, consisted of noble and fire-opals. One large stone with superb pink flames was especially beautiful.

A remarkable fire-opal was brought home from Mexico by Alexander von Humboldt, and is still preserved in the Berlin Mineralogical Museum.

The Spanish historians, in their marvellous stories of the wonders seen in Mexico at the time of the Conquest, describe the image of the mystic deity, Quetzalcoatl—God of the Air—on the great pyramid of Cholula, as wearing a mitre waving with plumes of fire, an effect which is supposed to have been produced by masses of mosaics of fire-opal. A well-known Mexican opal is the one sold in the collection formed by Henry Philip Hope.[1] It was a Mexican fire-opal, or sun-opal, as it was called, carved with the head of the great Mexican Sun God, and is believed to have been taken from a Persian temple. It has been known since the sixteenth century, and brought £262 at the sale of the Hope jewels in London in 1886. With the fire-opal is also

[1] Catalogue Hope Collection, plate xxxi, fig. 3, p. 3 (London, 1839).

found the variety known as hydrophane. In this form the opal is generally white or dull yellow in color, but when it is wet, it becomes transparent, often brilliant in color. This variety often absorbs almost an equal bulk of water before it is fully saturated. It has no value as a gem, although often an object of very great beauty.

A beautiful variety of opal agate is found in the State of Jalisco. Pink, yellow, and green, especially the softer shades, occur and are blended and veined in the most pleasing manner. It exists in considerable quantity, and is valued as a decorative stone for metal work or jewelry.

Obsidian is abundant on the hill of the Navajas in Pachuca, in Tulancingo, in Ucareo, State of Michoacan, in Penjamo, and on the landed property of Pateo, belonging to the same State. In Magdalena village, in the State of Jalisco, in Cardereita Mendez, in the State of Queretaro, and in many other parts of Mexico it is found in a variety of colors, such as golden, silvery, black, bluish, greenish, or reddish. The included crystals which the obsidian contains often give this mineral a double color, the one black, the other chatoyant, either yellow, greenish-gray, or white, and always at right angles to the black. This stone holds an important place in the archæology of Mexico. Obsidian was most extensively used in Mexico, before the empire of the Aztecs succumbed to the Spanish invaders. The old obsidian mines are still to be seen on the Cerro de Navajas, or " Hill of Knives," which is situated in a northeasterly direction from the City of Mexico, at some distance from the Indian town Atotonilco el Grande. These mines provided the ancient population of Mexico with vast quantities of the much-prized stone, of which they made double-edged knives, arrows, and spear-heads, mirrors, skilfully executed masks, and ornaments of various kinds. Humboldt speaks of the Hill of Knives.[1] For a precise description we are indebted to Edward B. Tylor,[2] who visited that interesting locality in 1856, while traveling through Mexico in company with Mr. Christy. Besides many facts relating to the archæology and ethnology of Mexico, this writer furnishes the best ob-

[1] Essai politique sur la Nouvelle-Espagne, Vol. 3, p. 122.
[2] Anahuac : or, Mexico and the Mexicans, Ancient and Modern (London, 1861).

servations on obsidian from that country. Describing the mines, he says: "Some of the trachytic porphyry which forms the substance of the hills had happened to have cooled, under suitable conditions, from the molten state into a sort of slag, or volcanic glass, which is the obsidian in question; and in places, this vitreous lava, from one layer, having flowed over another which was already cool, it became regularly stratified. The mines were mere walls, not very deep, with horizontal workings into the obsidian, where it was very good and in thick layers. Round about were heaps of fragments, hundreds of tons of them; and it is clear, from the shape of these, that some of the manufacturing was done on the spot. There had been great numbers of pits worked, and it was from these little mines—minillas, as they are called—that we first got an idea how important an element this obsidian was in the old Aztec civilization. In excursions made since, we traveled over whole districts in the plains where fragments of these arrows and knives were to be found literally at every step, mixed with fragments of pottery, and here and there a little clay idol."

From the center of the State of Ohio to the country of the Shoshones, as well as the Rio Gila, and the mines in Mexico, the straight distances are almost equal, measuring about seventeen hundred English miles; indeed the Mexican mines are a little nearer to Ohio than the other districts. It would be idle, therefore, to speculate from which of these localities the obsidian found in Ohio and Tennessee was derived. The number of articles of this stone that have been met with east of the Mississippi is so exceedingly small that its technical significance hardly deserves any consideration. Two large obsidian knives, about 18 inches long, found in Mexico and of Mexican origin, and almost identical in appearance, are marvels for their fine chipping. They are to be seen, one in the United States National Museum at Washington (see Illustration), and one in the Trocadéro Collection in Paris. Lip-ornaments, mirrors, and other objects are to be found in the United States National Museum, in the National Museum in City of Mexico, the Trocadéro Museum at Paris, the Archæological Collection of the British Museum, London, and M. Goupil's collection at Paris. A number of the finest known mirror and engraved plaques of obsidian are in the Trocadéro Museum. A square one from Texcoco, measuring 9 ½ x 8 ½ x 1½ inches (24 x 21 ½ x 3 centimeters), and a round one, convex on one side, from Oaxaca, 6½ inches (16 centimeters) in diameter (see Fig. 16), are both wonderful pieces of primitive stone work. The one possessing the greatest archæological interest is the square plaque described by the director, Dr. E. Hamy,[1] on which is the inscription "Ypanquetza-litzli 4 acatl" (9th December, 1483), the date of the laying of the first stone of the Great Temple of Mexico. The polished carved figures are exceedingly interesting. (See Illustration.)

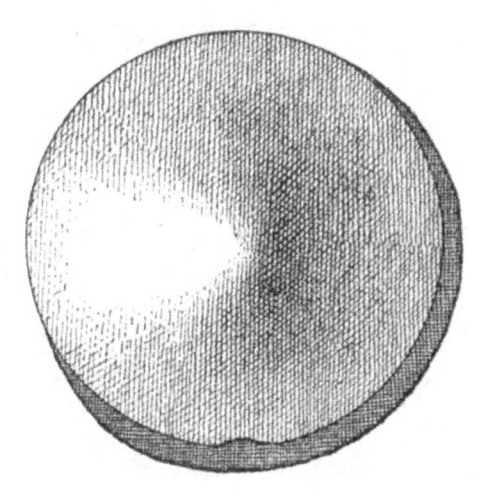

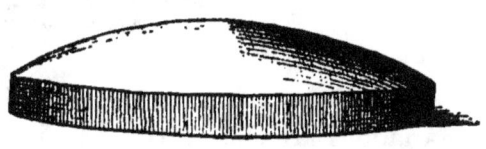

FIG. 16.
OBSIDIAN MIRROR, FROM OAXACA, MEXICO.
PINARD COLLECTION, TROCADÉRO MUSEUM.

The richly mottled red and black, brown and black, and yellow and black obsidian, called marekanite, is found in large quantities in the State of Jalisco, generally in sufficiently large masses to be useful as a decorative stone, since it admits of polish. Associated with it in quantity is pearlite, or sphærulite, which shows reddish-brown spherules in a gray matrix. Pitchstone exists in quantity with it.

Pyrite which is really a mixture of pyrite in cubes and marcasite in plates, as determined by Dr. Alexis A. Julien[2]—was worked by the ancient Aztecs into mirrors and other objects. The mirrors were generally semicircular on one side, and polished flat on the other side, and the polish is often still preserved.

[1] Revue d' Ethnographie, Vol. 2, p. 193, 1883.
[2] Ann. N. Y. Acad. Sci., Vol. 3, p. 365, 1886 and Vol. 4, p. 125, 1887.

The rounded side was often curiously carved and decorated. (See Fig. 17.) They also carved pyrite into other objects, notably a human head 2 inches high, in which were inserted eyes of white chalcedony, now in the Blake Collection in the United States National Museum, and a number of mirrors, now in the Trocadéro Museum, Paris. (See Fig. 17.)

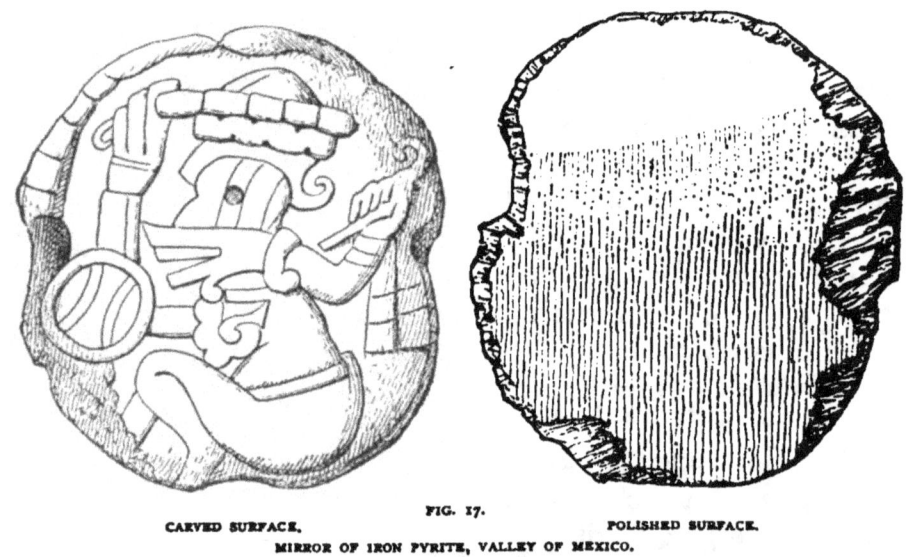

CARVED SURFACE.

FIG. 17.

POLISHED SURFACE.

MIRROR OF IRON PYRITE, VALLEY OF MEXICO.
PINARD COLLECTION, TROCADÉRO MUSEUM.

Mexican onyx, so called, is really an aragonite. Prof. Mariano Barcena, of the Mexican Commission to the World's Fair held in Philadelphia during 1876, has recently published an account of its occurrence and chemical character.[1] The principal deposits are located near the town of Tecali in the State of Pueblo. It is essentially a carbonate of calcium, containing small quantities of the oxides of iron and manganese, to which are due the variegated colors for which the rock is so much admired. The specific gravity, 2·9, shows that it is aragonite. It was extensively used by the ancient Mexicans, specimens of whose handiwork we still have preserved in our museums in the form of masks, idols, and a variety of objects. The softness of the material (it can be readily carved with a knife) has tempted some of the modern residents of Mex-

[1] The Rocks Known as Mexican Onyx, Proc. Acad. Nat. Sci., Phila., Vol. 28, p. 166, 1876.

ico to imitate the ancient objects, to meet the demand of visitors. to that country. This material is entirely stalagmatic in its for-mation, and yellow-brown and red oxides of iron have been de-posited between the layers. It is generally cut across the layers, which gives it a beautiful veined appearance. When it is cut in the same direction as the deposition, the botryoidal structure is well shown, the mineral being so translucent that the markings. resemble colored clouds. It is one of the most beautiful orna-mental stones of any age, and has been used extensively for or-namental purposes in Europe as well as in the United States, where it was first introduced about 1876, when it brought about ten times its present price. The natives in the vicinity of Pueblo sell large quantities of this material, made into trays, crucifixes, reliquaries, inkstands, penholders, paper-folders, and paper-weights, in the form of single fruits or bunches of fruit, fish, or other natural objects, which are copied, not only with regard to form, but often with remarkable skill in the utilization of the col-ors in the stone. So great is the variety of tints of color in which the material is found that there is scarcely a limit to its possibilities for such purposes. Bernardino de Sahagun refers to iztac chalchihuitl, white or fine green, and quite transparent, obtained from quarries in the vicinity of Tecalco, which Dr. Daniel G. Brinton [1] believes to be the modern Tecali; and the de-scription and locality answer so well to those of our so-called Mexican onyx that there can scarcely be a doubt that this was referred to by Sahagun as iztac chalchihuitl.

In the summer of 1888, William Cooper, of Esperanza, discovered in the volcano of Zempoaetepetl, in southern Mex-ico, a deposit of a beautiful mineral, which has received the trade name of mosaic agate; this is really the so-called Mexican onyx. It is an aragonite, with the difference, however, that the latter is always veined or stratified, whereas the new material is a brecciated or "ruin aragonite." The original formation has evidently been entirely broken up, the fragments having been cemented together and the crevices all filled in with a new depo-sition of aragonite, showing conclusively that a deposit of Mexi-can onyx had been fractured by some disturbance, possibly vol-

[1] Science, Vol. 12, p. 168, Oct. 5, 1888.

canic, and that a subsequent deposition of the material cemented
it into its present form. Like aragonite, it is susceptible of a
high polish, the difference between the two being that in the
onyx the straight bands of color of the aragonite are broken and
disseminated throughout the mass, making its general effect even
more pleasing and brilliant than that of the latter. It can easily
be cut into thin slabs, and makes beautiful tops for ornamental
tables and bureaus. It is often cut into solid columns and used
for pedestals for busts or statuary.

Specimens of a very remarkable amber have occasionally
been brought by travelers, for the last fifteen or twenty years,
from some locality in southern Mexico. The only information
gained concerning it is that it is brought to the coast by natives,
who say that it occurs in the interior so plentifully that it is used
by them for making fires. The color of this amber is a rich
golden-yellow, and when viewed in different positions, it exhibits
a remarkable fluorescence, similar to that of uranine, which it also
resembles in color. A specimen in the possession of Martius T.
Lynde measures 4 x 3 x 2 inches, is perfectly transparent, and is
even more beautiful than the famous so-called opalescent or green
amber found in Catania, Sicily. This material would be extremely
valuable for use in the arts.

Amber was formerly used as incense by the Aztecs, and
fragments have been found on the altars of ancient temples, also
in the Catholic churches in early Mexico.

CHAPTER XV.

Aboriginal Lapidarian Work in North America.

THIS chapter may seem out of place in a treatise on precious and ornamental stones, yet the chipping of an arrow-point, the grinding and polishing of a groove in an axe-head, the drilling of a bead or tube or an ear-ornament, all are done by the application of the same lapidarian methods that are practised to-day by cutters of agates or precious stones. The cutter of to-day, with a hammer, chips into shape the crystal or piece of agate before it is ground; and there is little difference between the ancient method of drilling and that of the present. The stone bead of ancient time was drilled from both ends, the drill holes often overlapping, or not meeting as neatly as by the modern method of drilling from one end.

The old way of drilling is still practised in the East, where the primitive bow-drill is used by lapidaries to-day precisely as it has been used by savage tribes in all quarters of the globe, though producing at different periods different qualities of work. Nowhere was its use better understood than in ancient Greece and Rome, where by its means were engraved the wonderful intaglios and cameos which now grace our museums, and which have never been surpassed in any period of the world's history. For the special use of gem engraving, the bow-drill has been replaced by a horizontal lathe, which, however, does not allow the freedom of touch or softness of feeling which artists attained by the use of the bow-drill. The instrument known as the dental drill is really an improved form of bow-drill, working much more rapidly. An S. S. White dental engine, provided with a suitable series of drill-points, answers every purpose, and has been found especially useful in exposing fossils and minerals when covered with rock, the objects being opened with great rapidity, with little danger of injury. As shown by the author in a paper on a new method of engraving cameos and intaglios,[1] an artist could be so trained to the use of this improved bow-drill as to attain the same softness and feeling developed by the old lapidarian masters.

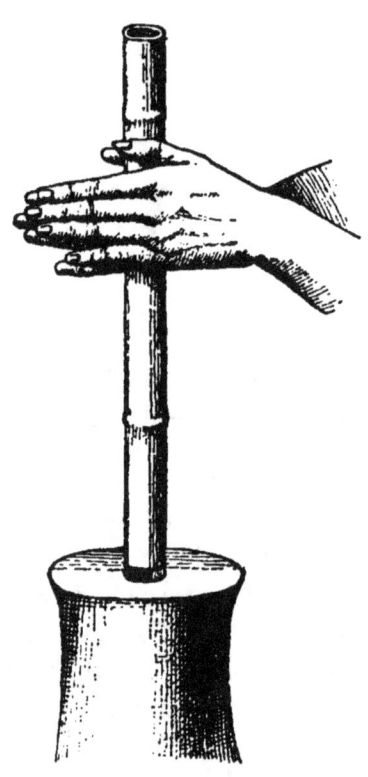

FIG. 18.
PRIMITIVE METHOD OF DRILLING A HARD STONE WITH A REED AND SAND.

In the ancient specimens of work, tubes from which a core had been drilled out by means of a reed and sand, revolved by the hand (see Fig. 18), were done as neatly as anything can be done, the reason being that the object was entirely drilled from end to end. This method of drilling is still practised, except that the hollow reed is replaced by the diamond or steel drill. When a valuable stone is being drilled, a sheet of steel or a thin iron tube is substituted for it. The polishing and grinding now is done on rapidly revolving disks, horizontal or lay wheels, as they are called, whereas, formerly, the slow process of rubbing with the hand on board or leather was perhaps resorted to. No lapidary can do finer work than that shown by the obsidian objects from Mexico (see Illustration), the labrets, and the ear-ornaments, which are even more highly polished, though no portion of the circle is thicker than $\frac{1}{32}$ of an inch. An obsidian coyote head in the Blake Collection in

[1] Trans. N. Y. Acad. Sci., Vol. 3, p. 105, June, 1884, also Jeweler's Circular, June, 1884.

the United States National Museum is a large ornament 6 inches across, highly polished, and bored through the center. The spear-points and hoes from East St. Louis and other parts of Missouri and Illinois, and beautiful sacrificial knives— notably the immense knife, 18 inches in length, in the Blake Collection of the United States National Museum (see Illustration), and the one in the Ethnological Museum at the Trocadéro in Paris—show the greatest skill in chipping.

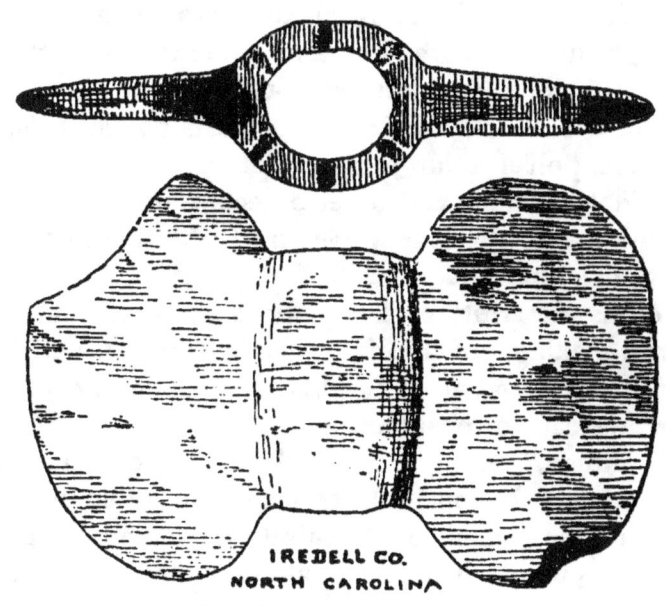

FIG. 19.
BANNER-STONE OF FERRUGINOUS QUARTZ.

Many of the aboriginal stone objects found in North Amer-ica and elsewhere are marvels of lapidarian skill in chipping, drilling, grinding, and polishing. Few lapidaries could duplicate the arrow-points of obsidian from New Mexico, or those of jasper, agate, agatized wood, and other minerals found along the Willamette River, Oregon. No lapidary could drill a hard stone object truer than some of the banner-stones (see Fig. 19), tubes, and other objects made of quartz, greenstone, and granite that have been found in North Carolina, Georgia, and Tennessee, or make anything more graceful in form and general outline than are some of the quartz discordal stones found in these same States. These latter objects are often from 4 to 6 inches, and occasionally 7 inches, in diameter, ground in the center until they are of the thinness of paper and almost transparent, and the great regularity of the two sides would almost suggest that they had been turned in a lathe. This may have been accomplished by mounting a log in the side of a tree so that it would revolve, and cementing the stones with pitch to the end of the log, as a lapidary would do to-day at Oberstein, Germany, or by allowing the shaft of the lathe to protrude through the side of the log, and cementing the stone to be turned on this. The Egyptian wood-turner at work in the Rue du Caire, at the World's Fair held in Paris during 1889, might, with his lathe, polish a large ornament of jade for jadeite, like the masks, idols, tablets, and other objects found in Mexico and Central America, or the jade knives from Alaska, in the United States National Museum.

Numerous descriptions have appeared of the chipping—or rather arrow-making—of aboriginal lapidarians. Caleb Lyon describes a California Indian of the Shasta tribe, whom he had seen making arrow-heads of obsidian.

"The Indian," he says, " seated himself on the floor, and placing a stone anvil upon his knee, which was of compact talcose slate, with one blow of his agate chisel he separated the obsidian pebble into two parts, then giving another blow to the fractured side he split off a slab a fourth of an inch in thickness. Holding the piece against the anvil with thumb and finger of his left hand, he commenced a series of continuous blows, every one of which chipped off fragments of the brittle substance. It gradually assumed the required shape. After finishing the base of the arrow-head (the whole being only a little over an inch in length) he began striking gentler blows, every one of which he expected would break it into pieces. Yet such was their adroit application, his skill and dexterity, that in little over an hour he produced a perfect obsidian arrow-head. Among them arrow-making is a distinct trade or profession, which many attempt, but in which few attain excellence."[1]

[1] Bulletin of the American Ethnological Society, vol. 1, p. 39, New York, 1861.

Another method of arrow-making practised by the California tribes is mentioned by Edward E. Chever in an article[1] in which is illustrated the implement used in the process. "The arrow-head," he says, "is held in the left hand, while the nick in the side of the tool is used as a nipper to chip off small fragments. This operation is very curious, both the holder and the striker singing, and the strokes of the mallet, given exactly in time

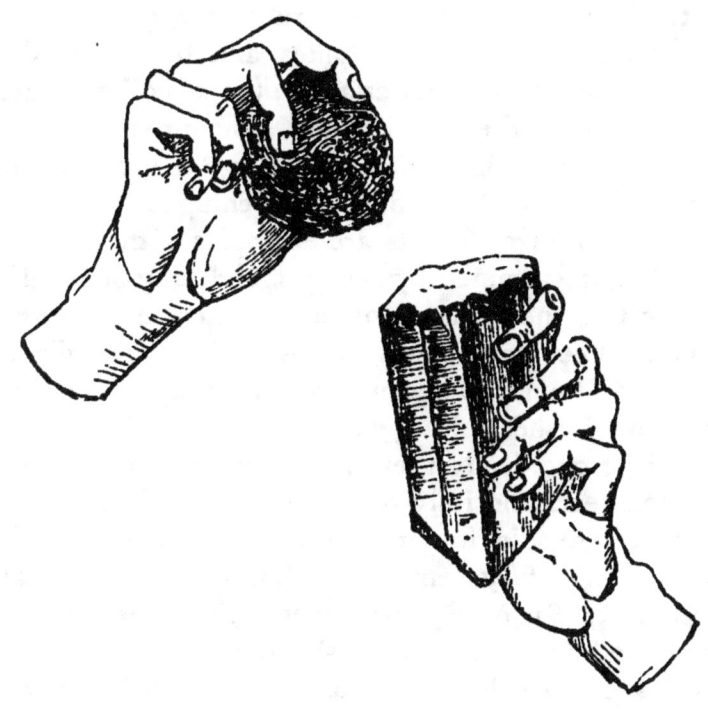

FIG. 20.
PRIMITIVE METHOD OF CLIPPING FLINT.

with the music, and with a sharp and rebounding blow, in which, the Indians tell us, is the great medicine (or mystery) of the operation. Every tribe has its factory in which these arrow-heads are made, and in those only certain adepts are able or allowed to make them for the use of the tribe."

Arrow-heads of glass, flint, obsidian, or similar substances

[1] American Naturalist, vol. 4, p. 139, May, 1870.

have been made by a number of our modern archæologists, and
a series is now in the United States National Museum.
The process consists, first, in chipping off a thick, suitable flake,
then pressing against one of the sides a bone object (nothing
better than the handle of a tooth-brush) until a small nick is
made in the side. Each pressure makes a nick, and the flake is
constantly reversed so that the nicks are alike on both sides.
With a little practise, any one can make a fairly good arrow-
point. Dr. F. Capitan, of Paris, exhibited in the liberal arts section of the World's Fair held in Paris during 1889, in connection
with a collection of flint cores, flakes, and fragments illustrating
the manipulation of the material in the palæolithic age, models
of two wax hands, the left holding a hammer stone, the right
the flint core (see Fig. 20), thus giving a graphic illustration
of the manner in which the flints were chipped.

The Oregon arrow-points are examples of the highest degree
in stone chipping attained by savages, and they often afford gem
material so that the demand for them as articles of jewelry is not
surprising. They are not now made by Oregon Indians and are
only sparingly found. They may be picked up in certain districts,
after a heavy freshet. Large quantities were formerly found in
Oregon City; one dealer there is said to have sold 40,000 during
the past ten years and fully 50,000 have been found there in that
time, which were sold for at least $6,000. Originally their price
was from five to fifty cents each, but the present scarcity has
increased it to $1, or $5 for exceptionally fine ones, which are
usually made of rock crystal; flesh-colored, red, yellow-brown, or
mottled jasper; obsidian; variously colored chalcedony; or agatized wood. They are sold principally in the East, as scarcely
any are used in Oregon for jewelry. Many thousands are in the
collections of Prof. Othniel C. Marsh of Yale College, New
Haven, and of James Terry at the American Museum of Natural
History in New York City, and they can be seen in all of our better collections. In North Carolina, South Carolina, and Georgia,
beautiful arrow-points of pellucid rock crystal or transparent and
smoky quartz are occasionally found. Some of these are two
inches long. A beautiful rock-crystal knife, which was found at
Wind River, Ariz., was exhibited at the World's Fair held in

Paris during 1889. In this connection see also three papers by Charles Rau, "Drilling in Stone without the Use of Metals, Report of the Smithsonian Institution for 1868," "Ancient Aboriginal Trade in North America, Report of the Smithsonian Institution for 1872"; and "The Stock in Trade of an Aboriginal Lapidary, Report of the Smithsonian Institution for 1878." (See also Chapter on Mexico, page 284.)

CHAPTER XVI.

Definitions, Imports, and Production, Values, Cutting of Diamonds and Other Stones,
Watch Jewels, Collections of Gems, Minerals, and Jade, Uses of Precious
and Ornamental Stones for the Ornamentation of Silver,
and Furniture and for Interior Decoration,
Trilobite Ornaments.

WHAT is a precious stone? The answer to this question is not easy, for the value of a particular kind of stone is often due in great measure to the caprice of fashion, or to some adventitious circumstance of time or place; and some stones that are to-day of small value have, during certain periods in the past, almost displaced the diamond or the ruby in public estimation. Beauty of color, hardness, and rarity are the essential qualities which entitle a mineral to be called precious. Strictly speaking, the only precious stones are the diamond, ruby, sapphire, and emerald, though the term is often extended to the opal, notwithstanding its lack of hardness, and to the pearl, which is not a mineral but strictly an animal product.

Popularly, a gem is a precious or semi-precious stone, especially when cut or polished for ornamental purposes. Mineralogically, the term designates a class or family of minerals hard enough to scratch quartz, without metallic lustre, but generally brilliant and beautiful, and includes the semi-precious or fancy stones (called pierres de fantaisie by the French), such as the chrysoberyls, alexandrites, tourmalines, spinel, and topaz. Archæologically, the term is restricted to engraved stones, such as intaglios and cameos. The term jewel is applied to a gem only after it has been mounted.

The epithet phenomenal has been applied in this work to stones which exhibit a play of color, like the opal, moonstone, sunstone, and labrador spar; or which change their color by artificial light, like the alexandrite; or show a line or band or bands of light, as the line in the cat's-eye and the star in the ruby or sapphire asteria.

Public interest in semi-precious stones has increased greatly during the last ten years. Formerly jewelers sold only diamonds, rubies, sapphires, emeralds, pearls, garnets, and agates, but at present it is not unusual to have almost any of the mineralogical gems, such as zircon, asteria or star sapphire or star ruby, tourmaline, spinel, or titanite, called for, not only by collectors, but by the public, whose taste has advanced in the matter of precious stones as well as in the fine arts. Ten years ago $100,000 was an unusual amount for even the wealthiest to have invested in diamonds: to-day there are a number of families each owning diamonds to the value of half a million dollars. Ear-rings worth from $5,000 to $8,000 a pair excite no wonder to-day: formerly, they were seldom seen. Of the French crown jewels sold in Paris, May, 1886, more than one-third, aggregating over $500,000 in value, came to the United States.

Three diamonds are owned in this country weighing $55\frac{1}{4}$, 77, and $125\frac{5}{8}$ carats respectively; the latter, known as the Tiffany diamond, is the handsomest large yellow diamond ever found. A number of necklaces worth over $100,000 each are owned in the United States, and one necklace, worth $320,000, was recently sold at the death of its owner. Among other diamonds worthy of note that are owned in the United States are four of the Mazarins[1] from the French crown jewels, and a diamond that belonged to the Empress Catharine of Russia. Besides these, there is one single stone weighing $25\frac{11}{32}$ carats, valued at $45,000; a ruby of $9\frac{5}{8}$ carats, worth over $33,000, and a number

---

[1] In the inventory of the French crown jewels, 1791, mention is made of "the tenth Mazarin." Cardinal Mazarin had fostered the diamond-cutting industry in Paris, and the diamonds called "Mazarins" are supposed either to have belonged to him originally or to have been recut under his direction. All of the ten Mazarins, however, were not found.

of rubies worth over $10,000 ; white pearls worth from $5,000 to $11,000 each ; black pearls worth over $6,000 each ; pearl necklaces worth from $20,000 to $100,000 each ; an emerald worth over $12,000 ; and half a dozen families at least own jewels which rank in value, as regards intrinsic worth, with those belonging to some of the royal and imperial families of Europe, differing only, perhaps, in quality ; as perfect stones were bought by the Americans, whereas the regal jewels depend to some extent for their value on historic association.

The expression first water, when applied to a diamond, denotes that it is free from all trace of color, blemish, flaw, or other imperfection, and that its brilliancy is perfect. It is, however, frequently applied to stones not quite perfect, but the best that the dealer has, and they may be of only second quality. It is almost impossible to value a diamond by its weight only. Color, brilliancy, cutting, and the general perfection of the stone have all to be taken into account. Of two stones, both flawless, and of the same weight, one may be worth $600, and the other $12,000. Exceptional stones often bring unusual prices, while off-colored stones sell for $60 to $100 a carat, regardless of size. The poor qualities have depreciated so much in value that some are worth only from one-tenth to one-fourth what they were worth twenty years ago. This is specially true of large stones of the second or third quality. To show the variation of diamonds in value, the following may be instanced. Four of the Mazarins were appraised in 1792 by the commission appointed at that time, and the price at which they were estimated, and the price paid for them in 1887 by dealers, to be re-sold, were as follows :

WEIGHT (CARATS) [1]	APPRAISAL	PRICE REALIZED
28 $\frac{7}{16}$	$50,000	$31,000
24 $\frac{11}{16}$	40,000	25,600
22 $\frac{1}{4}$	32,000	16,200

Owing to the absorption, by the DeBeers Mines of South Africa, of nearly all the African mines, to such an extent that these mines produce twenty-nine thirtieths of all the diamonds mined to-day, cut diamonds advanced fully one quarter in price

[1] The carat used was the international carat. See p. 14.

during the last months of 1889; and as mining must now be carried to greater depths, involving a higher cost, the price is likely to be maintained, if not advanced.

IMPORTS OF DIAMONDS INTO THE UNITED STATES.—From the customs import-lists, after deducting the approximate value of cut stones other than the diamond, we find that import duty was paid on about $120,000,000 worth of cut diamonds in the last twenty-four years, of which $90,000,000 worth were imported during the last twelve years. In 1868 $1,000,000 worth were imported, and about $1,200,000 worth in 1867, but about $11,000,000 in 1888, and the same amount in 1889, or ten times as many in the latter year as twenty years previous, showing the increase of wealth and the great popularity of the diamond among Americans, the previous figures representing the import prices, exclusive of mounting or dealers' profits. A single firm at present sells yearly more than the annual import of 1867.

Diamond dust worth $464,905 has been imported since 1878, $289,430 worth from 1868 to 1878, and in 1869 to 1871 only $228 worth; but the first year after the opening of the Kimberley Mines, $80,707 worth was imported, showing one of the great benefits the arts received from the opening of the great South African diamond mines.

In 1878 the importations of uncut diamonds amounted to $63,270, in 1887 to $262,357, showing that four times as many diamonds were cut in 1887 as in 1878, though the importations were falling off. The total for the decade was $2,728,214, while in 1883 there were imported $443,996 worth, in 1888 $322,356 worth, and in 1889 $191,341. The falling off in importation is partly because in the years since 1882 a number of jewelers, who had opened diamond-cutting establishments, either gave up or sold that branch of business; for, in spite of the protective duty of 10 per cent. on cut stones, cutting cannot be profitably carried on unless on a scale large enough to enable a partner to reside in London, the great market for rough diamonds, in order to take advantage of every fluctuation in the market, and purchase large parcels, to be cut immediately and converted into cash as fast as they are sold.

DIAMOND-CUTTING.—This industry is now carried on in the

IMPORTS

Diamonds and other precious stones imported and entered for consumption in the United States, 1867 to 1889 inclusive

Fiscal years ending June 30—	Glaziers'.	Dust.	Rough or uncut.	Diamonds and other stones not set.	Set in gold or other metal.	Total.
1867	$906	......	....	$1,317,420	$291	$1,318,617
1868	484	......	....	1,060,544	1,465	1,062,493
1869	445	$140	....	1,997,282	23	1,997,890
1870	9,372	71	....	1,768,324	1,504	1,779,271
1871	976	17	....	2,349,482	256	2,350,731
1872	2,386	89,707	....	2,939,155	2,400	3,033,648
1873	......	40,424	$176,426	2,917,216	326	3,134,392
1874	......	68,621	144,629	2,158,172	114	2,371,536
1875	......	32,518	211,920	3,234,319	......	3,478,757
1876	......	20,678	186,404	2,409,516	45	2,616,643
1877	......	45,264	78,033	2,110,215	1,734	2,235,246
1878	......	36,409	63,270	2,970,469	1,025	3,071,173
1879	......	18,889	104,158	3,841,335	538	3,964,920
1880	......	49,360	129,207	6,690,912	765	6,870,244
1881	......	51,409	233,596	8,320,315	1,307	8,606,627
1882	......	92,853	449,513	8,377,200	3,205	8,922,771
1883	......	82,628	443,996	7,598,176	(a)2,081	8,126,881
1884	22,208	37,121	367,816	8,712,315	......	9,139,460
1885	11,526	30,426	371,679	5,628,916	......	6,042,547
1886	8,949	32,316	302,822	7,915,660	......	8,259,747
1887	9,027	33,498	262,357	10,526,998	......	10,831,880
1888	9,571	37,657	322,356	10,473,329	......	10,842,913
1889	......	......	191,341	11,466,708	......	......
Total	$75,850	$800,006	$4,039,523	$16,784,978	$17,079	$121,717,436

a Not specified since 1883.

Imports of substances not included in the foregoing table, 1868 to 1888 inclusive

Fiscal years ending June 30—	Unmanufactured agates	Bookbinders' and other manufactured agates.	Carnelian.	Brazilian pebbles.	Amber.	Amber beads.	Unmanufactured coral.	Manufactured coral.	Unmanufactured meerschaum.	Total.
1868	......	......	......	......	......	......	......	$62,270	......	$62,270
1869	......	$70	$269	......	$427	......	......	22,417	$6,407	29,590
1870	......	......	766	......	1,433	......	......	18,975	3,998	25,172
1871	......	1	661	......	180	......	......	37,877	698	39,417
1872	......	529	207	......	2,426	......	$83	59,598	2,194	65,037
1873	$151	1,310	......	$1,237	1,534	$595	230	63,805	5,608	74,470
1874	177	1,524	......	......	1,448	1,057	527	28,152	270	33,155
1875	520	5,165	......	57	7,169	715	1,278	33,567	2,902	51,373
1876	293	1,567	......	......	15,502	187	109	33,559	21,939	73,156
1877	579	1,904	(a)69	......	17,307	329	718	28,650	9,304	58,860
1878	82	404	......	76	13,215	1,119	1,252	12,667	16,308	45,123
1879	138	364	......	......	17,821	203	147	11,327	19,088	49,088
1880	57	2,346	......	......	36,860	2,317	62	5,492	30,849	77,983
1881	486	1,700	......	5	42,400	1,102	89	2,501	72,754	121,037
1882	901	5,084	......	111	72,479	4,174	1,474	669	56,118	141,010
1883	14	2,895	......	......	40,166	3,472	681	(b)1,303	58,885	107,416
1884	......	6,100	......	3,496	56,301	4,692	158	......	43,169	113,916
1885	124	......	......	6,541	21,722	3,242	659	......	42,590	74,878
1886	284	......	......	17,379	27,215	5,665	219	......	23,417	74,179
1887	12	1,247	......	35,291	34,238	10,011	307	......	35,478	116,584
1888	24	......	......	24,330	31,922	18,777	59,350	......	21,013	155,416
Total	$3,842	$32,210	$1,972	$88,523	$441,765	$57,657	$67,343	$422,829	$472,989	$1,589,130

a Not separately classified since 1877.

b Not specified since 1883.

United States much more extensively than ever before, but it has not always proved profitable. London is the great market for diamonds in the rough, and diamonds are sold so soon after arrival there, and the competition of the cutters of continental Europe is so keen, that Americans, who have not such ready access to the market, cannot always make a profitable purchase ; moreover, the recent consolidation of the larger mines has placed the control of the price of rough diamonds in a few hands. The trade, therefore, has in many cases been given up, and among the successful dealers the standard of merit has been raised until to-day the finest cutting is done in the United States. A large part of the work done here consists in cutting fine material, in recutting old stones that were valued in Europe for weight only, or in improving modern work, and these branches are generally profitable. But even with a protective duty of 10 per cent. on cut gems (a higher rate would encourage smugglers), it is improbable that the work of the great foreign cutting-centres can be rivalled in this country, since the demand is for fine material, and large parcels of rough stones seldom yield more than 10 per cent. of the best quality.

Henry D. Morse, of Boston, was the first to cut diamonds in this country, and the best cutters in the United States to-day received their training under him. Educating in this art young Americans, both men and women, was not his greatest work, for he showed that diamond-cutting, which had so long been monopolized by the Hollanders, was degenerating in their hands into a mere mechanical trade. He studied the diamond scientifically, and taught his pupils that mathematical precision in cutting greatly enhanced the beauty and consequently the value of the gem ; and his artistic sense, sound judgment, and keen perception enabled him to carry the art to a degree of perfection not often attained. His treatment of the diamond gave a great impetus to the industry both here and abroad, shops being opened, both in this country and in London, in consequence of his success. In his shop a machine for cutting diamonds was invented, that did away in a great measure with the tediousness and inaccuracy of the old manual process. Thanks to his labors, there are now in the United States some of the best cutters·in the world,—men who can treat the diamond as it should be treated in order to develop its greatest beauty. The fact that so many fine stones have been recut here led to a great improvement abroad in cutting, especially in the French Jura and in Switzerland, where both sexes are now employed at that trade; and, as a result, diamonds sold to-day are better than those of twenty years ago. Mr. Morse, above all others, has shown that diamond-cutting is an art, and not an industry.

There are at present about twelve diamond-cutting establishments in this country, employing from one to fifty men each, in all about a hundred, at salaries ranging from $20 to $50 a week. Most of the cutting done is of a high class, some shops being almost entirely employed in recutting stones previously cut abroad. Ten years ago nearly all the diamonds used in the United States were purchased through brokers or importers: now, owing to the marvelous growth of the diamond business here, and the facilities for transatlantic travel, some of the large retail houses buy their diamonds direct in the European markets, and in more than one instance have established branches or agencies abroad.

AGATE CUTTING.—In cutting large surfaces of hard materials, such as agate, jasper, and quartz, no better work has been done than that of the Drake Company of Sioux Falls, South Dak. Agate-cutting, as already stated, has been carried on for over three hundred years in the Oberstein district in Germany. But little attention, however, has been paid to the cutting of large masses, because no agates are found over a foot in diameter, and the banding is not such as to offer any inducement for polishing. Perfected methods for sawing and polishing such material have resulted from experiments recently conducted[1] by the Drake Company. They have undertaken the preparation of agatized wood for the market, and have succeeded in producing a large number of columns from 8 to 12 inches wide, and 2 to 3 feet high, cut transversely across the tree, so that the heart is visible on two sides, with radiations in all directions; they also cut sections measuring respectively 13, 17, 24, and 25 inches in diameter, and so highly polished that when turned with the back

[1] Eng. and Min. Jour., Vol. 45, p. 214, March 24, 1888.

to the light, they form perfect mirrors. All the specimens are brilliant in color and rival any work ever done in hard material. One of the finest sections of an agatized trunk was sent to New York in the autumn of 1888. It measured 40½ by 34 inches on the top, was 36 inches high, and weighed 2½ tons. The top was four months undergoing the process of grinding down and polishing; it is a deep, rich red, yellow, black, mottled and variegated, and beautifully polished. This is probably the finest piece of hard-stone polishing that has been done in the United States. The company has removed from the forest 180 tons of material, and 20 tons of sections have been ground down to show its characteristic beauty. The process is briefly as follows: The faces of the rough sections are irregular, and must be worn down to a smooth surface. To accomplish this they are set in circular form in what is known as the "Drake Beds," about ten feet in diameter, composed of various-sized sections of the material set by the use of a spirit level, in order to secure an even face. They are then cemented together, and large slabs of Sioux Falls quartzite are attached to two arms of a powerful vertical shaft. These large, flat stones, which are almost as hard as the silicified wood and extend the full length of the arms, are revolved about the bed by a stream of water, with crushed quartzite reduced to the size of a pea. The silicified wood, being tougher than quartzite, soon wears grooves in these large stones, which are frequently reversed, and sometimes discarded for new ones. This initial stage of the work continues for nearly forty hours, when the quartzite stones are replaced by large sections of the silicified trees, which have been previously worn upon the bed, and these are revolved sometimes for one week, sometimes for two weeks, and fed with sand of quartzite until, by abrasion rather than cutting, a face is disclosed on the bed, which, for the first time, indicates the true colorings and quality of the material. From these beds, each of which requires about thirty horse-power when doing the best work, the specimens are taken up, and rebedded on a car thirty feet long and eight feet wide. The success of the operation depends upon the exactness of face of the different pieces. This car moves by cogs and concentric rings, the outermost of which is six feet in diameter, revolving at forty revolutions, and here is continued the sand quartzite feed, in order to wear down any irregularity of resetting upon the car. This operation usually lasts for two days, when the bed is cleansed, and diminutive globules of chilled shot-iron are rolled under the rings. Then follows treatment with emery, beginning with the coarser grade and ending with the finer. After a week of this work, the bed is thoroughly washed, the rings removed, and large wheels, made from blocks of bass-wood clamped together, presenting a rough surface by being set across the grain of the wood, are placed in position. The speed, both in the movement of the car and of the wheel, is now increased, and tin oxide is used to burnish the surface, which is brought to a mirror-like finish by means of tripoli, fed to felt-covered wheels, that are revolved at the rate of 300 revolutions a minute.

The cutting and carving of rock crystal now done in the United States, even the cutting of crystal balls, vases, cups, and vials, is equal to work produced anywhere, as the vials, bonbon boxes, and clock exhibited at the Paris Exposition in 1889 by Messrs. Tiffany & Co. fully demonstrated. Much of the cutting of precious stones, such as ruby, sapphire, emerald, and garnet, is of the highest order. Sards, bloodstones, and other cheap agates are often cut abroad to a uniform size for mounting, because it costs less to fit the stone to the mounting than the mounting to the stone, and such stones as are found here are generally cut in this country.

WATCH JEWELS.—About 1,200,000 watches with jeweled works are annually manufactured in the United States, requiring about 12,-000,000 jewels, seven to twenty-one for each watch ; of these 5,000,-000 are ruby and sapphire, and 7,000,000 are garnet jewels, valued at over $300,000. Most of them are imported, but the Waltham Company does its own cutting, employing about 200 hands. About 15,000 carats of diamond in the form of bort, are used annually in slitting and drilling these jewels. Nearly all the ruby, sapphire, and garnet used for jewels are imported, but it is hoped that American materials will soon be used. To be of value for this purpose, the material must not only be flawless, but also be of some decided shade of red or blue, and of a hardness greater than that of quartz.

Estimated production of precious stones in the United States from 1883 to 1885

Species	1883. Value of stones found and sold as specimens and curiosities, occasionally polished to beautify or show structure.	1883. Value of stones found and sold to be cut into gems.	1883. Total.	1884. Value of stones found and sold as specimens and curiosities, occasionally polished to beautify or show structure.	1884. Value of stones found and sold to be cut into gems.	1884. Total.	1885. Value of stones found and sold as specimens and curiosities, occasionally polished to beautify or show structure.	1885. Value of stones found and sold to be cut into gems.	1885. Total.
Diamond	$500	.....	.....	$250	$800	$800	.....	$500	$500
Sapphire gems	100	$2,000	$2,200	25	1,500	1,750	.....	$500	.....
Chrysoberyl	1,000	.....	100	200	.....	25	$1,000	250	1,250
Topaz	200	300	1,000	300	300	500	250	500	750
Beryl	500	.....	500	.....	400	700	.....	.....	3,200
Emerald	.....	500	500	.....	.....	.....	.....	.....	2,500
Hiddenite	100	.....	600	.....	.....	.....	100	100	600
Tourmaline	.....	.....	.....	1,500	500	2,000	500	100	600
Smoky quartz	2,500	7,500	10,000	2,000	10,000	12,000	2,000	5,000	7,000
Quartz	10,000	1,500	11,500	10,000	1,500	11,500	10,000	1,500	11,500
Silicified wood	5,000	.....	5,000	10,000	500	10,500	5,000	1,500	6,500
Garnet	1,000	5,000	6,000	1,000	3,000	4,000	200	2,500	2,700
Anthracite	.....	2,500	2,500	.....	.....	2,500	2,500	.....	2,500
Pyrite	1,500	500	2,000	2,000	1,000	3,000	1,500	500	2,000
Amazonstone	3,500	250	3,750	2,500	250	2,750	2,500	250	2,750
Catlinite (pipestone)	10,000	.....	10,000	10,000	.....	10,000	10,000	10,000	10,000
Arrow points	1,000	.....	1,000	1,000	.....	1,000	.....	2,500	2,500
Trilobites	500	.....	500	500	500	500	.....	1,000	1,000
Sagenitic rutile	500	500	1,000	500	500	1,000	.....	250	250
Hornblende in quartz	500	100	600	500	100	600	.....	300	300
Thomsonite	250	500	750	250	500	750	250	300	750
Diopside	200	100	300	.....	.....	.....	100	100	100
Agate	1,000	500	1,500	4,000	500	4,500	1,000	.....	2,000
Chlorastrolite	500	1,000	1,500	500	1,000	1,500	.....	1,000	.....
Turquoise	1,500	500	2,000	1,500	500	2,000	1,500	2,000	3,500
Moss agate	1,000	2,000	3,000	1,000	2,000	3,000	500	2,000	2,500
Amethyst	2,000	250	2,250	2,000	250	2,250	2,000	100	3,100
Jasper	2,000	500	2,500	2,000	500	2,500	.....	.....	.....
Sunstone	250	200	450	250	200	450	250	.....	350
Fossil coral	500	250	750	500	250	750	.....	100	.....
Rutile	.....	.....	.....	.....	.....	.....	750	.....	750
Total	47,300	26,450	73,750	54,275	28,550	82,825	39,300	24,850	69,850
Gold quartz	40,000	75,000	115,000	40,000	100,000	140,000	40,000	100,000	140,000

Estimated production of precious stones in the United States from 1886 to 1888

Species.	1886. Value of stones found and sold as specimens and curiosities, occasionally polished to beautify or show structure.	1886. Value of stones found and sold to be cut into gems.	1886. Total.	1887. Value of stones found and sold as specimens and curiosities, occasionally polished to beautify or show structure.	1887. Value of stones found and sold to be cut into gems.	1887. Total.	1888. Value of stones found and sold as specimens and curiosities, occasionally polished to beautify or show structure.	1888. Value of stones found and sold to be cut into gems.	1888. Value.
Diamond		$60	$60		$500	$500		$500	$500
Sapphire gems	$250	500	750	1,500	500	2,000	500	100	600
Topaz	1,000		1,000	500	3,000	3,500	300	500	800
Beryl		5,500	5,500	500			650		650
Phenacite							100		100
Emerald	3,000	200	3,200						
Hiddenite	3,500	1,000	4,500						
Tourmaline	3,500	2,000	5,500	300	200	500			
Smoky quartz	2,000	5,000	7,000	1,500	3,000	4,500	1,000	3,000	4,000
Quartz	10,000	1,500	11,500	10,000	1,500	11,500	10,000	1,150	11,150
Silicified wood	500	1,000	1,500	35,000	1,000	36,000	1,000	15,000	16,000
Garnet	1,250	2,000	3,250	2,500	1,000	3,500	2,000	1,500	3,500
Anthracite		2,500	2,500	2,500	500	3,000	1,500		1,500
Pyrite	1,500	500	2,000	2,000	500	2,500	2,500		2,500
Amazonstone	2,000	250	2,250	1,500	200	1,700	1,500	200	1,700
Catlinite (pipestone)	10,000		10,000	5,000		5,000	5,000		5,000
Arrow points		2,500	2,500		1,500	1,500	1,500		1,500
Trilobites	1,000		1,000	500		500	500		500
Sagenitic rutile	1,750		1,750						
Hornblende in quartz	200		200	100		100			
Thomsonite	100	300	400	250	500	750	300	200	500
Diopside				50		50			
Agate	1,000	1,000	2,000	3,000	1,000	4,000	3,000	1,000	4,000
Chlorastrolite	500	500	1,000	300	500	800		800	800
Turquoise	1,000	2,000	3,000	1,000	1,500	2,500	1,500	1,500	3,000
Moss agate	1,000	1,000	2,000	200	750	950	200	750	950
Amethyst	2,000	100	2,100	2,000	100	2,100	2,200	300	2,500
Jasper				50	100	150	100		100
Sunstone	200	100	300						
Fossil coral	1,000		1,000	1,500	500	2,000	2,500	500	3,000
Rutile	750		750						
Total	49,000	29,510	78,510	70,650	17,950	88,600	37,650	27,200	64,850
Gold quartz			40,000			75,000		40,000	40,000

NOTE.—These tables are taken from the volume on the Mineral Resources of the United States for 1888, edited by Dr. David T. Day.

COLLECTIONS.—One of the first collections of precious stones formed in the United States was that begun early in the century by J. R. Cox of the University of Pennsylvania, whose cabinet in 1860 passed into the possession of Prof. Joseph Leidy, of Philadelphia, who continued adding to it until 1880. At that time it comprised 221 lots, and was then considered the finest collection in the United States, the specimens all having been chosen with great care and scientific accuracy. The cabinet was offered for sale, but failed to find a purchaser and was disposed of to a dealer, who soon scattered the fruits of over half a century's patient gathering among his customers.

Dr. Lewis Feuchtwanger, of New York, was, in 1838, an authority on precious stones, and his book[1] was one of the first to be written on the subject in the United States. During his long residence in this country, he made an interesting and valuable collection of minerals, fossils, and gems, many of the latter being fine specimens, but it has not been sold, and is in the possession of his daughters. Moving has greatly damaged this collection, which originally included nearly 900 specimens; it was placed on sale in 1874, and for a time was deposited in the American Museum of Natural History, New York.

A remarkable and most interesting collection was made by Dr. Isaac Lea, of Philadelphia, who died in December, 1886. This eminent scientist for the last thirty years of his long life devoted much time to the study of microscopic inclusions in gems and minerals. The cabinet bequeathed by him to his daughter contains thousands of specimens of rubies, sapphires chrysoberyls, tourmalines, garnets, quartz, and other stones, each specimen labelled, and generally accompanied by a drawing, showing the interesting inclusions. His extensive bibliography includes several papers on inclusions in precious stones.[2]

The finest known collection of precious stones, and the finest collection of those found in the United States, is the one,

---

[1] A Treatise on Gems, in Reference to their Practical and Scientific Value. A useful guide for the jeweler, lapidary, artist, amateur, mineralogist, and chemist. Accompanied by a description of the most interesting American gems and ornamental and architectural materials. (New York, 1838. Subsequent enlarged editions appeared in 1859 and in 1872.

[2] Proc. Acad. Nat. Sci. of Philadelphia, vol. 21 pp. 4 and 119, 1869, and vol. 28, p. 98, 1876.

presented by Mr. J. Pierpont Morgan to the American Museum of Natural History, Central Park, N. Y.   This collection, which contains over one thousand specimens of all the obtainable precious and ornamental stones native to the United States, and some of the finest known examples of foreign stones, was prepared with the assistance of the author by Messrs. Tiffany & Co., and was exhibited, under his charge, at the World's Fair held in Paris in 1889, before a larger number of people than were ever before gathered within a given time at any one place.   (See Colored Plate 1, Figs. E and H ; Plate 3, Figs F and K ; Plate 5, Fig. D ; Plate 6, Fig. C ; Plate 7, Fig. B.)   It occupied a circular case in the center of the American Section, and, with the collection of pearls found in North America, gained the award of two gold medals.   The collection, of which a catalogue was published,[1] was a central point of attraction in the exhibit, was visited by leading scientists, lapidaries, stone-workers, and decorators, as well as the general public.

Among benefits resulting from the New Orleans Exposition held 1884–1885 was the appropriation to the United States Museum to perfect its exhibit there.   This money was expended by Prof. Frank W. Clarke, the curator in mineralogy, in the purchase of a very complete series of precious and ornamental stones, many of which are of great value from an educational point of view.   Since the Exposition, numerous fine specimens have been added by purchase and donation, notably the 171 diamonds and 150 pearls presented by the Imâm of Muscat to President Van Buren, all of which are of good quality.   The collection now numbers about a thousand specimens, and includes examples of almost every known variety of precious stones, many of them being remarkably good specimens.

At the Metropolitan Museum of Art, Central Park, New York, are the Curium gems, brought by Gen. Luigi P. di Cesnola from the island of Cyprus.   Some of these were described in his volume on Cyprus,[2] and are on exhibition with the Cesnola Collection.

[1] Catalogue of a Collection of Precious and Ornamental Stones of North America, p. 32, 8vo, New York, 1889.

[2] Researches and Discoveries in Cyprus. (New York, 1878.)

A full description of them was prepared by the Rev. C. W. King, of Trinity College, England, the greatest of all writers on engraved gems; this has never been published, but Mr. King's numerous writings mark an epoch in the study of this branch of archæology.   His collection of antique gems, numbering 331 pieces, is the summary of Mr. King's vast knowledge, and none has ever been more thoroughly studied.[1]   It was sent to the United States for sale in 1881, and in October, through the friendly mediation of Gaston L. Feuardent, it was purchased and presented to the Metropolitan Museum of Art by John Taylor Johnston, then president of the Museum. Near it is deposited the Sommerville Collection.   Maxville Sommerville, during thirty-two years passed in Europe, Asia, and Africa, collected cameos, intaglios, seals, and other historical gems, and as a result of his liberal expenditure of time and money is to-day the owner of one of the most unique and valuable collections of engraved gems in the world.   It numbers over 1,500 objects, including specimens of Egyptian, Persian, Babylonian, Etruscan, Greek, Roman, Aztec, and Mexican glyptic or jewel-carving art of singular excellence, affording a panoramic view of the achievements of civilized man in this direction. Descriptive of his remarkable collection, Mr. Sommerville has just published an illustrated catalogue.[2]   It is hoped that the Metropolitan Museum of Art, New York, will become the permanent owner of the collection.   .

Of greater antiquity, and of great archæological value, because representing a period before gems were cut in the form of intaglios, is the collection of the Rev. W. Hayes Ward, consisting of 300 Babylonian, Persian, and other cylinders.   Two hundred, collected by himself in Babylon and its vicinity, during the Catharine Wolfe Exploration, were sold to the Museum at a nominal figure.   Since that time, he has collected 100 more cylinders, many of which date from 2500 B. C. to 300 B. C., and are made of lapis lazuli (the sapphire of the ancients), agate, carnelian, hematite, and chalcedony.

[1] See Antique Gems and Rings, by C. W. King, 2 vols.   (London, 1872.)
[2] Engraved Gems : Their History and an Elaborate View of Their Place in Art. 4to, pp. 777, plates 104.   (Philadelphia, 1889.)   Printed by the Author.

The private collection of Clarence S. Bement, of Philadelphia, Pa., numbering 12,000 specimens of choice and carefully selected minerals, is the finest in the United States, and is exceeded in magnitude and excellence by only one or two collections of foreign museums. The high standard of all the specimens is due to the fact that Mr. Bement purchased from more than a dozen collections, one the largest—not his own—in the country. The collection is remarkable for its magnificent series, all in their natural state, of emeralds and sapphires, from North Carolina; its Colorado and Mexican topaz; its very fine series of garnets from Chester County, Pa., and other American localities; its sets of rutiles from Graves Mountain, Ga. Magnet Cove, Ark., Alexander County, N. C., and Vermont. It also contains a unique series of quartz specimens from every American locality; brown, black, and white tourmaline from northern New York, and the green, red, and blue varieties from Maine; and some of the finest known crystals of green microcline (amazonstone) from Pike's Peak, Col., and Amelia County, Va. In fact, nearly all the gem minerals, both American and foreign, are fully represented in this cabinet in their native state, although Mr. Bement says he is not a gem collector.[1]

Dr. Augustus C. Hamlin, of Bangor, Me., owns a collection, the nucleus of which was formed in 1826 by Elijah J. Hamlin. It includes nearly all the precious stones found at Mount Mica and other tourmaline localities in Maine, and contains several thousand crystals of every possible shade of color from white to pink, red, blue, green, yellow, to black, including some of the finest known specimens of rubellites, achroites, and other varieties of tourmalines, also some fine foreign gems. Dr. Hamlin has published two works on precious stones.[2]

Frederick Stearns, of Detroit, and Thomas T. Bouvé, of Boston, Mass., have excellent collections of precious stones, which, while not specially valuable, are still full and representative as regards species and varieties. Augustus Lowell, of Boston,

---

[1] See Prof. Gerhard von Rath's descriptive article on this collection in the Verhandlungen Des Naturh. Vereins d. Preuss. Rheinl. u. Westf., 1884, pages 295–304, of which an abstract by the author was published in the Jeweler's Circular for January, 1886.

[2] The Tourmaline. (Boston, 1875.) Leisure Hours Among the Gems. (Boston, 1884.)

has some superb fancy colored diamonds, sapphires and other interesting precious stones.

Judge Henry Hilton, of New York City, owns the finest collection of colored diamonds in the United States, ninety-seven in number, including many shades of brown, yellow, green, pink, and other colors. Mrs. T. N. C. Lowe, of Norristown, Pa., has an extensive collection of precious and fancy stones, and Mrs. M. J. Chase, of Philadelphia, Pa., some exceptionally choice and rare specimens in her cabinet.

Among the fine collections containing interesting gem specimens, to some of which reference has been made, may be mentioned the magnificent cabinet of Yale University, formed in the early part of the century, which contains the well-known Gibbs collection; the Tenney tourmalines; and in the same building, Peabody Museum, the private collection of Prof. George J. Brush; the Harvard University collection at Cambridge, Mass.; the collection of the School of Mines, Columbia College, New York City; that of Cornell University, at Ithaca, N. Y., which contains the Silliman Cabinet; of the University of Michigan, at Ann Arbor, containing the collection of the late Baron Lederer, one of the finest and most complete private collections known; incorporated with it is the famous Blum Collection;[1] that of Johns Hopkins University, Baltimore; of the Academy of Natural Sciences of Philadelphia, and in the building with it, the magnificent cabinet of the late William S. Vaux. At Hamilton College, Clinton, Oneida County, N. Y., is the Root Collection, and also the collection of Prof. Albert H. Chester; at Union College, Schenectady, N. Y., the collection of the late Charles M. Wheatley; at Amherst College, Amherst, Mass., the collection of Charles U. Shepard and others. The State Cabinet at Albany, N. Y., contains the Emmons Collection among others, as well as a neat case of precious and ornamental stones. The cabinet of Prof. Thomas Egleston of New York, is one of the finest crystallographical collections in the United States, and contains many unique and choice crystals of Russian and other gem minerals. The Canfield Collection, formed by the father of the

[1] Dr. Blum, of Heidelberg, was the author of works on precious stones, minerals, and pseudomorphs.

present owners over forty years ago, and containing many of the finest New Jersey and southern New York specimens, in addition to others equally choice, is at Dover, N. J., the William W. Jefferis collection at West Chester, Pa., the mineralogical collection of the University of Minnesota at Minneapolis, and the cabinet of the State Mining Bureau at San Francisco, Cal.

The foreign museums which contain the best American specimens are the British Museum in London (the finest mineralogical collection in the world), the Austrian Imperial Mineral Cabinet at Vienna; and the collections of the Jardin des Plantes and of the École des Mines in Paris. During the last ten years, the disposition to collect jade and other hard, carved stone objects has greatly increased, especially in the United States, owing to the stimulus given by the World's Fairs, at Philadelphia, Paris, and Amsterdam, and the breaking up by sale of many of the large collections. In December, 1889, a number of fine objects was furnished by American collectors to fill four large cases for a Loan Exhibition at the Union League Club, New York City. The value of carved jades, outside of China and India, cannot be less than $2,000,000. In the United States there are, perhaps, less than a dozen buyers, who have purchased $500,000 worth of this material. Many of the pieces are among the finest known, such as the private seal and other objects of the Emperor of China, taken at the sacking of the summer palace by the Chinese themselves, after it had been looted by the British and the French. The pieces brought by Tien Pau to Paris included some of the finest work that ever left China : they were intended for the Amsterdam Exposition. The choicest specimens of the Wells, Guthrie, Michael, and Hamilton Palace collections are now owned in the United States ; and experienced agents have been frequently sent to India and China to secure the finest objects as they presented themselves. Even during the year 1889, after the famine in China, a buyer securing a number of objects of priceless value.

Jade collectors may be divided into those who collect oriental jade, and those who collect archæological jade.

Among the principal collectors in this country are Heber R.

Bishop, Brayton Ives, William C. Oastler, John Harper, Samuel P. Avery, Charles Stewart Smith, Edward G. Low, Thomas B. Clarke, James W. Ellsworth, and James A. Garland, of New York; Samuel M. Nickerson and Potter Palmer, of Chicago; William T. Walters, of Baltimore; Frederick Ames, Dr. Bigelow, and Quincy Shaw, of Boston. There is a good collection at the Peabody Museum, Yale College, gathered and bequeathed by Dr. S. Wells Williams, formerly Secretary of Legation at Pekin, and author of the standard work on China, " The Middle Kingdom." Of foreign collectors who have a notable quantity of jade objects, there are Alfred Morrison, of London; Messrs. Bing and Gentian and Vicomte de Samalle, of Paris. The Louvre and the Musée de Fontainebleau contain some specimens of great interest, and the South Kensington Museum has quite a large and valuable collection.

Explorations in Alaska have brought to light the fact that jade was used by the natives of Alaska for making implements; almost conclusive proof, also, has been offered to show that it is found, not only as boulders, but in place. The United States National Museum at Washington; the Emmons Collection, and that of James Terry in the American Museum of Natural History; the Everett Collection; the Peabody Collection, at Cambridge, Mass.; the collections in the Canadian Geological Survey at Ottawa, and the Peter Redpath Museum, McGill College, Montreal, Can.; the Dresden Collection; the Freiberg Collection, at Baden; and others, including the writer's own, contain several hundred objects made from this very interesting material found in Alaska and British Columbia.

For nearly ten years fresh-water pearls, jade, rock crystal, rhodonite, and other stones have been used in the decoration of high-class silverware and some examples were shown at the World's Fair, held in Paris during 1889.

Taste in household decoration in the United States has of late attained a high standard, and any new idea that has been applied elsewhere is at once made use of. Minerals as yet have been only slightly utilized because they have not been thoroughly understood, and because of the absence of any accepted method of so applying them as to avoid inappropriateness. Ruskin, the eminent art critic, approves of the application of precious stones to the decoration of fine furniture, for, as he says, "furniture can be made to last indefinitely, and hence is worthy of the highest artistic effects." Instances of the use of precious stones for decorative purposes are more common in Europe and the East than on this side of the Atlantic. The famous peacock throne of India, looted by Nadir Shah, the Persian conqueror, in the eighteenth century, is estimated to have contained millions of dollars' worth of precious stones. Even now the altars of the Catholic and Greek churches throughout the world are often gorgeously decorated. The new palace in Potsdam, built by Frederick the Great, after the Seven Years' War, to show that the financial resources of Prussia were not exhausted, contains an apartment the walls of which are covered with minerals and precious stones, offering, perhaps, the most unique example of this style of decoration. The pair of rosewood pedestals with silver panels made for Mrs. Mary Jane Morgan, at a cost of over $2,000, were greatly improved in appearance by the application of a number of pieces of red, jaspery agate from Texas, cut en cabochon. Gold quartz has been used with pleasing effect in fine furniture and small ornaments, especially in California. Its rich colors, its hardness, and the beautiful polish of which it is susceptible give to the agatized wood of Arizona many advantages for inlaid work, and judging from the reception it met with at the World's Fair of 1889, it will probably soon be used extensively in furniture and interior decoration. Other cheap and ornamental stones, such as jasper, turquoise, rose-quartz, and amazonstone, might be introduced with advantage into inlaid work on clocks, mantels, and fine furniture. The employment of rock crystal for hand-glasses, crystal balls, and similar articles is treated in the chapter on quartz. One of the new departures in the United States in the uses made of common stones has been the introduction of the so-called Scotch jewelry; the designs were greatly improved, and native gem stones were used to such an extent that this jewelry found a ready sale, displacing many of the cheaper varieties of gold and silver pins. Among the minerals that have been so employed are agate, moss agate, jasper of all colors, rhodonite, pyrite, labradorite, and moonstone. The designs used are crowns, knots, thistles, shepherds' crooks, nails, horse-shoes, crescents, daggers, keys, spears, umbrellas, and many like shapes. In 1880, thousands of so-called mineral clocks, each in a plain wooden case, usually in the form of a house, completely covered with specimens, about an inch square, of pyrite, galenite, amazon-stone, ores from celebrated mines, and other Colorado minerals, were made. The minerals are glued on, each bearing a number referring to a list of the minerals on the back of the case. The interior consists of common Connecticut clock-work. Fully $15,000 worth have been annually disposed of. This form of decoration has also been applied to paper-weights, inkstands, and a large number of objects.

The fossils known as trilobites, which are found in various parts of the United States, are used, when fossilized, or curled up into proper forms, as charms, scarf-pins, and other ornaments. Most of those employed for such purposes are procured from the vicinity of Cincinnati, Ohio, and from near Covington, Ky. The species of trilobite used is principally Calymene senaria, which is generally found curled up, evidently in dying, and therefore appears either round or slightly oval in form, making a very suitable charm or an ornament for a scarf-pin. They vary in size from ¼ inch to 2 inches in diameter, and are sold at from twenty-five cents to $5.00 each, according to beauty and per-fection. The casts of the Calymene senaria, variety blumen-bachii, if entirely flattened out and perfect in form, have been worn as scarf-pins. As they are pure limestone, the surface is generally covered with thousands of brilliant microscopic crystals of calcite, that glitter beautifully in the sunlight.

INDEX.

ERRATA.

Page 20, line 13, *for* twined *read* twinned.
Page 29, line 5, *for* 1 1-5 carats *read* 1½ carats.
Page 35, line 14, *for* Onaka *read* Unaka.
Page 44, line 23, *for* Shorting *read* Shooting.
Page 51, line 28, *insert* and *after* pyrite.
Page 68, in heading of fifth analysis, *for* Turnbull *read* Trumbull.
Page 69, line 26, *for* Mountains *read* Mountain.
Page 74, line 27, *for* covered *read* uncovered.
Page 74, line 29, *for* dips *read* strikes.
Page 75, line 5, *for* or *read* of.
Page 75, line 6, *for* rubellites *read* rubellite.
Page 79, line 4, *for* the *read* this.
Page 83, lines 12 and 19, *for* melonite *read* melanite.
Page 101, last line, *for* is *read* are.
Page 104, line 4, *for* microline *read* microcline.
Page 108, line 20, *for* Ash *read* Ashe.
Page 114, line 21, *for* Jefferies *read* Jefferis.
Page 116, line 13, *for* Anteros *read* Antero.
Page 116, line 16, *for* plegmatic *read* pegmatitic.
Page 131, lines 24 and 33, *for* Uraguay *read* Uruguay.
Page 134, line 24, *after* Richmond *insert* Indiana.
Page 137, line 24, *omit* Cretaceous.
Page 141, line 5 from bottom, *for* representations *read* representatives.
Page 142, line 5 from bottom, *remove* comma *after* only.
Page 145, line 30, *after* bluish green *insert* coatings.
Page 150, analysis note 2, *for* Zoizite *read* Zoisite.
Page 155, head of analyses, *for* Sandford *read* Sanford.
Page 157, line 2, *for* pleochrism *read* pleochroism.
Page 166, line 3 from bottom, *for* Bythurst *read* Bathurst.

Page 168, line 20, *for* Sante *read* Santa.
Page 168, line 22, *for* sperolite *read* spherolitic.
Page 169, line 16, *for* Squire *read* Squier.
Page 173, line 5, *for* A like B *read* B like A.
Page 176, line 3, *for* 10 to 6 *read* 10 by 6.
Page 179, line 9, *for* breccilated *read* brecciated.
Page 182, lines 1 and 11, *for* Marias *read* Marais.
Page 187, line 20, *for* Hartford *read* Harford.
Page 193, line 11, *for* popular *read* poplar.
Page 198, line 20, *for* ashes *read* oxide.
Page 201, line 6, *for* Cresswicks *read* Crosswicks.
Page 203, line 11, *after* species *insert* of insects.
Page 206, foot-note 4, *for* Missouri *read* Mission.
Page 230, line 26, *for* Olentangg *read* Olentangy.
Page 248, line 26, *for* Jefferies *read* Jeffries.
Page 250, line 19, *for* Caciques *read* Cacique.
Page 250, foot-note 1, *for* Charlestown *read* Charleston.
Page 253, line 9, *for* Jefferies *read* Jeffries.
Page 254, line 17, *for* costalus *read* costatus.
Page 254, line 8 from bottom, *for* Techa *read* Teche.
Page 259, line 15, *for* was *read* were.
Page 266, line 23, *for* and *read* as.
Page 276, line 8, *for* goedes *read* geodes.
Page 279, line 9 from bottom, *for* 1-5 inch *read* 3-5 inch.
Page 280, line 6, *before* an *insert* as.
Page 284, line 7 from bottom, *for* matamorphic *read* metamorphic.
Page 288, line 5, *for* quinzite *read* quincite.
Page 288, line 11 from bottom, *for* rocks *read* rock.
Page 298, line 8, *for* walls *read* wells.
Page 306, line 1, *for* discordal *read* discoidal.

GOLD RUSH BOOKS

OREGON, USA

www.GoldMiningBooks.com

Books On Mining

Visit: www.goldminingbooks.com to order your copies or ask your favorite book seller to offer them.

Mining Books by Kerby Jackson

Gold Dust: Stories From Oregon's Mining Years - Oregon mining historian and prospector, Kerby Jackson, brings you a treasure trove of seventeen stories on Southern Oregon's rich history of gold prospecting, the prospectors and their discoveries, and the breathtaking areas they settled in and made homes. 5" X 8", 98 ppgs. Retail Price: $11.99

The Golden Trail: More Stories From Oregon's Mining Years - In his follow-up to "Gold Dust: Stories of Oregon's Mining Years", this time around, Jackson brings us twelve tales from Oregon's Gold Rush, including the story about the first gold strike on Canyon Creek in Grant County, about the old timers who found gold by the pail full at the Victor Mine near Galice, how Iradel Bray discovered a rich ledge of gold on the Coquille River during the height of the Rogue River War, a tale of two elderly miners on the hunt for a lost mine in the Cascade Mountains, details about the discovery of the famous Armstrong Nugget and others. 5" X 8", 70 ppgs. Retail Price: $10.99

Oregon Mining Books

Geology and Mineral Resources of Josephine County, Oregon - Unavailable since the 1970's, this important publication was originally compiled by the Oregon Department of Geology and Mineral Industries and includes important details on the economic geology and mineral resources of this important mining area in South Western Oregon. Included are notes on the history, geology and development of important mines, as well as insights into the mining of gold, copper, nickel, limestone, chromium and other minerals found in large quantities in Josephine County, Oregon. 8.5" X 11", 54 ppgs. Retail Price: $9.99

Mines and Prospects of the Mount Reuben Mining District - Unavailable since 1947, this important publication was originally compiled by geologist Elton Youngberg of the Oregon Department of Geology and Mineral Industries and includes detailed descriptions, histories and the geology of the Mount Reuben Mining District in Josephine County, Oregon. Included are notes on the history, geology, development and assay statistics, as well as underground maps of all the major mines and prospects in the vicinity of this much neglected mining district. 8.5" X 11", 48 ppgs. Retail Price: $9.99

The Granite Mining District - Notes on the history, geology and development of important mines in the well known Granite Mining District which is located in Grant County, Oregon. Some of the mines discussed include the Ajax, Blue Ribbon, Buffalo, Continental, Cougar-Independence, Magnolia, New York, Standard and the Tillicum. Also included are many rare maps pertaining to the mines in the area. 8.5" X 11", 48 ppgs. Retail Price: $9.99

Ore Deposits of the Takilma and Waldo Mining Districts of Josephine County, Oregon - The Waldo and Takilma mining districts are most notable for the fact that the earliest large scale mining of placer gold and copper in Oregon took place in these two areas. Included are details about some of the earliest large gold mines in the state such as the Llano de Oro, High Gravel, Cameron, Platerica, Deep Gravel and others, as well as copper mines such as the famous Queen of Bronze mine, the Waldo, Lily and Cowboy mines. This volume also includes six maps and 20 original illustrations. 8.5" X 11", 74 ppgs. Retail Price: $9.99

Metal Mines of Douglas, Coos and Curry Counties, Oregon - Oregon mining historian Kerby Jackson introduces us to a classic work on Oregon's mining history in this important re-issue of Bulletin 14C Volume 1, otherwise known as the Douglas, Coos & Curry Counties, Oregon Metal Mines Handbook. Unavailable since 1940, this important publication was originally compiled by the Oregon Department of Geology and Mineral Industries includes detailed descriptions, histories and the geology of over 250 metallic mineral mines and prospects in this rugged area of South West Oregon. 8.5" X 11", 158 ppgs. Retail Price: $19.99

Metal Mines of Jackson County, Oregon - Unavailable since 1943, this important publication was originally compiled by the Oregon Department of Geology and Mineral Industries includes detailed descriptions, histories and the geology of over 450 metallic mineral mines and prospects in Jackson County, Oregon. Included are such famous gold mining areas as Gold Hill, Jacksonville, Sterling and the Upper Applegate. **8.5" X 11", 220 ppgs. Retail Price: $24.99**

Metal Mines of Josephine County, Oregon - Oregon mining historian Kerby Jackson introduces us to a classic work on Oregon's mining history in this important re-issue of Bulletin 14C, otherwise known as the Josephine County, Oregon Metal Mines Handbook. Unavailable since 1952, this important publication was originally compiled by the Oregon Department of Geology and Mineral Industries includes detailed descriptions, histories and the geology of over 500 metallic mineral mines and prospects in Josephine County, Oregon. **8.5" X 11", 250 ppgs. Retail Price: $24.99**

Metal Mines of North East Oregon - Oregon mining historian Kerby Jackson introduces us to a classic work on Oregon's mining history in this important re-issue of Bulletin 14A and 14B, otherwise known as the North East Oregon Metal Mines Handbook. Unavailable since 1941, this important publication was originally compiled by the Oregon Department of Geology and Mineral Industries and includes detailed descriptions, histories and the geology of over 750 metallic mineral mines and prospects in North Eastern Oregon. **8.5" X 11", 310 ppgs. Retail Price: $29.99**

Metal Mines of North West Oregon - Oregon mining historian Kerby Jackson introduces us to a classic work on Oregon's mining history in this important re-issue of Bulletin 14D, otherwise known as the North West Oregon Metal Mines Handbook. Unavailable since 1951, this important publication was originally compiled by the Oregon Department of Geology and Mineral Industries and includes detailed descriptions, histories and the geology of over 250 metallic mineral mines and prospects in North Western Oregon. **8.5" X 11", 182 ppgs. Retail Price: $19.99**

Mines and Prospects of Oregon - Mining historian Kerby Jackson introduces us to a classic mining work by the Oregon Bureau of Mines in this important re-issue of The Handbook of Mines and Prospects of Oregon. Unavailable since 1916, this publication includes important insights into hundreds of gold, silver, copper, coal, limestone and other mines that operated in the State of Oregon around the turn of the 19th Century. Included are not only geological details on early mines throughout Oregon, but also insights into their history, production, locations and in some cases, also included are rare maps of their underground workings. **8.5" X 11", 314 ppgs. Retail Price: $24.99**

Lode Gold of the Klamath Mountains of Northern California and South West Oregon
(See California Mining Books)

Mineral Resources of South West Oregon - Unavailable since 1914, this publication includes important insights into dozens of mines that once operated in South West Oregon, including the famous gold fields of Josephine and Jackson Counties, as well as the Coal Mines of Coos County. Included are not only geological details on early mines throughout South West Oregon, but also insights into their history, production and locations. **8.5" X 11", 154 ppgs. Retail Price: $11.99**

Chromite Mining in The Klamath Mountains of California and Oregon
(See California Mining Books)

Southern Oregon Mineral Wealth - Unavailable since 1904, this rare publication provides a unique snapshot into the mines that were operating in the area at the time. Included are not only geological details on early mines throughout South West Oregon, but also insights into their history, production and locations. Some of the mining areas include Grave Creek, Greenback, Wolf Creek, Jump Off Joe Creek, Granite Hill, Galice, Mount Reuben, Gold Hill, Galls Creek, Kane Creek, Sardine Creek, Birdseye Creek, Evans Creek, Foots Creek, Jacksonville, Ashland, the Applegate River, Waldo, Kerby and the Illinois River, Althouse and Sucker Creek, as well as insights into local copper mining and other topics. **8.5" X 11", 64 ppgs. Retail Price: $8.99**

Geology and Ore Deposits of the Takilma and Waldo Mining Districts - Unavailable since the 1933, this publication was originally compiled by the United States Geological Survey and includes details on gold and copper mining in the Takilma and Waldo Districts of Josephine County, Oregon. The Waldo and Takilma mining districts are most notable for the fact that the earliest large scale mining of placer gold and copper in Oregon took place in these two areas. Included in this report are details about some of the earliest large gold mines in the state such as the Llano de Oro, High Gravel, Cameron, Platerica, Deep Gravel and others, as well as copper mines such as the famous Queen of Bronze mine, the Waldo, Lily and Cowboy mines. In addition to geological examinations, insights are also provided into the production, day to day operations and early histories of these mines, as well as calculations of known mineral reserves in the area. This volume also includes six maps and 20 original illustrations. **8.5" X 11", 74 ppgs. Retail Price: $9.99**

Gold Mines of Oregon - Oregon mining historian Kerby Jackson introduces us to a classic work on Oregon's mining history in this important re-issue of Bulletin 61, otherwise known as "Gold and Silver In Oregon". Unavailable since 1968, this important publication was originally compiled by geologists Howard C. Brooks and Len Ramp of the Oregon Department of Geology and Mineral Industries and includes detailed descriptions, histories and the geology of over 450 gold mines Oregon. Included are notes on the history, geology and gold production statistics of all the major mining areas in Oregon including the Klamath Mountains, the Blue Mountains and the North Cascades. While gold is where you find it, as every miner knows, the path to success is to prospect for gold where it was previously found. **8.5" X 11", 344 ppgs. Retail Price: $24.99**

Mines and Mineral Resources of Curry County Oregon - Originally published in 1916, this important publication on Oregon Mining has not been available for nearly a century. Included are rare insights into the history, production and locations of dozens of gold mines in Curry County, Oregon, as well as detailed information on important Oregon mining districts in that area such as those at Agness, Bald Face Creek, Mule Creek, Boulder Creek, China Diggings, Collier Creek, Elk River, Gold Beach, Rock Creek, Sixes River and elsewhere. Particular attention is especially paid to the famous beach gold deposits of this portion of the Oregon Coast. **8.5" X 11", 140 ppgs. Retail Price: $11.99**

Chromite Mining in South West Oregon - Originally published in 1961, this important publication on Oregon Mining has not been available for nearly a century. Included are rare insights into the history, production and locations of nearly 300 chromite mines in South Western Oregon. **8.5" X 11", 184 ppgs. Retail Price: $14.99**

Mineral Resources of Douglas County Oregon - Originally published in 1972, this important publication on Oregon Mining has not been available for nearly forty years. Included are rare insights into the geology, history, production and locations of numerous gold mines and other mining properties in Douglas County, Oregon. **8.5" X 11", 124 ppgs. Retail Price: $11.99**

Mineral Resources of Coos County Oregon - Originally published in 1972, this important publication on Oregon Mining has not been available for nearly forty years. Included are rare insights into the geology, history, production and locations of numerous gold mines and other mining properties in Coos County, Oregon. **8.5" X 11", 100 ppgs. Retail Price: $11.99**

Mineral Resources of Lane County Oregon - Originally published in 1938, this important publication on Oregon Mining has not been available for nearly seventy five years. Included are extremely rare insights into the geology and mines of Lane County, Oregon, in particular in the Bohemia, Blue River, Oakridge, Black Butte and Winberry Mining Districts. **8.5" X 11", 82 ppgs. Retail Price: $9.99**

Mineral Resources of the Upper Chetco River of Oregon: Including the Kalmiopsis Wilderness - Originally published in 1975, this important publication on Oregon Mining has not been available for nearly forty years. Withdrawn under the 1872 Mining Act since 1984, real insight into the minerals resources and mines of the Upper Chetco River has long been unavailable due to the remoteness of the area. Despite this, the decades of battle between property owners and environmental extremists over the last private mining inholding in the area has continued to pique the interest of those interested in mining and other forms of natural resource use. Gold mining began in the area in the 1850's and has a rich history in this geographic area, even if the facts surrounding it are little known. Included are twenty two rare photographs, as well as insights into the Becca and Morning Mine, the Emmly Mine (also known as Emily Camp), the Frazier Mine, the Golden Dream or Higgins Mine, Hustis Mine, Peck Mine and others. **8.5" X 11", 64 ppgs. Retail Price: $8.99**

Gold Dredging in Oregon - Originally published in 1939, this important publication on Oregon Mining has not been available for nearly seventy five years. Included are extremely rare insights into the history and day to day operations of the dragline and bucketline gold dredges that once worked the placer gold fields of South West and North East Oregon in decades gone by. Also included are details into the areas that were worked by gold dredges in Josephine, Jackson, Baker and Grant counties, as well as the economic factors that impacted this mining method. This volume also offers a unique look into the values of river bottom land in relation to both farming and mining, in how farm lands were mined, re-soiled and reclamated after the dredges worked them. Featured are hard to find maps of the gold dredge fields, as well as rare photographs from a bygone era. **8.5" X 11", 86 ppgs. Retail Price: $8.99**

Quick Silver Mining in Oregon - Originally published in 1963, this important publication on Oregon Mining has not been available for over fifty years. This publication includes details into the history and production of Elemental Mercury or Quicksilver in the State of Oregon. **8.5" X 11", 238 ppgs. Retail Price: $15.99**

Mines of the Greenhorn Mining District of Grant County Oregon - Originally published in 1948, this important publication on Oregon Mining has not been available for over sixty five years. In this publication are rare insights into the mines of the famous Greenhorn Mining District of Grant County, Oregon, especially the famous Morning Mine. Also included are details on the Tempest, Tiger, Bi-Metallic, Windsor, Psyche, Big Johnny, Snow Creek, Banzette and Paramount Mines, as well as prospects in the vicinities in the famous mining areas of Mormon Basin, Vinegar Basin and Desolation Creek. Included are hard to find mine maps and dozens of rare photographs from the bygone era of Grant County's rich mining history. **8.5" X 11", 72 ppgs. Retail Price: $9.99**

Geology of the Wallowa Mountains of Oregon: Part I (Volume 1) - Originally published in 1938, this important publication on Oregon Mining has not been available for nearly seventy five years. Included are details on the geology of this unique portion of North Eastern Oregon. This is the first part of a two book series on the area. Accompanying the text are rare photographs and historic maps.**8.5" X 11", 92 ppgs. Retail Price: $9.99**

Geology of the Wallowa Mountains of Oregon: Part II (Volume 2) - Originally published in 1938, this important publication on Oregon Mining has not been available for nearly seventy five years. Included are details on the geology of this unique portion of North Eastern Oregon. This is the first part of a two book series on the area. Accompanying the text are rare photographs and historic maps.**8.5" X 11", 94 ppgs. Retail Price: $9.99**

Field Identification of Minerals For Oregon Prospectors - Originally published in 1940, this important publication on Oregon Mining has not been available for nearly seventy five years. Included in this volume is an easy system for testing and identifying a wide range of minerals that might be found by prospectors, geologists and rockhounds in the State of Oregon, as well as in other locales. Topics include how to put together your own field testing kit and how to conduct rudimentary tests in the field. This volume is written in a clear and concise way to make it useful even for beginners. **8.5" X 11", 158 ppgs. Retail Price: $14.99**

The Bohemia Mining District of Oregon - Originally published in 1900, this important publication on Oregon Mining has not been available for over a century. Included in this volume are important insights into the famous Bohemia Mining District of Oregon, including the histories and locations of important gold mines in the area such as the Ophir Mine, Clarence, Acturas, Peek-a-boo, White Swan, Combination Mine, the Musick Mine, The California, White Ghost, The Mystery, Wall Street, Vesuvius, Story, Lizzie Bullock, Delta, Elsie Dora, Golden Slipper, Broadway, Champion Mine, Knott, Noonday, Helena, White Wings, Riverside and others. Also included are notes on the nearby Blue River Mining District. **8.5" X 11", 58 ppgs. Retail Price: $9.99**

The Gold Fields of Eastern Oregon - Unavailable since 1900, this publication was originally compiled by the Baker City Chamber of Commerce Offering important insights into the gold mining history of Eastern Oregon, "The Gold Fields of Eastern Oregon" sheds a rare light on many of the gold mines that were operating at the turn of the 19th Century in Baker County and Grant County in North Eastern Oregon. Some of the areas featured include the Cable Cove District, Baisely-Elhorn, Granite, Red Boy, Bonanza, Susanville, Sparta, Virtue, Vaughn, Sumpter, Burnt River, Rye Valley and other mining districts. Included is basic information on not only many gold mines that are well known to those interested in Eastern Oregon mining history, but also many mines and prospects which have been mostly lost to the passage of time. Accompanying are numerous rare photos **8.5" X 11", 78 ppgs. Retail Price: $10.99**

Gold Mining in Eastern Oregon - Originally published in 1938, this important publication on Oregon Mining has not been available for over a century. Included in this volume are important insights into the famous mining districts of Eastern Oregon during the late 1930's. Particular attention is given to those gold mines with milling and concentrating facilities in the Greenhorn, Red Boy, Alamo, Bonanza, Granite, Cable Cove, Cracker Creek, Virtue, Keating, Medical Springs, Sanger, Sparta, Chicken Creek, Mormon Basin, Connor Creek, Cornucopia and the Bull Run Mining Districts. Some of the mines featured include the Ben Harrison, North Pole-Columbia, Highland Maxwell, Baisley-Elkhorn, White Swan, Balm Creek, Twin Baby, Gem of Sparta, New Deal, Gleason, Gifford-Johnson, Cornucopia, Record, Bull Run, Orion and others. Of particular interest are the mill flow sheets and descriptions of milling operations of these mines. **8.5" X 11", 68 ppgs. Retail Price: $8.99**

The Gold Belt of the Blue Mountains of Oregon - Originally published in 1901, this important publication on Oregon Mining has not been available for over a century. Included in this volume are rare insights into the gold deposits of the Blue Mountains of North East Oregon, including the history of their early discovery and early production. Extensive details are offered on this important mining area's mineralogy and economic geology, as well as insights into nearby gold placers, silver deposits and copper deposits. Featured are the Elkhorn and Rock Creek mining districts, the Pocahontas district, Auburn and Minersville districts, Sumpter and Cracker Creek, Cable Cove, the Camp Carson district, Granite, Alamo, Greenhorn, Robinsonville, the Upper Burnt River Valley and Bonanza districts, Susanville, Quartzburg, Canyon Creek, Virtue, the Copper Butte district, the North Powder River, Sparta, Eagle Creek, Cornucopia, Pine Creek, Lower Powder River, the Upper Snake River Canyon, Rye Valley, Lower Burnt River Valley, Mormon Basin, the Malheur and Clarks Creek districts, Sutton Creek and others. Of particular interest are important details on numerous gold mines and prospects in these mining districts, including their locations, histories, geology and other important information, as well as information on silver, copper and fire opal deposits. **8.5" X 11", 250 ppgs. Retail Price: $24.99**

<u>Mining in the Cascades Range of Oregon</u> - Originally published in 1938, this important publication on Oregon Mining has not been available for over seventy five years. Included in this volume are rare insights into the gold mines and other types of metal mines in the Cascades Mountain Range of Oregon. Some of the important mining areas covered include the famous Bohemia Mining District, the North Santiam Mining District, Quartzville Mining District, Blue River Mining District, Fall Creek Mining District, Oakridge District, Zinc District, Buzzard-Al Sarena District, Grand Cove, Climax District and Barron Mining District. Of particular interest are important details on over 100 mines and prospects in these mining districts, including their locations, histories, geology and other important information. **8.5" X 11", 170 ppgs. Retail Price: $14.99**

<u>Beach Gold Placers of the Oregon Coast</u> - Originally published in 1934, this important publication on Oregon Mining has not been available for over 80 years. Included in this volume are rare insights into the beach gold deposits of the State of Oregon, including their locations, occurance, composition and geology. Of particular interest is information on placer platinum in Oregon's rich beach deposits. Also included are the locations and other information on some famous Oregon beach mines, including the Pioneer, Eagle, Chickamin, Iowa and beach placer mines north of the mouth of the Rogue River. **8.5" X 11", 60 ppgs. Retail Price: $8.99**

Idaho Mining Books

<u>Gold in Idaho</u> - Unavailable since the 1940's, this publication was originally compiled by the Idaho Bureau of Mines and includes details on gold mining in Idaho. Included is not only raw data on gold production in Idaho, but also valuable insight into where gold may be found in Idaho, as well as practical information on the gold bearing rocks and other geological features that will assist those looking for placer and lode gold in the State of Idaho. This volume also includes thirteen gold maps that greatly enhance the practical usability of the information contained in this small book detailing where to find gold in Idaho. **8.5" X 11", 72 ppgs. Retail Price: $9.99**

<u>Geology of the Couer D'Alene Mining District of Idaho</u> - Unavailable since 1961, this publication was originally compiled by the Idaho Bureau of Mines and Geology and includes details on the mining of gold, silver and other minerals in the famous Coeur D'Alene Mining District in Northern Idaho. Included are details on the early history of the Coeur D'Alene Mining District, local tectonic settings, ore deposit features, information on the mineral belts of the Osburn Fault, as well as detailed information on the famous Bunker Hill Mine, the Dayrock Mine, Galena Mine, Lucky Friday Mine and the infamous Sunshine Mine. This volume also includes sixteen hard to find maps. **8.5" X 11", 70 ppgs. Retail Price: $9.99**

<u>The Gold Camps and Silver Cities of Idaho</u> - Originally published in 1963, this important publication on Idaho Mining has not been available for nearly fifty years. Included are rare insights into the history of Idaho's Gold Rush, as well as the mad craze for silver in the Idaho Panhandle. Documented in fine detail are the early mining excitements at Boise Basin, at South Boise, in the Owyhees, at Deadwood, Long Valley, Stanley Basin and Robinson Bar, at Atlanta, on the famous Boise River, Volcano, Little Smokey, Banner, Boise Ridge, Hailey, Leesburg, Lemhi, Pearl, at South Mountain, Shoup and Ulysses, Yellow Jacket and Loon Creek. The story follows with the appearance of Chinese miners at the new mining camps on the Snake River, Black Pine, Yankee Fork, Bay Horse, Clayton, Heath, Seven Devils, Gibbonsville, Vienna and Sawtooth City. Also included are special sections on the Idaho Lead and Silver mines of the late 1800's, as well as the mining discoveries of the early 1900's that paved the way for Idaho's modern mining and mineral industry. Lavishly illustrated with rare historic photos, this volume provides a one of a kind documentary into Idaho's mining history that is sure to be enjoyed by not only modern miners and prospectors who still scour the hills in search of nature's treasures, but also those enjoy history and tromping through overgrown ghost towns and long abandoned mining camps. **8.5" X 11", 186 ppgs. Retail Price: $14.99**

<u>Ore Deposits and Mining in North Western Custer County Idaho</u> - Unavailable since 1913, this important publication was originally published by the Us Department of the Interior and has been unavailable for a century. Included are fine details on the geology, geography, gold placers and gold and silver bearing quartz veins of the mining region of North West Custer County, Idaho. Of particular interest is a rare look at the mines and prospects of the region, including those such as the Ramshorn Mine, SkyLark, Riverview, Excelsior, Beardsley, Pacific, Hoosier, Silver Brick, Forest Rose and dozens of others in the Bay Horse Mining District. Also covered are the mines of the Yankee Fork District such as the Lucky Boy, Badger, Black, Enterprise, Charles Dickens, Morrison, Golden Sunbeam, Montana, Golden Gate and others, as well as those in the Loon Mining District. **8.5" X 11", 126 ppgs. Retail Price: $12.99**

Gold Rush To Idaho - Unavailable since 1963, this important publication was originally published by the Idaho Bureau of Mines and has been unavailable for 50 years. "Gold Rush To Idaho" revisits the earliest years of the discovery of gold in Idaho Territory and introduces us to the conditions that the pioneer gold seekers met when they blazed a trail through the wilderness of Idaho's mountains and discovered the precious yellow metal at Oro Fino and Pierce. Subsequent rushes followed at places like Elk City, Newsome, Clearwater Station, Florence, Warrens and elsewhere. Of particular interest is a rare look at the hardships that the first miners in Idaho met with during their day to day existences and their attempts to bring law and order to their mining camps. 8.5" X 11", 88 ppgs. **Retail Price: $9.99**

The Geology and Mines of Northern Idaho and North Western Montana - Unavailable since 1909, this important publication was originally published by the Us Department of the Interior and has been unavailable for a century. Included are fine details on the geology and geography of the mining regions of Northern Idaho and North Western Montana. Of particular interest is a rare look at the mines and prospects of the region, including those in the Pine Creek Mining District, Lake Pend Oreille district, Troy Mining District, Sylvanite District, Cabinet Mining District, Prospect Mining District and the Missoula Valley. Some of the mines featured include the Iron Mountain, Silver Butte, Snowshoe, Grouse Mountain Mine and others. 8.5" X 11", 142 ppgs. **Retail Price: $12.99**

Mining in the Alturas Quadrangle of Blaine County Idaho - Unavailable since 1922, this important publication was originally published by the Idaho Bureau of Mines and has been unavailable for ninety years. Topics include the geology, rock formations and the formation of ore deposits in this important mining area of Idaho. Of particular focus is information on the local geology, quartz veins and ore deposits of this portion of Idaho. Included are hard to find details, including the descriptions and locations of numerous gold and silver mines in the area including the Silver King, Pilgrim, Columbia, Lone Jack, Sunbeam, Pride of the West, Lucky Boy, Scotia, Atlanta, Beaver-Bidwell and others mines and prospects. 8.5" X 11", 56 ppgs. **Retail Price: $8.99**

Mining in Lemhi County Idaho - Originally published in 1913, this important book on Idaho Mining has not been available to miners for over a century. Included are rare insights into hundreds of gold, silver, copper and other mines in this famous Idaho mining area. Details include the locations, geology, history, production and other facts of the mines of this region, not only gold and silver hardrock mines, but also gold placer mines, lead-silver deposits, copper mines, cobalt-nickel deposits, tungsten and tin mines . It is lavishly illustrated with hard to find photos of the period and rare mining maps. Some of the vicinities featured include the Nicholia Mining District, Spring Mountain District, Texas District, Blue Wing District, Junction District, McDevitt District, Pratt Creek, Eldorado District, Kirtley Creek, Carmen Creek, Gibbonsville, Indian Creek, Mineral Hill District, Mackinaw, Eureka District, Blackbird District, YellowJacket District, Gravel Range District, Junction District, Parker Mountain and other mining districts. 8.5" X 11", 226 ppgs. **Retail Price: $19.99**

Utah Mining Books

Fluorite in Utah - Unavailable since 1954, this publication was originally compiled by the USGS, State of Utah and U.S. Atomic Energy Commission and details the mining of fluorspar, also known as fluorite in the State of Utah. Included are details on the geology and history of fluorspar (fluorite) mining in Utah, including details on where this unique gem mineral may be found in the State of Utah. 8.5" X 11", 60 ppgs. **Retail Price: $8.99**

California Mining Books

The Tertiary Gravels of the Sierra Nevada of California - Mining historian Kerby Jackson introduces us to a classic mining work by Waldemar Lindgren in this important re-issue of The Tertiary Gravels of the Sierra Nevada of California. Unavailable since 1911, this publication includes details on the gold bearing ancient river channels of the famous Sierra Nevada region of California. 8.5" X 11", 282 ppgs. **Retail Price: $19.99**

The Mother Lode Mining Region of California - Unavailable since 1900, this publication includes details on the gold mines of California's famous Mother Lode gold mining area. Included are details on the geology, history and important gold mines of the region, as well as insights into historic mining methods, mine timbering, mining machinery, mining bell signals and other details on how these mines operated. Also included are insights into the gold mines of the California Mother Lode that were in operation during the first sixty years of California's mining history. 8.5" X 11", 176 ppgs. **Retail Price: $14.99**

Lode Gold of the Klamath Mountains of Northern California and South West Oregon - Unavailable since 1971, this publication was originally compiled by Preston E. Hotz and includes details on the lode mining districts of Oregon and California's Klamath Mountains. Included are details on the geology, history and important lode mines of the French Gulch, Deadwood, Whiskeytown, Shasta, Redding, Muletown, South Fork, Old Diggings, Dog Creek (Delta), Bully Choop (Indian Creek), Harrison Gulch, Hayfork, Minersville, Trinity Center, Canyon Creek, East Fork, New River, Denny, Liberty (Black Bear), Cecilville, Callahan, Yreka, Fort Jones and Happy Camp mining districts in California, as well as the Ashland, Rogue River, Applegate, Illinois River, Takilma, Greenback, Galice, Silver Peak, Myrtle Creek and Mule Creek districts of South Western Oregon. Also included are insights into the mineralization and other characteristics of this important mining region. 8.5" X 11", 100 ppgs. **Retail Price: $10.99**

Mines and Mineral Resources of Shasta County, Siskiyou County, Trinity County: California - Unavailable since 1915, this publication was originally compiled by the California State Mining Bureau and includes details on the gold mines of this area of Northern California. Also included are insights into the mineralization and other characteristics of this important mining region, as well as the location of historic gold mines. **8.5" X 11", 204 ppgs. Retail Price: $19.99**

Geology of the Yreka Quadrangle, Siskiyou County, California - Unavailable since 1977, this publication was originally compiled by Preston E. Hotz and includes details on the geology of the Yreka Quadrangle of Siskiyou County, California. Also included are insights into the mineralization and other characteristics of this important mining region. **8.5" X 11", 78 ppgs. Retail Price: $7.99**

Mines of San Diego and Imperial Counties, California - Originally published in 1914, this important publication on California Mining has not been available for a century. This publication includes important information on the early gold mines of San Diego and Imperial County, which were some of the first gold fields mined in California by early Spanish and Mexican miners before the 49ers came on the scene. Included are not only details on early mining methods in the area, production statistics and geological information, but also the location of the early gold mines that helped make California "The Golden State". Also included are details on the mining of other minerals such as silver, lead, zinc, manganese, tungsten, vanadium, asbestos, barite, borax, cement, clay, dolomite, fluospar, gem stones, graphite, marble, salines, petroleum, stronium, talc and others. **8.5" X 11", 116 ppgs. Retail Price: $12.99**

Mines of Sierra County, California - Unavailable since 1920, this publication was originally compiled by the California State Mining Bureau and includes details on the gold mines of Sierra County, California. Also included are insights into the mineralization and other characteristics of this important mining region, as well as the location of historic gold mines. **8.5" X 11", 156 ppgs. Retail Price: $19.99**

Mines of Plumas County, California - Unavailable since 1918, this publication was originally compiled by the California State Mining Bureau and includes details on the gold mines of Plumas County, California. Also included are insights into the mineralization and other characteristics of this important mining region, as well as the location of historic gold mines. **8.5" X 11", 200 ppgs. Retail Price: $19.99**

Mines of El Dorado, Placer, Sacramento and Yuba Counties, California - Originally published in 1917, this important publication on California Mining has not been available for nearly a century. This publication includes important information on the early gold mines of El Dorado County, Placer County, Sacramento County and Yuba County, which were some of the first gold fields mined by the Forty-Niners during the California Gold Rush. Included are not only details on early mining methods in the area, production statistics and geological information, but also the location of the early gold mines that helped make California "The Golden State". Also included are insights into the early mining of chrome, copper and other minerals in this important mining area. **8.5" X 11", 204 ppgs. Retail Price: $19.99**

Mines of Los Angeles, Orange and Riverside Counties, California - Originally published in 1917, this important publication on California Mining has not been available for nearly a century. This publication includes important information on the early gold mines of Los Angeles County, Orange County and Riverside County, which were some of the first gold fields mined in California by early Spanish and Mexican miners before the 49ers came on the scene. Included are not only details on early mining methods in the area, production statistics and geological information, but also the location of the early gold mines that helped make California "The Golden State". **8.5" X 11", 146 ppgs. Retail Price: $12.99**

Mines of San Bernadino and Tulare Counties, California - Originally published in 1917, this important publication on California Mining has not been available for nearly a century. This publication includes important information on the early gold mines of San Bernadino and Tulare County, which were some of the first gold fields mined in California by early Spanish and Mexican miners before the 49ers came on the scene. Included are not only details on early mining methods in the area, production statistics and geological information, but also the location of the early gold mines that helped make California "The Golden State". Also included are details on the mining of other minerals such as copper, iron, lead, zinc, manganese, tungsten, vanadium, asbestos, barite, borax, cement, clay, dolomite, fluospar, gem stones, graphite, marble, salines, petroleum, stronium, talc and others. **8.5" X 11", 200 ppgs. Retail Price: $19.99**

Chromite Mining in The Klamath Mountains of California and Oregon - Unavailable since 1919, this publication was originally compiled by J.S. Diller of the United States Department of Geological Survey and includes details on the chromite mines of this area of Northern California and Southern Oregon. Also included are insights into the mineralization and other characteristics of this important mining region, as well as the location of historic mines. Also included are insights into chromite mining in Eastern Oregon and Montana. **8.5" X 11", 98 ppgs. Retail Price: $9.99**

Mines and Mining in Amador, Calaveras and Tuolumne Counties, California - Unavailable since 1915, this publication was originally compiled by William Tucker and includes details on the mines and mineral resources of this important California mining area. Included are details on the geology, history and important gold mines of the region, as well as insights into other local mineral resources such as asbestos, clay, copper, talc, limestone and others. Also included are insights into the mineralization and other characteristics of this important portion of California's Mother Lode mining region. **8.5" X 11", 198 ppgs. Retail Price: $14.99**

The Cerro Gordo Mining District of Inyo County California - Unavailable since 1963, this publication was originally compiled by the United States Department of Interior. Included are insights into the mineralization and other characteristics of this important mining region of Southern California. Topics include the mining of gold and silver in this important mining district in Inyo County, California, including details on the history, production and locations of the Cerro Gordo Mine, the Morning Star Mine, Estelle Tunnel, Charles Lease Tunnel, Ignacio, Hart, Crosscut Tunnel, Sunset, Upper Newtown, Newtown, Ella, Perseverance, Newsboy, Belmont and other silver and gold mines in the Cerro Gordo Mining District. This volume also includes important insights into the fossil record, geologic formations, faults and other aspects of economic geology in this California mining district. **8.5" X 11", 104 ppgs. Retail Price: $10.99**

Mining in Butte, Lassen, Modoc, Sutter and Tehama Counties of California - Unavailable since 1917, this publication was originally compiled by the United States Department of Interior. Included are insights into the mineralization and other characteristics of this important mining region of California. Topics include the mining of asbestos, chromite, gold, diamonds and manganese in Butte County, the mining of gold and copper in the Hayden Hill and Diamond Mountain mining districts of Lassen County, the mining of coal, salt, copper and gold in the High Grade and Winters mining districts of Modoc County, gold mining in Sutter County and the mining of gold, chromite, manganese and copper in Tehama County. This volume also includes the production records and locations of numerous mines in this important mining region. **8.5" X 11", 114 ppgs. Retail Price: $11.99**

Mines of Trinity County California - Originally published in 1965, this important publication on California Mining has not been available for nearly fifty years. This publication includes important information on mines and mining in Trinity County, California, as well insights into the mineralization and geology of this important mining area in Northern California. Included are extensive details on hardrock and placer gold mines and prospects, including charts showing the locations of these historic mines.. **8.5" X 11", 144 ppgs. Retail Price: $12.99**

Mines of Kern County California - Originally published in 1962, this important publication on California Mining has not been available for nearly fifty years. This publication includes important information on mines and mining in Kern County, California, as well insights into the mineralization and geology of this important mining area in California. Included are extensive details on hardrock and placer gold mines and prospects, including charts showing the locations of these historic mines. **8.5" X 11", 398 ppgs. Retail Price: $24.99**

Mines of Calaveras County California - Originally published in 1962, this important publication on California Mining has not been available for nearly fifty years. This publication includes important information on mines and mining in Calaveras County, California, as well insights into the mineralization and geology of this important mining area in Northern California. Included are extensive details on hardrock and placer gold mines and prospects, including charts showing the locations of these historic mines. **8.5" X 11", 236 ppgs. Retail Price: $19.99**

Lode Gold Mining in Grass Valley California - Unavailable since 1940, this publication was originally compiled by the United States Department of Interior. Included are insights into the gold mineralization and other characteristics of this important mining region of Nevada County, California. This volume also includes important insights into the geologic formations, faults and other aspects of economic geology in this California mining district. Of particular interest are the fine details on many hardrock gold mines in the area, including their locations, histories, development and mineralization. Some of the mines featured include the Gold Hill Mine, Massachusetts Hill, Boundary, Peabody, Golden Center, North Star, Omaha, Lone Jack, Homeward Bound, Hartery, Wisconsin, Allison Ranch, Phoenix, Kate Hayes, W.Y.O.D., Empire, Rich Hill, Daisy Hill, Orleans, Sultana, Centennial, Conlin, Ben Franklin, Crown Point and many others. **8.5" X 11", 148 ppgs. Retail Price: $12.99**

Lode Mining in the Alleghany District of Sierra County California - Unavailable since 1913, this publication was originally compiled by the United States Department of Interior. Included are insights into the mineralization and other characteristics of this important mining region of Sierra County. Included are details on the history, production and locations of numerous hardrock gold mines in this famous California area, including the Tightner Mine, Minnie D., Osceola, Eldorado, Twenty One, Sherman, Kenton, Oriental, Rainbow, Plumbago, Irelan, Gold Canyon, North Fork, Federal, Kate Hardy and others. This volume also includes important insights into the fossil record, geologic formations, faults and other aspects of economic geology in this California mining district. **8.5" X 11", 48 ppgs. Retail Price: $7.99**

Six Months In The Gold Mines During The California Gold Rush - Unavailable since 1850, this important work is a first hand account of one "49'ers" personal experience during the great California Gold Rush, shedding important light on one of the most exciting periods in the history of not only California, but also the world. Compiled from journals written between 1847 and 1849 by E. Gould Buffum, a native of New York, "Six Months In The Gold Mines During The California Gold Rush" offers a rare look into the day to day lives of the people who came to California to work in her gold mines when the state was still a great frontier. 8.5" X 11", 290 ppgs. Retail Price: $19.99

Quartz Mines of the Grass Valley Mining District of California - Unavailable since 1867, this important publication has not been available since those days. This rare publication offers a short dissertation on the early hardrock mines in this important mining district in the California Mother Lode region between the 1850's and 1860's. Also included are hard to find details on the mineralization and locations of these mines, as well as how they were operated in those day. 8.5" X 11", 44 ppgs. Retail Price: $8.99

Alaska Mining Books

Ore Deposits of the Willow Creek Mining District, Alaska - Unavailable since 1954, this hard to find publication includes valuable insights into the Willow Creek Mining District near Hatcher Pass in Alaska. The publication includes insights into the history, geology and locations of the well known mines in the area, including the Gold Cord, Independence, Fern, Mabel, Lonesome, Snowbird, Schroff-O'Neil, High Grade, Marion Twin, Thorpe, Webfoot, Kelly-Willow, Lane, Holland and others. 8.5" X 11", 96 ppgs. Retail Price: $9.99

The Juneau Gold Belt of Alaska - Unavailable since 1906, this hard to find publication includes valuable insights into the gold mines around Juneau, Alaska. The publication includes important details into the history, geology and locations of the well known gold mines and prospects in the area, including those around Windham Bay, Holkham Bay, Port Snettisham, on Grindstone and Rhine Creeks, Gold Creek, Douglas Island, Salmon Creek, Lemon Creek, Nugget Creek, from the Mendenhall River to Berners Bay, McGinnis Creek, Montana Creek, Peterson Creek, Windfall Creek, the Eagle River, Yankee Basin, Yankee Curve, Kowee Creek and elsewhere. Not only are gold placer mines included, but also hardrock gold mines. 8.5" X 11", 224 ppgs. Retail Price: $19.99

Arizona Mining Books

Mines and Mining in Northern Yuma County Arizona - Originally published in 1911, this important publication on Arizona Mining has not been available for over a hundred years. Included are rare insights into the gold, silver, copper and quicksilver mines of Yuma County, Arizona together with hard to find maps and photographs. Some of the mines and mining districts featured include the Planet Copper Mine, Mineral Hill, the Clara Consolidated Mine, Viati Mine, Copper Basin prospect, Bowman Mine, Quartz King, Billy Mack, Carnation, the Wardwell and Osbourne, Valensuella Copper, the Mariquita, Colonial Mine, the French American, the New York-Plomosa, Guadalupe, Lead Camp, Mudersbach Copper Camp, Yellow Bird, the Arizona Northern (Salome Strike), Bonanza (Harqua Hala), Golden Eagle, Hercules, Socorro and others. 8.5" X 11", 144 ppgs. Retail Price: $11.99

The Aravaipa and Stanley Mining Districts of Graham County Arizona - Originally published in 1925, this important publication on Arizona Mining has not been available for nearly ninety years. Included are rare insights into the gold and silver mines of these two important mining districts, together with hard to find maps. 8.5" X 11", 140 ppgs. Retail Price: $11.99

Gold in the Gold Basin and Lost Basin Mining Districts of Mohave County, Arizona - This volume contains rare insights into the geology and gold mineralization of the Gold Basin and Lost Basin Mining Districts of Mohave County, Arizona that will be of benefit to miners and prospectors. Also included is a significant body of information on the gold mines and prospects of this portion of Arizona. This volume is lavishly illustrated with rare photos and mining maps. 8.5" X 11", 188 ppgs. Retail Price: $19.99

Mines of the Jerome and Bradshaw Mountains of Arizona - This important publication on Arizona Mining has not been available for ninety years. This volume contains rare insights into the geology and ore deposits of the Jerome and Bradshaw Mountains of Arizona that will be of benefit to miners and prospectors who work those areas. Included is a significant body of information on the mines and prospects of the Verde, Black Hills, Cherry Creek, Prescott, Walker, Groom Creek, Hassayampa, Bigbug, Turkey Creek, Agua Fria, Black Canyon, Peck, Tiger, Pine Grove, Bradshaw, Tintop, Humbug and Castle Creek Mining Districts. This volume is lavishly illustrated with rare photos and mining maps. 8.5" X 11", 218 ppgs. Retail Price: $19.99

The Ajo Mining District of Pima County Arizona - This important publication on Arizona Mining has not been available for nearly seventy years. This volume contains rare insights into the geology and mineralization of the Ajo Mining District in Pima County, Arizona and in particular the famous New Cornelia Mine. 8.5" X 11", 126 ppgs. Retail Price: $11.99

Mining in the Santa Rita and Patagonia Mountains of Arizona - Originally published in 1915, this important publication on Arizona Mining has not been available for nearly a century. Included are rare insights into hundreds of gold, silver, copper and other mines in this famous Arizona mining area. Details include the locations, geology, history, production and other facts of the mines of this region. **8.5" X 11", 394 ppgs. Retail Price: $24.99**

Mining in the Bisbee Quadrangle of Arizona - Originally published in 1906, this important publication on Arizona Mining has not been available for nearly a century. Included are rare insights into hundreds of gold, silver, copper and other mines in this famous Arizona mining area. Details include the locations, geology, history, production and other facts of the mines of this important mining region. **8.5" X 11", 188 ppgs. Retail Price: $14.99**

Montana Mining Books

A History of Butte Montana: The World's Greatest Mining Camp - First published in 1900 by H.C. Freeman, this important publication sheds a bright light on one of the most important mining areas in the history of The West. Together with his insights, as well as rare photographs of the periods, Harry Freeman describes Butte and its vicinity from its early beginnings, right up to its flush years when copper flowed from its mines like a river. At the time of publication, Butte, Montana was known worldwide as "The Richest Mining Spot On Earth" and produced not only vast amounts of copper, but also silver, gold and other metals from its mines. Freeman illustrates, with great detail, the most important mines in the vicinity of Butte, providing rare details on their owners, their history and most importantly, how the mines operated and how their treasures were extracted. Of particular interest are the dozens of rare photographs that depict mines such as the famous Anaconda, the Silver Bow, the Smoke House, Moose, Paulin, Buffalo, Little Minah, the Mountain Consolidated, West Greyrock, Cora, the Green Mountain, Diamond, Bell, Parnell, the Neversweat, Nipper, Original and many others. **8.5" X 11", 142 ppgs. Retail Price: $12.99**

The Butte Mining District of Montana - This important publication on Montana Mining has not been available for over a century. Included are rare insights into the gold, copper and silver mines of Butte, Montana together with hard to find maps and photographs. Some of the topics include the early history of gold, silver and copper mining in the Butte area, insight into the geology of its mining areas, the local distribution of gold, silver and copper ores, as well their composition and how to identify them. Also included are detailed facts about the mines in the Butte Mining District, including the famous Anaconda Mine, Gagnon, Parrot, Blue Vein, Moscow, Poulin, Stella, Buffalo, Green Mountain, Wake Up Jim, the Diamond-Bell Group, Mountain Consolidated, East Greyrock, West Greyrock, Snowball, Corra, Speculator, Adirondack, Miners Union, the Jessie-Edith May Group, Otisco, Iduna, Colorado, Lizzie, Cambers, Anderson, Hesperus, Preferencia and dozens of others. **8.5" X 11", 298 ppgs. Retail Price: $24.99**

Mines of the Helena Mining Region of Montana - This important publication on Montana Mining has not been available for over a century. Included are rare insights into the gold, copper and silver mines of the vicinity of Helena, Montana, including the Marysville Mining District, Elliston Mining District, Rimini Mining District, Helena Mining District, Clancy Mining District, Wickes Mining District, Boulder and Basin Mining Districts and the Elkhorn Mining District. Some of the topics include the early history of gold, silver and copper mining in the Helena area, insight into the geology of its mining areas, the local distribution of gold, silver and copper ores, as well their composition and how to identify them. Also included are detailed facts, history, geology and locations of over one hundred gold, silver and copper mines in the area . **8.5" X 11", 162 ppgs, Retail Price: $14.99**

Mines and Geology of the Garnet Range of Montana - This important publication on Montana Mining has not been available for over a century. Included are rare insights into the gold, copper and silver mines of the vicinity of this important mining area of Montana. Some of the topics include the early history of gold, silver and copper mining in the Garnet Mountains, insight into the geology of its mining areas, the local distribution of gold, silver and copper ores, as well their composition and how to identify them. Also included are detailed facts, history, geology and locations of numerous gold, silver and copper mines in the area . **8.5" X 11", 100 ppgs, Retail Price: $11.99**

Mines and Geology of the Philipsburg Quadrangle of Montana - This important publication on Montana Mining has not been available for over a century. Included are rare insights into the gold, copper and silver mines of the vicinity of this important mining area of Montana. Some of the topics include the early history of gold, silver and copper mining in the Philipsburg Quadrangle, insight into the geology of its mining areas, the local distribution of gold, silver and copper ores, as well their composition and how to identify them. Also included are detailed facts, history, geology and locations of over one hundred gold, silver and copper mines in the area **8.5" X 11", 290 ppgs, Retail Price: $24.99**

Geology of the Marysville Mining District of Montana - Included are rare insights into the mining geology of the Marysville Mining District. Some of the topics include the early history of gold, silver and copper mining in the area, insight into the geology of its mining areas, the local distribution of gold, silver and copper ores, as well their composition and how to identify them. Also included are detailed facts, history, geology and locations of gold, silver and copper mines in the area **8.5" X 11", 198 ppgs, Retail Price: $19.99**

The Geology and Mines of Northern Idaho and North Western Montana

See listing under Idaho.

Nevada Mining Books

The Bull Frog Mining District of Nevada - Unavailable since 1910, this publication was originally compiled by the United States Department of Interior. This volume also includes important insights into the geologic formations, faults and other aspects of economic geology in this Nevada mining district. Of particular interest are the fine details on many mines in the area, including their locations, histories, development and mineralization. Some of the mines featured include the National Bank Mine, Providence, Gibraltor, Tramps, Denver, Original Bullfrog, Gold Bar, Mayflower, Homestake-King and other mines and prospects. **8.5″ X 11″, 152 ppgs, Retail Price: $14.99**

History of the Comstock Lode - Unavailable since 1876, this publication was originally released by John Wiley & Sons. This volume also includes important insights into the famous Comstock Lode of Nevada that represented the first major silver discovery in the United States. During its spectacular run, the Comstock produced over 192 million ounces of silver and 8.2 million ounces of gold. Not only did the Comstock result in one of the largest mining rushes in history and yield immense fortunes for its owners, but it made important contributions to the development of the State of Nevada, as well as neighboring California. Included here are important details on not only the early development and history of the Comstock, but also rare early insight into its mines, ore and its geology.**8.5″ X 11″, 244 ppgs, Retail Price: $19.99**

Colorado Mining Books

Ores of The Leadville Mining District - Unavailable since 1926, this publication was originally compiled by the United States Department of Interior. This volume also includes important insights into the ores and mineralization of the Leadville Mining District in Colorado. Topics include historic ore prospecting methods, local geology, insights into ore veins and stockworks, the local trend and distribution of ore channels, reverse faults, shattered rock above replacement ore bodies, mineral enrichment in oxidized and sulphide zones and more. **8.5″ X 11″, 66 ppgs, Retail Price: $8.99**

Mining in Colorado - Unavailable since 1926, this publication was originally compiled by the United States Department of Interior. This volume also includes important insights into the mining history of Colorado from its early beginnings in the 1850's right up to the mid 1920's. Not only is Colorado's gold mining heritage included, but also its silver, copper, lead and zinc mining industry. Each mining area is treated separately, detailing the development of Colorado's mines on a county by county basis. **8.5″ X 11″, 284 ppgs, Retail Price: $19.99**

Gold Mining in Gilpin County Colorado - Unavailable since 1876, this publication was originally compiled by the Register Steam Printing House of Central City, Colorado. A rare glimpse at the gold mining history and early mines of Gilpin County, Colorado from their first discovery in the 1850's up to the "flush years" of the mid 1870's. Of particular interest is the history of the discovery of gold in Gilpin County and details about the men who made those first strikes. Special focus is given to the early gold mines and first mining districts of the area, many of which are not detailed in other books on Colorado's gold mining history. **8.5″ X 11″, 156 ppgs, Retail Price: $12.99**

Mining in the Gold Brick Mining District of Colorado - Important insights into the history of the Gold Brick Mining District, as well as its local geography and economic geology. Also included are the histories and locations of historic mines in this important Colorado Mining District, including the Cortland, Carter, Raymond, Gold Links, Sacramento, Bassick, Sandy Hook, Chronicle, Grand Prize, Chloride, Granite Mountain, Lucille, Gray Mountain, Hilltop, Maggie Mitchell, Silver Islet, Revenue, Roosevelt, Carbonate King and others. In addition to hardrock mining, are also included are details on gold placer mining in this portion of Colorado. **8.5″ X 11″, 140 ppgs, Retail Price: $12.99**

Washington Mining Books

The Republic Mining District of Washington - Unavailable since 1910, this important publication was originally published by the Washington Geologic Survey and has been unavailable for a century. Topics include the geology, rock formations and the formation of ore deposits in this important mining area of Washington State. Also included are hard to find details on the geology, history and locations of dozens of mines in the area. Some of the mines featured include the New Republic Mine, Ben Hur, Morning Glory, the South Republic Mine, Quilp, Surprise, Black Tail, Lone Pine, San Poil, Mountain Lion, Tom Thumb, Elcaliph and many others. **8.5″ X 11″, 94 ppgs, Retail Price: $10.99**

The Myers Creek and Nighthawk Mining Districts of Washington - Unavailable since 1911, this important publication was originally published by the Washington Geologic Survey and has been unavailable for a century. Topics include the geology, rock formations and the formation of ore deposits in these important mining areas of Washington State. Also included are hard to find details on the geology, history and locations of dozens of mines in the area. Some of the mines featured include the Grant Mine, Monterey, Nip and Tuck, Myers Creek, Number Nine, Neutral, Rainbow, Aztec, Crystal Butte, Apex, Butcher Boy, Molson, Mad River, Olentangy, Delate, Kelsey, Golden Chariot, Okanogan, Ohio, Forty-Ninth Parallel, Nighthawk, Favorite, Little Chopaka, Summit, Number One, California, Peerless, Caaba, Prize Group, Ruby, Mountain Sheep, Golden Zone, Rich Bar, Similkameen, Kimberly, Triune, Hiawatha, Trinity, Hornsilver, Maquae, Bellevue, Bullfrog, Palmer Lake, Ivanhoe, Copper World and many others.
    8.5" X 11", 136 ppgs, Retail Price: $12.99

The Blewett Mining District of Washington - Unavailable since 1911, this important publication was originally published by the Washington Geologic Survey and has been unavailable for a century. Topics include the geology, rock formations and the formation of ore deposits in this important mining area of Washington State. Also included are hard to find details on the geology, history and locations of dozens of mines in the area. Some of the mines featured include the Washington Meteor, Alta Vista, Pole Pick, Blinn, North Star, Golden Eagle, Tip Top, Wilder, Golden Guinea, Lucky Queen, Blue Bell, Prospect, Homestake, Lone Rock, Johnson, and others. **8.5" X 11", 134 ppgs, Retail Price: $12.99**

Silver Mining In Washington - Unavailable since 1955, this important publication was originally published by the Washington Geologic Survey. Featured are the hard to find locations and details pertaining to Washington's silver mines. **8.5" X 11", 180 ppgs, Retail Price: $15.99**

The Mines of Snohomish County Washington - Unavailable since 1942, this important publication was originally published by the Washington Geologic Survey and has been unavailable for seventy years. Featured are details on a large number of gold, silver, copper, lead and other metallic mineral mines. Included are the locations of each historic mine, along with information on the commodity produced. **8.5" X 11", 98 ppgs, Retail Price: $10.99**

The Mines of Chelan County Washington - Unavailable since 1943, this important publication was originally published by the Washington Geologic Survey and has been unavailable for seventy years. Featured are details on a large number of gold, silver, copper, lead and other metallic mineral mines. Included are the locations of each historic mine, along with information on the commodity. **8.5" X 11", 88 ppgs, Retail Price: $9.99**

Metal Mines of Washington - Unavailable since 1921, this important publication was originally published by the Washington Geologic Survey and has been unavailable for nearly ninety years. Widely considered a masterpiece on the Washington Mining Industry, "Metal Mines of Washington" sheds light on the important details of Washington's early mining years. Featured are details on hundreds of gold, silver, copper, lead and other metallic mineral mines. Included are hard to find details on the mineral resources of this state, as well as the locations of historic mines. Lavishly illustrated with maps and historic photos and complete with a glossary to explain any technical terms found in the text, this is one of the most important works on mining in the State of Washington. No prospector or miner should be without it if they are interested in mining in Washington. **8.5" X 11", 396 ppgs, Retail Price: $24.99**

Gem Stones In Washington - Unavailable since 1949, this important publication was originally published by the Washington Geologic Survey and has been unavailable since first published. Included are details on where to find naturally occurring gem stones in the State of Washington, including quartz crystal, amethyst, smoky quartz, milky quartz, agates, bloodstone, carnelian, chert, flint, jasper, onyx, petrified wood, opal, fire opal, hyalite and others. **8.5" X 11", 54 ppgs, Retail Price: $8.99**

The Covada Mining District of Washington - Unavailable since 1913, this important publication was originally published by the Washington Geologic Survey and has been unavailable for a century. Topics include the geology, rock formations and the formation of ore deposits in this important mining area of Washington State. Also included are hard to find details on the geology, history and locations of dozens of mines in the area. Some of the mines featured include the Admiral, Advance, Algonkian, Big Bug, Big Chief, Big Joker, Black Hawk, Black Tail, Black Thorn, Captain, Cherokee Strip, Colorado, Dan Patch, Dead Shot, Etta, Good Ore, Greasy Run, Great Scott, Idora, IXL, Jay Bird, Kentucky Bell, King Solomon, Laurel, Laura S, Little Jay, Meteor, Neglected, Northern Light, Old Nell, Plymouth Rock, Polaris, Quandary, Reserve, Shoo Fly, Silver Plume, Three Pines, Vernie, White Rose and dozens of others. **8.5" X 11", 114 ppgs, Retail Price: $10.99**

The Index Mining District of Washington - Unavailable since 1912, this important publication was originally published by the Washington Geologic Survey and has been unavailable for a century. Topics include the geology, rock formations and the formation of ore deposits in this important mining area of Washington State. Also included are hard to find details on the geology, history and locations of dozens of mines in the area. Some of the mines featured include the Sunset, Non-Pareil, Ethel Consolidated, Kittaning, Merchant, Homestead, Co-operative, Lost Creek, Uncle Sam, Calumet, Florence-Rae, Bitter Creek, Index Peacock, Gunn Peak, Helena, North Star, Buckeye. Copper Bell, Red Cross and others. **8.5" X 11", 114 ppgs, Retail Price: $11.99**

Mining & Mineral Resources of Stevens County Washington - Unavailable since 1920, this important publication was originally published by the Washington Geologic Survey and has been unavailable for a century. Topics include the geology, rock formations and the formation of ore deposits in these important mining areas of Washington State. Also included are hard to find details on the geology, history and locations of hundreds of mines in the area. **8.5" X 11", 372 ppgs, Retail Price: $24.99**

The Mines and Geology of the Loomis Quadrangle Okanogan County, Washington - Unavailable since 1972, this important publication was originally published by the Washington Geologic Survey and has been unavailable for a century. Topics include the geology, rock formations and the formation of ore deposits in this important mining area of Washington State. Also included are hard to find details on the geology, history and locations of dozens of gold, copper, silver and other mines in the area. **8.5" X 11", 150 ppgs, Retail Price: $12.99**

The Conconully Mining District of Okanogan County Washington - Unavailable since 1973, this important publication was originally published by the Washington Geologic Survey and has been unavailable for a century. Topics include the geology, rock formations and the formation of ore deposits in this important mining area of Washington State, which also includes Salmon Creek, Blue Lake and Galena. Also included are hard to find details on the geology, mining history and locations of dozens of mines in the area. Some of the mines include Arlington, Fourth of July, Sonny Boy, First Thought, Last Chance, War Eagle-Peacock, Wheeler, Mohawk, Lone Star, Woo Loo Moo Loo, Keystone, Hughes, Plant-Callahan, Johnny Boy, Leuena, Gubser, John Arthur, Tough Nut, Homestake, Key and many others **8.5" X 11", 68 ppgs, Retail Price: $8.99**

Wyoming Mining Books

Mining in the Laramie Basin of Wyoming - Unavailable since 1909, this publication was originally compiled by the United States Department of Interior. Also included are insights into the mineralization and other characteristics of this important mining region, especially in regards to coal, limestone, gypsum, bentonite clay, cement, sand, clay and copper. **8.5" X 11", 104 ppgs, Retail Price: $11.99**

New Mexico Mining Books

The Mogollon Mining District of New Mexico - Unavailable since 1927, this important publication was originally published by the US Department of Interior and has been unavailable for 80 years. Topics include the geology, rock formations and the formation of ore deposits in this important mining area in New Mexico. Of particular focus is information on the history and production of the ore deposits in this area, their form and structure, vein filling, their paragenesis, origins and ore shoots, as well as oxidation and supergene enrichment. Also included are hard to find details, including the descriptions and locations of numerous gold, silver and other types of mines, including the Eureka, Pacific, South Alpine, Great Western, Enterprise, Buffalo, Mountain View, Floride, Gold Dust, Last Chance, Deadwood, Confidence, Maud S., Deep Down, Little Fanney, Trilby, Johnson, Alberta, Comet, Golden Eagle, Cooney, Queen, the Iron Crown, Eberle, Clifton, Andrew Jackson mine, Mascot and others. **8.5" X 11", 144 ppgs, Retail Price: $12.99**

The Percha Mining District of Kingston New Mexico - Unavailable since 1883, this important publication was originally published by the Kingston Tribune and has been unavailable for over one hundred and thirty five years. Having been written during the earliest years of gold and silver mining in the Percha Mining District, unlike other books on the subject, this work offers the unique perspective of having actually been written while the early mining history of this area was still being made. In fact, the work was written so early in the development of this area that many of the notable mines in the Percha District were less than a few years old and were still being operated by their original discoverers with the same enthusiasm as when they were first located. Included are hard to find details on the very earliest gold and silver mines of this important mining district near Kingston in Sierra County, New Mexico. **8.5" X 11", 68 ppgs, Retail Price: $9.99**

East Coast Mining Books

The Gold Fields of the Southern Appalachians - Unavailable since 1895, this important publication was originally published by the US Department of Interior and has been unavailable for nearly 120 years. Topics include the geology, rock formations and the formation of ore deposits in this important mining area of the American South. Of particular focus is information on the history and statistics of the ore deposits in this area, their form and structure and veins. Also included are details on the placer gold deposits of the region. The gold fields of the Georgian Belt, Carolinian Belt and the South Mountain Mining District of North Carolina are all treated in descriptive detail. Included are hard to find details, including the descriptions and locations of numerous gold mines in Georgia, North Carolina and elsewhere in the American South. Also included are details on the gold belts of the British Maritime Provinces and the Green Mountains. **8.5" X 11", 104 ppgs, Retail Price: $9.99**

Gold Rush Tales Series

<u>Millions in Siskiyou County Gold</u> - In this first volume of the "Gold Rush Tales" series, leading mining historian and editor Kerby Jackson, introduces us to the story of how millions of dollars worth of gold was discovered in Siskiyou County during the California Gold Rush. Lavishly illustrated with photos from the 19th Century, this hard to find information was first published in 1897 and sheds important light onto the gold rush era in Siskiyou County, California and the experiences of the men who dug for the gold and actually found it. **8.5" X 11", 82 ppgs, Retail Price: $9.99**

<u>The California Rand in the Days of '49</u> - In this second volume of the "Gold Rush Tales" series, leading mining historian and editor Kerby Jackson, introduces us to four tales from the California Gold Rush. Lavishly illustrated with photos from the 19th Century, this hard to find information was first published in 1890's and includes the stories of "California's Rand", details about Chinese miners, how one early miner named Baker struck it rich and also the story of Alphonzo Bowers, who invented the first hydraulic gold dredge. **8.5" X 11", 54 ppgs, Retail Price: $9.99**

More Mining Books

<u>Prospecting and Developing A Small Mine</u> - Topics covered include the classification of varying ores, how to take a proper ore sample, the proper reduction of ore samples, alluvial sampling, how to understand geology as it is applied to prospecting and mining, prospecting procedures, methods of ore treatment, the application of drilling and blasting in a small mine and other topics that the small scale miner will find of benefit. **8.5" X 11", 112 ppgs, Retail Price: $11.99**

<u>Timbering For Small Underground Mines</u> - Topics covered include the selection of caps and posts, the treatment of mine timbers, how to install mine timbers, repairing damaged timbers, use of drift supports, headboards, squeeze sets, ore chute construction, mine cribbing, square set timbering methods, the use of steel and concrete sets and other topics that the small underground miner will find of benefit. This volume also includes twenty eight illustrations depicting the proper construction of mine timbering and support systems that greatly enhance the practical usability of the information contained in this small book. **8.5" X 11", 88 ppgs. Retail Price: $10.99**

<u>Timbering and Mining</u> - A classic mining publication on Hard Rock Mining by W.H. Storms. Unavailable since 1909, this rare publication provides an in depth look at American methods of underground mine timbering and mining methods. Topics include the selection and preservation of mine timbers, drifting and drift sets, driving in running ground, structural steel in mine workings, timbering drifts in gravel mines, timbering methods for driving shafts, positioning drill holes in shafts, timbering stations at shafts, drainage, mining large ore bodies by means of open cuts or by the "Glory Hole" system, stoping out ore in flat or low lying veins, use of the "Caving System", stoping in swelling ground, how to stope out large ore bodies, Square Set timbering on the Comstock and its modifications by California miners, the construction of ore chutes, stoping ore bodies by use of the "Block System", how to work dangerous ground, information on the "Delprat System" of stoping without mine timbers, construction and use of headframes and much more. This volume provides a reference into not only practical methods of mining and timbering that may be employed in narrow vein mining by small miners today, but also rare insights into how mines were being worked at the turn of the 19th Century. **8.5" X 11", 288 ppgs. Retail Price: $24.99**

<u>A Study of Ore Deposits For The Practical Miner</u> - Mining historian Kerby Jackson introduces us to a classic mining publication on ore deposits by J.P. Wallace. First published in 1908, it has been unavailable for over a century. Included are important insights into the properties of minerals and their identification, on the occurrence and origin of gold, on gold alloys, insights into gold bearing sulfides such as pyrites and arsenopyrites, on gold bearing vanadium, gold and silver tellurides, lead and mercury tellurides, on silver ores, platinum and iridium, mercury ores, copper ores, lead ores, zinc ores, iron ores, chromium ores, manganese ores, nickel ores, tin ores, tungsten ores and others. Also included are facts regarding rock forming minerals, their composition and occurrences, on igneous, sedimentary, metamorphic and intrusive rocks, as well as how they are geologically disturbed by dikes, flows and faults, as well as the effects of these geologic actions and why they are important to the miner. Written specifically with the common miner and prospector in mind, the book will help to unlock the earth's hidden wealth for you and is written in a simple and concise language that anyone can understand. **8.5" X 11", 366 ppgs. Retail Price: $24.99**

<u>Mine Drainage</u> - Unavailable since 1896, this rare publication provides an in depth look at American methods of underground mine drainage and mining pump systems. This volume provides a reference into not only practical methods of mining drainage that may be employed in narrow vein mining by small miners today, but also rare insights into how mines were being worked at the turn of the 19th Century. **8.5" X 11", 218 ppgs. Retail Price: $24.99**

Fire Assaying Gold, Silver and Lead Ores - Unavailable since 1907, this important publication was originally published by the Mining and Scientific Press and was designed to introduce miners and prospectors of gold, silver and lead to the art of fire assaying. Topics include the fire assaying of ores and products containing gold, silver and lead; the sampling and preparation of ore for an assay; care of the assay office, assay furnaces; crucibles and scorifiers; assay balances; metallic ores; scorification assays; cupelling; parting' crucible assays, the roasting of ores and more. This classic provides a time honored method of assaying put forward in a clear, concise and easy to understand language that will make it a benefit to even beginners. **8.5" X 11", 96 ppgs. Retail Price: $11.99**

Methods of Mine Timbering - Originally published in 1896, this important publication on mining engineering has not been available for nearly a century. Included are rare insights into historical methods of timbering structural support that were used in underground metal mines during the California that still have a practical application for the small scale hardrock miner of today. **8.5" X 11", 94 ppgs. Retail Price: $10.99**

The Enrichment of Copper Sulfide Ores - First published in 1913, it has been unavailable for over a century. Topics include the definition and types of ore enrichment, the oxidation of copper ores, the precipitation of metallic sulfides. Also included are the results of dozens of lab experiments pertaining to the enrichment of sulfide ores that will be of interest to the practical hard rock mine operator in his efforts to release the metallic bounty from his mine's ore. **8.5" X 11", 92 ppgs. Retail Price: $9.99**

A Study of Magmatic Sulfide Ores - Unavailable since 1914, this rare publication provides an in depth look at magmatic sulfide ores. Some of the topics included are the definition and classification of magmatic ores, descriptions of some magmatic sulfide ore deposits known at the time of publication including copper and nickel bearing pyrrohitic ore bodies, chalcopyrite-bornite deposits, pyritic deposits, magnetite-ileminite deposits, chromite deposits and magmatic iron ore deposits. Also included are details on how to recognize these types of ore deposits while prospecting for valuable hardrock minerals. **8.5" X 11", 138 ppgs. Retail Price: $11.99**

The Cyanide Process of Gold Recovery - Unavailable since 1894 and released under the name "The Cyanide Process: Its Practical Application and Economical Results", this rare publication provides an in depth look at the early use of cyanide leaching for gold recovery from hardrock mine ores. This volume provides a reference into the early development and use of cyanide leaching to recover gold. **8.5" X 11", 162 ppgs. Retail Price: $14.99**

California Gold Milling Practices - Unavailable since 1895 and released under the name "California Gold Practices", this rare publication provides an in depth look at early methods of milling used to reduce gold ores in California during the late 19th century. This volume provides a reference into the early development and use of milling equipment during the earliest years of the California Gold Rush up to the age of the Industrial Revolution. Much of the information still applies today and will be of use to small scale miners engaging in hardrock mining. **8.5" X 11", 104 ppgs. Retail Price: $10.99**

Leaching Gold and Silver Ores With The Plattner and Kiss Processes - Mining historian Kerby Jackson introduces us to a classic mining publication on the evaluation and examination of mines and prospects by C.H. Aaron. First published in 1881, it has been unavailable for over a century and sheds important light on the leaching of gold and silver ores with the Plattner and Kiss processes. **8.5" X 11", 204 ppgs. Retail Price: $15.99**

The Metallurgy of Lead and the Desilverization of Base Bullion - First published in 1896, it has been unavailable for over a century and sheds important light on the the recovery of silver from lead based ores. Some of the topics include the properties of lead and some of its compounds, lead ores such as galenite, anglesite, cerussite and others, the distribution of lead ores throughout the United States and the sampling and assaying of lead ores. Also covered is the metallurgical treatment of lead ores, as well as the desilverization of lead by the Pattinson Process and the Parkes Process. Hofman's text has long been considered one of the most important early works on the recovery of silver from lead based ores. **8.5" X 11", 452 ppgs. Retail Price: $29.99**

Ore Sampling For Small Scale Miners - First published in 1916, it has been unavailable for over a century and sheds important light on historic methods of ore sampling in hardrock mines. Topics include how to take correct ore samples and the conditions that affect sampling, such as their subdivision and uniformity. Particular detail is given to methods of hand sampling ore bodies by grab sample, pipe sample and coning, as well as sampling by mechanical methods. Also given are insights into the screening, drying and grinding processes to achieve the most consistent sample results and much more. **8.5" X 11", 124 ppgs. Retail Price: $12.99**

The Extraction of Silver, Copper and Tin from Ores - First published in 1896, it has been unavailable for over a century and sheds important light on how historic miners recovered silver, copper and tin from their mining operations. The book is split into three sections, including a discussion on the Lixiviation of Silver Ores, the mining and treatment of copper ores as practiced at Tharsis, Spain and the smelting of tin as it was practiced by metallurgists at Pulo Brani, Singapore. Also included is an overview and analysis of these historic metal recovery methods that will be of benefit to those interested in the extraction of silver, copper and tin from small mines. **8.5" X 11", 118 ppgs. Retail Price: $14.99**

The Roasting of Gold and Silver Ores - First published in 1880, it has been unavailable for over a century and sheds important light on how historic miners recovered gold and silver rom their mining operations. Topics include details on the most important silver and free milling gold ores, methods of desulphurization of ores, methods of deoxidation, the chlorination of ores, methods and details on roasting gold and silver ores, notes on furnaces and more. Also included are details on numerous methods of gold and silver recovery, including the Ottokar Hofman's Process, the Patera Process, Kiss Process, Augustin Process, Ziervogel Process and others. **8.5" X 11", 178 ppgs. Retail Price: $19.99**

The Examination of Mines and Prospects - First published in 1912, it has been unavailable for over a century and sheds important light on how to examine and evaluate hardrock mines, prospects and lode mining claims. Sections include Mining Examinations, Structural Geology, Structural Features of Ore Deposits, Primary Ores and their Distribution, Types of Primary Ore Deposits, Primary Ore Shoots, The Primary Alteration of Wall Rocks, Alterations by Surface Agencies, Residual Ores and their Distribution, Secondary Ores and Ore Shoots and Vein Outcrops. This hard to find information is a must for those who are interested in owning a mine or who already own a lode mining claim and wish to succeed at quartz mining. **8.5" X 11", 250 ppgs. Retail Price: $19.99**

www.ingramcontent.com/pod-product-compliance
Lightning Source LLC
Chambersburg PA
CBHW080757180526
45168CB00006B/2242